Asynchronous Circuit Design

Asynchronous Circuit Design

Chris J. Myers

A Wiley-Interscience Publication
JOHN WILEY & SONS, INC.
New York / Chichester / Weinheim / Brisbane / Singapore / Toronto

This text is printed on acid-free paper. ∞

Copyright © 2001 by John Wiley & Sons, Inc. All rights reserved.

Published simultaneously in Canada.

No part of this publication may be reproduced, stored in a retrieval system or transmitted in any form or by any means, electronic, mechanical, photocopying, recording, scanning or otherwise, except as permitted under Sections 107 or 108 of the 1976 United States Copyright Act, without either the prior written permission of the Publisher, or authorization through payment of the appropriate per-copy fee to the Copyright Clearance Center, 222 Rosewood Drive, Danvers, MA 01923, (978) 750-8400, fax (978) 750-4744. Requests to the Publisher for permission should be addressed to the Permissions Department, John Wiley & Sons, Inc., 605 Third Avenue, New York, NY 10158-0012, (212) 850-6011, fax (212) 850-6008, E-Mail: PERMREQ @ WILEY.COM.

For ordering and customer service, call 1-800-CALL-WILEY.

Library of Congress Cataloging-in-Publication Data

Myers, Chris J., 1969–
  Asynchronous Circuit Design / Chris J. Myers
    p. cm.
  Includes bibliographical references and index.
  1. Asynchronous circuits—Design and construction. 2. Electronic circuit design. I. Title.

TK7868.A79 M94 2001
621.3815—dc21   2001024028

Printed in the United States of America

10 9 8 7 6 5 4 3 2 1

To Ching and John

Contents

	Preface	*xiii*
	Acknowledgments	*xvii*
1	Introduction	1
	1.1 Problem Specification	1
	1.2 Communication Channels	2
	1.3 Communication Protocols	4
	1.4 Graphical Representations	8
	1.5 Delay-Insensitive Circuits	10
	1.6 Huffman Circuits	13
	1.7 Muller Circuits	16
	1.8 Timed Circuits	17
	1.9 Verification	20
	1.10 Applications	20
	1.11 Let's Get Started	21
	1.12 Sources	21
	Problems	22
2	Communication Channels	23
	2.1 Basic Structure	24

	2.2	Structural Modeling in VHDL		27
	2.3	Control Structures		31
		2.3.1	Selection	31
		2.3.2	Repetition	32
	2.4	Deadlock		34
	2.5	Probe		35
	2.6	Parallel Communication		35
	2.7	Example: MiniMIPS		36
		2.7.1	VHDL Specification	38
		2.7.2	Optimized MiniMIPS	48
	2.8	Sources		52
		Problems		53
3	Communication Protocols			57
	3.1	Basic Structure		57
	3.2	Active and Passive Ports		61
	3.3	Handshaking Expansion		61
	3.4	Reshuffling		65
	3.5	State Variable Insertion		66
	3.6	Data Encoding		67
	3.7	Example: Two Wine Shops		71
	3.8	Syntax-Directed Translation		73
	3.9	Sources		80
		Problems		82
4	Graphical Representations			85
	4.1	Graph Basics		85
	4.2	Asynchronous Finite State Machines		88
		4.2.1	Finite State Machines and Flow Tables	88
		4.2.2	Burst-Mode State Machines	91
		4.2.3	Extended Burst-Mode State Machines	93
	4.3	Petri Nets		100
		4.3.1	Ordinary Petri Nets	100
		4.3.2	Signal Transition Graphs	111
	4.4	Timed Event/Level Structures		116
	4.5	Sources		120
		Problems		121

5	Huffman Circuits		131
	5.1	Solving Covering Problems	132
		5.1.1 Matrix Reduction Techniques	134
		5.1.2 Bounding	137
		5.1.3 Termination	137
		5.1.4 Branching	138
	5.2	State Minimization	140
		5.2.1 Finding the Compatible Pairs	141
		5.2.2 Finding the Maximal Compatibles	143
		5.2.3 Finding the Prime Compatibles	145
		5.2.4 Setting Up the Covering Problem	148
		5.2.5 Forming the Reduced Flow Table	154
	5.3	State Assignment	154
		5.3.1 Partition Theory and State Assignment	155
		5.3.2 Matrix Reduction Method	157
		5.3.3 Finding the Maximal Intersectibles	158
		5.3.4 Setting Up the Covering Problem	161
		5.3.5 Fed-Back Outputs as State Variables	163
	5.4	Hazard-Free Two-Level Logic Synthesis	165
		5.4.1 Two-Level Logic Minimization	165
		5.4.2 Prime Implicant Generation	166
		5.4.3 Prime Implicant Selection	168
		5.4.4 Combinational Hazards	169
	5.5	Extensions for MIC Operation	171
		5.5.1 Transition Cubes	172
		5.5.2 Function Hazards	172
		5.5.3 Combinational Hazards	173
		5.5.4 Burst-Mode Transitions	176
		5.5.5 Extended Burst-Mode Transitions	177
		5.5.6 State Minimization	180
		5.5.7 State Assignment	183
		5.5.8 Hazard-Free Two-Level Logic Synthesis	183
	5.6	Multilevel Logic Synthesis	188
	5.7	Technology Mapping	189
	5.8	Generalized C-Element Implementation	193
	5.9	Sequential Hazards	194
	5.10	Sources	196
		Problems	199

6 Muller Circuits — 207
- 6.1 Formal Definition of Speed Independence — 208
  - 6.1.1 Subclasses of Speed-Independent Circuits — 210
  - 6.1.2 Some Useful Definitions — 212
- 6.2 Complete State Coding — 216
  - 6.2.1 Transition Points and Insertion Points — 217
  - 6.2.2 State Graph Coloring — 219
  - 6.2.3 Insertion Point Cost Function — 220
  - 6.2.4 State Signal Insertion — 222
  - 6.2.5 Algorithm for Solving CSC Violations — 223
- 6.3 Hazard-Free Logic Synthesis — 223
  - 6.3.1 Atomic Gate Implementation — 225
  - 6.3.2 Generalized C-Element Implementation — 226
  - 6.3.3 Standard C-Implementation — 230
  - 6.3.4 The Single-Cube Algorithm — 238
- 6.4 Hazard-Free Decomposition — 243
  - 6.4.1 Insertion Points Revisited — 245
  - 6.4.2 Algorithm for Hazard-Free Decomposition — 246
- 6.5 Limitations of Speed-Independent Design — 248
- 6.6 Sources — 249
- Problems — 251

7 Timed Circuits — 259
- 7.1 Modeling Timing — 260
- 7.2 Regions — 262
- 7.3 Discrete time — 265
- 7.4 Zones — 267
- 7.5 POSET Timing — 280
- 7.6 Timed Circuits — 289
- 7.7 Sources — 292
- Problems — 293

8 Verification — 295
- 8.1 Protocol Verification — 296
  - 8.1.1 Linear-Time Temporal Logic — 296
  - 8.1.2 Time-Quantified Requirements — 300
- 8.2 Circuit Verification — 303
  - 8.2.1 Trace Structures — 303

	8.2.2 Composition	305
	8.2.3 Canonical Trace Structures	308
	8.2.4 Mirrors and Verification	310
	8.2.5 Strong Conformance	312
	8.2.6 Timed Trace Theory	314
8.3	Sources	315
	Problems	316

9 Applications 321
   9.1 Brief History of Asynchronous Circuit Design   322
   9.2 An Asynchronous Instruction-Length Decoder   325
   9.3 Performance Analysis   329
   9.4 Testing Asynchronous Circuits   330
   9.5 The Synchronization Problem   332
       9.5.1 Probability of Synchronization Failure   334
       9.5.2 Reducing the Probability of Failure   335
       9.5.3 Eliminating the Probability of Failure   336
       9.5.4 Arbitration   340
   9.6 The Future of Asynchronous Circuit Design   341
   9.7 Sources   342
       Problems   346

Appendix A VHDL Packages 347
   A.1 nondeterminism.vhd   347
   A.2 channel.vhd   348
   A.3 handshake.vhd   355

Appendix B Sets and Relations 359
   B.1 Basic Set Theory   360
   B.2 Relations   362

References 365

Index 393

Preface

> An important scientific innovation rarely makes its way by gradually winning over and converting its opponents: it rarely happens that Saul becomes Paul. What does happen is that its opponents gradually die out and that the growing generation is familiarized with the idea from the beginning.
> —Max Planck

> I must govern the clock, not be governed by it.
> —Golda Meir

> All pain disappears, it's the nature of my circuitry.
> —nine inch nails

In 1969, Stephen Unger published his classic textbook on asynchronous circuit design. This book presented a comprehensive look at the asynchronous design methods of the time. In the 30 years hence, there have been numerous technical publications and even a few books [37, 57, 120, 203, 224, 267, 363, 393], but there has not been another textbook. This book attempts to fill this void by providing an updated look at asynchronous circuit design in a form accessible to a student who simply has some background in digital logic design.

An asynchronous circuit is one in which synchronization is performed without a global clock. Asynchronous circuits have several advantages over their synchronous counterparts, including:

1. *Elimination of clock skew problems.* As systems become larger, increasing amounts of design effort is necessary to guarantee minimal skew in the arrival time of the clock signal at different parts of the chip. In an asynchronous circuit, skew in synchronization signals can be tolerated.

2. *Average-case performance.* In synchronous systems, the performance is dictated by worst-case conditions. The clock period must be set to be long enough to accommodate the slowest operation even though the average delay of the operation is often much shorter. In asynchronous circuits, the speed of the circuit is allowed to change dynamically, so the performance is governed by the average-case delay.

3. *Adaptivity to processing and environmental variations.* The delay of a VLSI circuit can vary significantly over different processing runs, supply voltages, and operating temperatures. Synchronous designs have their clock rate set to allow correct operation under some allowed variations. Due to their adaptive nature, asynchronous circuits operate correctly under all variations and simply speed up or slow down as necessary.

4. *Component modularity and reuse.* In an asynchronous system, components can be interfaced without the difficulties associated with synchronizing clocks in a synchronous system.

5. *Lower system power requirements.* Asynchronous circuits reduce synchronization power by not requiring additional clock drivers and buffers to limit clock skew. They also automatically power-down unused components. Finally, asynchronous circuits do not waste power due to spurious transitions.

6. *Reduced noise.* In a synchronous design, all activity is locked into a very precise frequency. The result is nearly all the energy is concentrated in very narrow spectral bands at the clock frequency and its harmonics. Therefore, there is substantial electrical noise at these frequencies. Activity in an asynchronous circuit is uncorrelated, resulting in a more distributed noise spectrum and a lower peak noise value.

Despite all these potential advantages, asynchronous design has seen limited usage to date. Although there are many reasons for this, perhaps the most serious is a lack of designers with experience in asynchronous design. This textbook is a direct attempt at addressing this problem by providing a means for graduate or even undergraduate courses to be created that teach modern asynchronous design methods. I have used it in a course which includes both undergraduates and graduates. Lectures and other material used in this and future courses will be made available on our Web site: http://www.async.elen.utah.edu/book/. This book may also be used for self-study by engineers who would like to learn about modern asynchronous design methods. Each chapter includes numerous problems for the student to try out his or her new skills.

The history of asynchronous design is quite long. Asynchronous design methods date back to the 1950s and to two people in particular: Huffman and Muller. Every asynchronous design methodology owes its roots to one of these two men. Huffman developed a design methodology for what is known today as *fundamental-mode* circuits [170]. Muller developed the theoretical underpinnings of *speed-independent* circuits [279]. Unger is a member of the "Huffman School," so his textbook focused primarily on fundamental-mode circuit design with only a brief treatment of Muller circuits. Although I am a student of the "Muller School," in this book we present both design methods with the hope that members of both schools will grow to understand each other better, perhaps even realizing that the differences are not that great.

Since the early days, asynchronous circuits have been used in many interesting applications. In the 1950s and 1960s at the University of Illinois, Muller and his colleagues used speed-independent circuits in the design of the ILLIAC and ILLIAC II computers [46]. In the early days, asynchronous design was also used in the MU-5 and Atlas mainframe computers. In the 1970s at Washington University in St. Louis, asynchronous macromodules were developed [87]. These modules could be plugged together to create numerous special-purpose computing engines. Also in the 1970s, asynchronous techniques were used at the University of Utah in the design of the first operational dataflow computer [102, 103] and at Evans and Sutherland in design of the first commercial graphics system.

Due to the advantages cited above, there has been a resurgence of interest in asynchronous design. There have been several recent successful design projects. In 1989, researchers at Caltech designed the first fully asynchronous microprocessor [251, 257, 258]. Since that time, numerous other researchers have produced asynchronous microprocessors of increasing complexity [10, 13, 76, 134, 135, 138, 191, 259, 288, 291, 324, 379, 406]. Commercially, asynchronous circuits have had some recent success. Myranet uses asynchronous circuits coupled with *pipeline synchronization* [348] in their router design. Philips has designed numerous asynchronous designs targeting low power [38, 136, 192, 193]. Perhaps the most notable accomplishment to come out of this group is an asynchronous 80C51 microcontroller, which is now used in a fully asynchronous pager being sold by Philips. Finally, the RAPPID project at Intel demonstrated that a fully asynchronous instruction-length decoder for the x86 instruction set could achieve a threefold improvement in speed and a twofold improvement in power compared with the existing synchronous design [141, 142, 143, 144, 330, 367].

In the time of Unger's text, there were perhaps only a handful of publications each year on asynchronous design. As shown in Figure 0.1, this rate of publication continued until about 1985, when there was a resurgence of interest in asynchronous circuit design [309]. Since 1985, the publication rate has grown to well over 100 technical publications per year. Therefore,

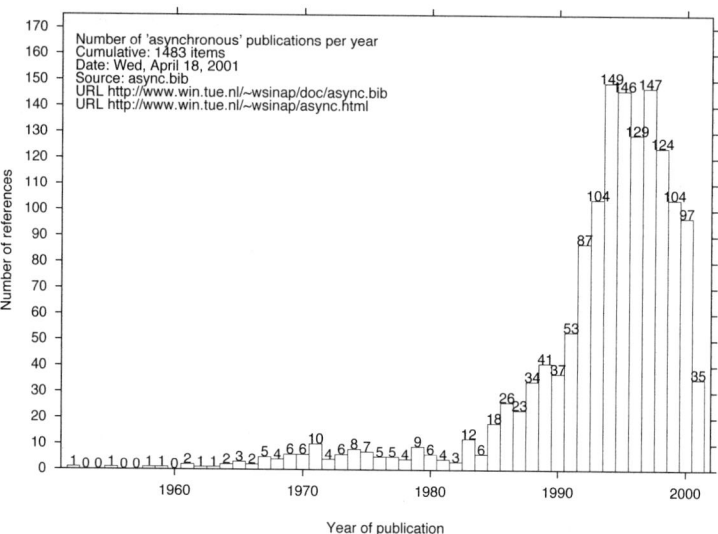

Fig. 0.1 Number of asynchronous publications per year.

although Unger did a superb job of surveying the field, this author has his work cut out for him. In the sources section at the end of each chapter, the interested reader is pointed to an extensive bibliography (over 400 entries) to probe deeper. Although an attempt has been made to give a flavor of the major design methodologies being developed and used, it is impossible even to reference every paper published on asynchronous design, as the number of entries in the asynchronous bibliography [309] now exceeds 1400. The interested reader should consult this bibliography and the proceedings from the recent symposiums on asynchronous circuits and systems [14, 15, 16, 17, 18, 19, 20].

The book is organized as follows. In Chapter 1 we introduce the asynchronous design problem through a small example illustrating the differences among the various timing models used. In Chapter 2 we introduce the concept of asynchronous communication and describe a methodology for specifying asynchronous designs using VHDL. In Chapter 3 we discuss various asynchronous protocols. In Chapter 4 we introduce graphical representations that are used for asynchronous design. In Chapter 5 we discuss Huffman circuits and in Chapter 6 we describe Muller circuits. In Chapter 7 we develop techniques for timing analysis and optimization which can lead to significant improvements in circuit quality. In Chapter 8 we introduce methods for the analysis and verification of asynchronous circuits. Finally, in Chapter 9 we give a brief discussion of issues in asynchronous application.

CHRIS J. MYERS

Salt Lake City, Utah

Acknowledgments

I am indebted to Alain Martin and Chuck Seitz of Caltech, who turned me onto asynchronous design as an undergraduate. I would also like to thank my graduate advisors, Teresa Meng and David Dill of Stanford University, who taught me alternative ways of looking at asynchronous design. My former officemate, Peter Beerel (USC), through numerous heated discussions throughout the years, has taught me much.

I would like to thank Erik Brunvand (Utah), Steve Nowick (Columbia), Peter Beerel (USC), Wendy Belluomini (IBM), Ganesh Gopalakrishnan (Utah), Ken Stevens (Intel), Charles Dike (Intel), Jim Frenzel (U. of Idaho), Steven Unger (Columbia), Dong-Ik Lee (KJIST), and Tomohiro Yoneda (Titech) for their comments and advice on earlier versions of this manuscript. I'm also grateful to the comments and ideas that I received from my graduate students: Brandon Bachman, Jie Dai, Hans Jacobson, Kip Killpack, Chris Krieger, Scott Little, Eric Mercer, Curt Nelson, Eric Peskin, Robert Thacker, and Hao Zheng. I would like to thank the students in my course on asynchronous circuit design in the spring of 2000 for putting up with the rough version of this text. I am grateful to Sanjin Piragic for drawing many of the figures in the book. Many other figures are due to `draw_astg` by Jordi Cortadella (UPC) and `dot` by Eleftherios Koutsofios and Stephen North (AT&T).

I would like especially to thank my family, Ching and John, for being patient with me while I wrote this book. Without their love and support, the book would not have been possible.

C.J.M.

Asynchronous Circuit Design

1
Introduction

Wine is bottled poetry.

—Robert Louis Stevenson

Wine gives courage and makes men more apt for passion.

—Ovid

I made wine out of raisins so I wouldn't have to wait for it to age.

—Steven Wright

In this chapter we use a simple example to give an informal introduction to many of the concepts and design methods that are covered in this book. Each of the topics in this chapter is addressed more formally in much more detail in subsequent chapters.

1.1 PROBLEM SPECIFICATION

In a small town in southern Utah, there's a little winery with a wine shop nearby. Being a small town in a county that thinks Prohibition still exists, there is only one wine patron. The wine shop has a single small shelf capable of holding only a single bottle of wine. Each hour, on the hour, the shopkeeper receives a freshly made bottle of wine from the winery which he places on the shelf. At half past each hour, the patron arrives to purchase the wine, making space for the next bottle of wine. Now, the patron has learned that it is very important to be right on time. When he has arrived early, he has found an empty shelf, making him quite irate. When he has arrived late, he has found that the shopkeeper drank the last bottle of wine to make room for the new bottle. The most frustrating experience was when he arrived at just the same time that the shopkeeper was placing the bottle on the shelf. In his excitement, he and the shopkeeper collided, sending the wine bottle to the floor, shattering it, so that no one got to partake of that lovely bottle of wine.

This synchronous method of wine shopping went on for some time, with all parties being quite happy. Then one day (in the mid-1980s), telephone service arrived in this town. This was a glorious invention which got the town really excited. The patron got a wonderful idea. He knew the winery could operate faster if only he had a way to purchase the wine faster. Therefore, he suggested to the shopkeeper, "Say, why don't you give me a call when the wine arrives?" This way he could avoid showing up too early, frustrating his fragile temperament. Shortly after the next hour, he received a call to pick up the wine. He was so excited that he ran over to the store. On his way out, he suggested to the shopkeeper, "Say, why don't you give the folks at the winery a call to tell them you have room for another bottle of wine?" This is exactly what the shopkeeper did; and wouldn't you know it, the wine patron got another call just 10 minutes later that a new bottle had arrived. This continued throughout the hour. Sometimes it would take 10 minutes to get a call while other times it would take as long as 20 minutes. There was even one time he got a call just five minutes after leaving the shop (fortunately, he lived very close by). At the end of the hour, he realized that he had drunk 5 bottles of wine in just one hour!

At this point, he was feeling a bit woozy, so he decided to take a little snooze. An hour later, he woke up suddenly quite upset. He realized that the phone had been ringing off the hook. "Oh my gosh, I forgot about the wine!" He rushed over, expecting to find that the shopkeeper had drunk several bottles of his wine, but to his dismay, he saw one bottle on the shelf with no empties laying around. He asked the shopkeeper, "Why did they stop delivering wine?" The shopkeeper said, "Well, when I did not call, they decided that they had better hold up delivery until I had a place on my shelf."

From that day forward, this asynchronous method of wine shopping became the accepted means of doing business. The winery was happy, as they sold more wine (on average). The shopkeeper's wife was happy, as the shopkeeper never had to drink the excess. The patron was extremely happy, as he could now get wine faster, and whenever he felt a bit overcome by his drinking indiscretion, he could rest easy knowing that he would not miss a single bottle of wine.

1.2 COMMUNICATION CHANNELS

One day a VLSI engineer stopped by this small town's wine shop, and he got to talking with the shopkeeper about his little business. Business was

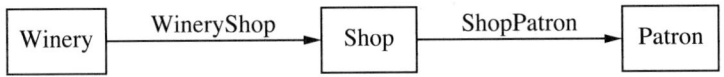

Fig. 1.1 Channels of communications.

good, but his wife kept bugging him to take that vacation to Maui he had been promising for years and years. He really did not know what to do, as he did not trust anyone to run his shop for him while he was away. Also, he was a little afraid that if he wasn't a little careful, the winery and patron might realize that they really did not need him and could deal with each other directly. He really could not afford that.

The VLSI engineer listened to him attentively, and when he was done announced, "I can solve all your problems. Let me design you a circuit!" At first, the shopkeeper was quite skeptical when he learned that this circuit would be powered by electricity (a new magical force that the locals had not completely accepted yet). The engineer announced, "It is really quite simple, actually." The engineer scribbled a little picture on his napkin (see Figure 1.1). This picture shows two *channels* of communication which must be kept synchronized. One is between the winery and the shop, and another is between the shop and the patron. When the winery receives a request from the shop over the *WineryShop communication channel*, the winery sends over a bottle of wine. This can be specified as follows:

```
Winery:process
begin
  send(WineryShop,bottle);
end process;
```

Note that the **process** statement implies that the *winery* loops forever sending bottles of wine. The wine patron, when requested by the shop over the *ShopPatron* communication channel, comes to receive a bottle of wine, which is specified as follows:

```
Patron:process
begin
  receive(ShopPatron,bag);
end process;
```

Now, what the shopkeeper does as the middleman (besides mark up the price) is provide a buffer for the wine to allow the winery to start preparing its next bottle of wine. This is specified as follows:

```
Shop:process
begin
  receive(WineryShop,shelf);
  send(ShopPatron,shelf);
end process;
```

1.3 COMMUNICATION PROTOCOLS

These three things together form a *specification*. The first two processes describe the types of folks the shopkeeper deals with (i.e., his *environment*). The last process describes the behavior of the shop.

After deriving a channel-level specification, it is then necessary to determine a *communication protocol* that implements the communication. For example, the shopkeeper calls the winery to "request" a new bottle of wine. After some time, the new bottle arrives, "acknowledging" the request. Once the bottle has been shelved safely, the shopkeeper can call the patron to "request" him to come purchase the wine. After some time, the patron arrives to purchase the wine, which "acknowledges" the request. This can be described as follows:

```
Shop: process
begin
  req_wine;    -- call winery
  ack_wine;    -- wine arrives
  req_patron;  -- call patron
  ack_patron;  -- patron buys wine
end process;
```

To build a VLSI circuit, it is necessary to assign signal wires to each of the four operations above. Two of the wires go to a device to place the appropriate phone call. These are called *outputs*. Another wire will come from a button that the wine delivery boy presses when he delivers the wine. Finally, the last wire comes from a button pressed by the patron. These two signals are *inputs*. Since this circuit is digital, these wires can only be in one of two states: either '0' (a low-voltage state) or '1' (a high-voltage state). Let us assume that the actions above are signaled by the corresponding wire changing to '1'. This can be described as follows:

```
Shop: process
begin
  assign(req_wine,'1');   -- call winery
  guard(ack_wine,'1');    -- wine arrives
  assign(req_patron,'1'); -- call patron
  guard(ack_patron,'1');  -- patron buys wine
end process;
```

The function `assign` used above sets a signal to a value. The function `guard` waits until a signal attains a given value. There is a problem with the specification given above in that when the second bottle of wine comes *req_wine* will already be '1'. Therefore, we need to reset these signals before looping back.

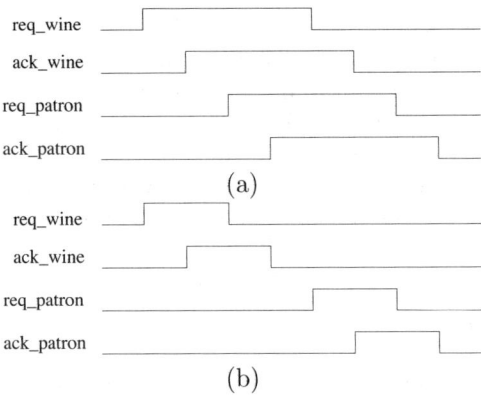

Fig. 1.2 (a) Waveform for a two-phase shop. (b) Waveform for a four-phase shop.

```
Shop_2Phase: process
begin
    assign(req_wine,'1');      -- call winery
    guard(ack_wine,'1');       -- wine arrives
    assign(req_patron,'1');    -- call patron
    guard(ack_patron,'1');     -- patron buys wine
    assign(req_wine,'0');      -- call winery
    guard(ack_wine,'0');       -- wine arrives
    assign(req_patron,'0');    -- call patron
    guard(ack_patron,'0');     -- patron buys wine
end process;
```

When *req_wine* changes from '0' to '1', a phone call is placed, and when it changes again from '1' to '0', another call is placed. We call this *transition signaling*. It is also known as *two-phase* or *two-cycle signaling*, for obvious reasons. A waveform showing the behavior of a two-phase shop is shown in Figure 1.2(a). Another alternative is given below.

```
Shop_4Phase: process
begin
    assign(req_wine,'1');      -- call winery
    guard(ack_wine,'1');       -- wine arrives
    assign(req_wine,'0');      -- reset req_wine
    guard(ack_wine,'0');       -- ack_wine resets
    assign(req_patron,'1');    -- call patron
    guard(ack_patron,'1');     -- patron buys wine
    assign(req_patron,'0');    -- reset req_patron
    guard(ack_patron,'0');     -- ack_patron resets
end process;
```

This protocol is called *level signaling* because a call is placed when the request signal is '1'. It is also called *four-phase* or *four-cycle signaling*, since it takes four transitions to complete. A waveform showing the behavior of a four-phase shop is shown in Figure 1.2(b). Although this protocol may appear to be a little more complex in that it requires twice as many transitions of signal wires, it often leads to simpler circuitry.

There are still more options. In the original protocol, the shop makes the calls to the winery and the patron. In other words, the shop is the *active* participant in both communications. The winery and the patron are the *passive* participants. They simply wait to be told when to act. Another alternative would be for the winery to be the active participant and call the shop when a bottle of wine is ready, as shown below.

```
Shop_PA: process
begin
  guard(req_wine,'1');    -- winery calls
  assign(ack_wine,'1');   -- wine is received
  guard(req_wine,'0');    -- req_wine resets
  assign(ack_wine,'0');   -- reset ack_wine
  assign(req_patron,'1'); -- call patron
  guard(ack_patron,'1');  -- patron buys wine
  assign(req_patron,'0'); -- reset req_patron
  guard(ack_patron,'0');  -- ack_patron resets
end process;
```

Similarly, the patron could be active as well and call when he has finished his last bottle of wine and requires another. Of course, in this case, the shopkeeper needs to install a second phone line.

```
Shop_PP: process
begin
  guard(req_wine,'1');    -- winery calls
  assign(ack_wine,'1');   -- wine is received
  guard(req_wine,'0');    -- req_wine resets
  assign(ack_wine,'0');   -- reset ack_wine
  guard(req_patron,'1');  -- patron calls
  assign(ack_patron,'1'); -- sells wine
  guard(req_patron,'0');  -- req_patron resets
  assign(ack_patron,'0'); -- reset ack_patron
end process;
```

Unfortunately, none of these specifications can be transformed into a circuit as is. Let's return to the initial four-phase protocol (i.e., the one labeled *Shop_4Phase*). Initially, all the signal wires are set to '0' and the circuit is supposed to call the winery to request a bottle of wine. After the wine has arrived and the signal *req_wine* and *ack_wine* have been reset, the state of the signal wires is again all '0'. The problem is that in this case the circuit must call the patron. In other words, when all signal wires are set to '0', the circuit is in a state of confusion. Should the winery or the patron be called at this point? We need to determine some way to clarify this. Considering again the initial four-phase protocol, this can be accomplished by *reshuffling* the order in which these signal wires change. Although it is important that the wine

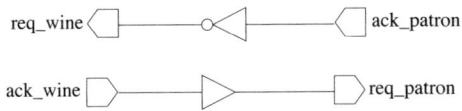

Fig. 1.3 Circuit for active/active shop.

arrives before the patron is called, exactly when the handshaking wires reset is less important. Rearranging the protocol as shown below allows the circuit always to be able to tell what to do. Also, the eager patron gets a call sooner. On top of that, it results in the very simple circuit shown in Figure 1.3.

```
Shop_AA_reshuffled: process
begin
    assign(req_wine,'1');      -- call winery
    guard(ack_wine,'1');       -- wine arrives
    assign(req_patron,'1');    -- call patron
    guard(ack_patron,'1');     -- patron buys wine
    assign(req_wine,'0');      -- reset req_wine
    guard(ack_wine,'0');       -- ack_wine resets
    assign(req_patron,'0');    -- reset req_patron
    guard(ack_patron,'0');     -- ack_patron resets
end process;
```

Alternatively, we could have reshuffled the protocol in which the shop passively waits for the winery to call but still actively calls the patron, as shown below. The resulting circuit is shown in Figure 1.4. The gate with a C in the middle is called a *Muller C-element*. When both of its inputs are '1', its output goes to '1'. Similarly, when both of its inputs are '0', its output goes to '0'. Otherwise, it retains its old value.

```
Shop_PA_reshuffled: process
begin
    guard(req_wine,'1');       -- winery calls
    assign(ack_wine,'1');      -- receives wine
    guard(ack_patron,'0');     -- ack_patron resets
    assign(req_patron,'1');    -- call patron
    guard(req_wine,'0');       -- req_wine resets
    assign(ack_wine,'0');      -- reset ack_wine
    guard(ack_patron,'1');     -- patron buys wine
    assign(req_patron,'0');    -- reset req_patron
end process;
```

Another curious thing about this protocol is that it waits for *ack_patron* to be '0', but it is '0' to begin with. A guard in which its expression is already satisfied simply passes straight through. We call that first guard *vacuous* because it does nothing. However, the second time around, *ack_patron* may actually have not reset at that point. Postponing this guard until this point

8   INTRODUCTION

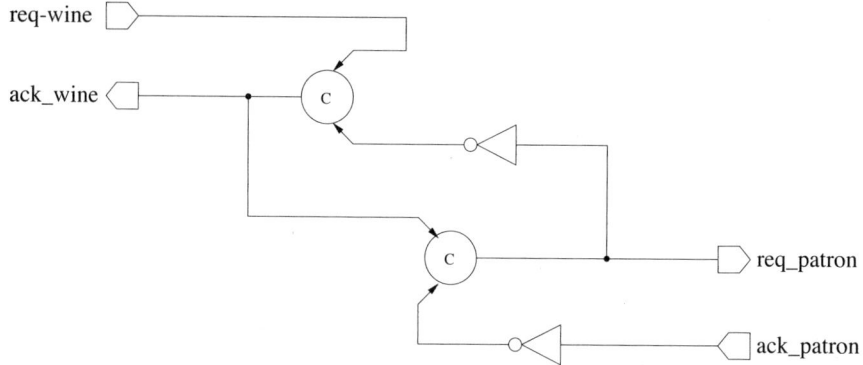

Fig. 1.4 Circuit for passive/active shop.

allows *ack_patron* to be reset concurrently with *req_wine* and *ack_wine* being set. This increased concurrency can potentially improve the performance.

1.4 GRAPHICAL REPRESENTATIONS

Before describing how these circuits are derived, let us first consider an alternative way of looking at these specifications using graphs. The first method is to use an *asynchronous finite state machine* (AFSM). As an example, consider the active/active protocol from above (see *Shop_AA_reshuffled*): It can be represented as an AFSM, as shown in Figure 1.5(a), or in a tabular form called a *Huffman flow table*, shown in Figure 1.5(b). In the state machine, each node in the graph represents a state, and each arc represents a state transition. The state transition is labeled with the value of the inputs needed to make a state transition (these are the numbers to the left of the "/"). The numbers to the right of the "/" represent what the outputs do during the state transition. Starting in state 0, if both *ack_wine* and *ack_patron* are '0', as is the case initially, the output *req_wine* is set to '1' and the machine moves into state 1. In state 1, the machine waits until *ack_wine* goes to '1', and then it sets *req_patron* to '1' and moves to state 2. The same behavior is illustrated in the Huffman flow table, in which the rows are the states and the columns are the input values (i.e., *ack_wine* and *ack_patron*). Each entry is labelled with the next state and next value of the outputs (i.e., *req_wine* and *req_patron*) for a given state and input combination. When the next state equals the current state, it is circled to indicate that it is stable.

Not all protocols can be described using an AFSM. The AFSM model assumes that inputs change followed by output and state changes in sequence. In the second design (see *Shop_PA_reshuffled*), however, inputs and outputs can change concurrently. For example, *req_wine* may be set to '0' while *req_patron* is being set to '1'. Instead, we can use a different graphical method called

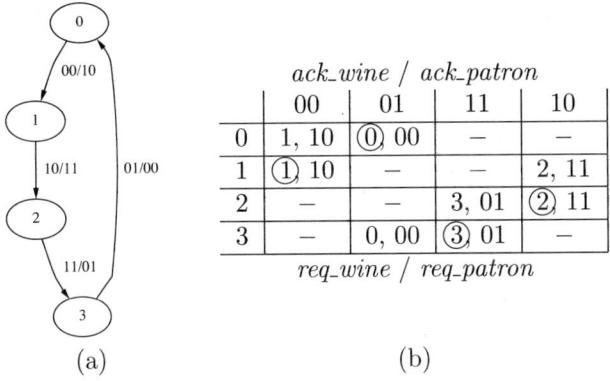

Fig. 1.5 (a) AFSM and (b) Huffman flow table for active/active shop (input/output vector is ⟨ack_wine, ack_patron⟩/⟨req_wine, req_patron⟩).

a *Petri net* (PN) to illustrate the behavior of the second design as shown in Figure 1.6(a).

In a Petri net, the nodes of the graph represent signal transitions. For example, *req_wine+* indicates that *req_wine* changes from '0' to '1'. Similarly, *req_wine−* indicates that *req_wine* changes from '1' to '0'. The arcs in this graph represent causal relationships between transitions. For example, the arc between *req_wine+* and *ack_wine+* indicates that *req_wine* must be set to '1' before *ack_wine* can be set to '1'. The little balls are called *tokens*, and a collection of tokens is called a *marking* of the Petri net. The initial marking is shown in Figure 1.6(a).

For a signal transition to occur, it must have tokens on all of its incoming arcs. Therefore, the only transition that may occur in the initial marking is *req_wine* may be set to '1'. After *req_wine* changes value, the tokens are removed from the incoming arcs and new tokens are placed on each outgoing arc. In this case, the token on the arc between *ack_wine−* and *req_wine+* would be removed, and a new token would be put on the arc between *req_wine+* and *ack_wine+*, as shown in Figure 1.6(b). In this marking, *ack_wine* can now be set to '1', and no other signal transition is possible. After *ack_wine* becomes '1', tokens are removed from its two incoming arcs and tokens are placed on its two outgoing arcs, as shown in Figure 1.6(c). In this new marking, there are two possible next signal transitions. Either *req_patron* will be set to '1' or *req_wine* will be set to '0'. These two signal transitions can occur in either order. The rest of the behavior of this circuit can be determined by similar analysis.

It takes quite a bit of practice to come up with a Petri-net model from any given word description. Another graphical model, called the *timed event/level* (TEL) *structure*, has a more direct correspondence with the word description. The TEL structure for the *Shop_PA_reshuffled* protocol is shown in Figure 1.7.

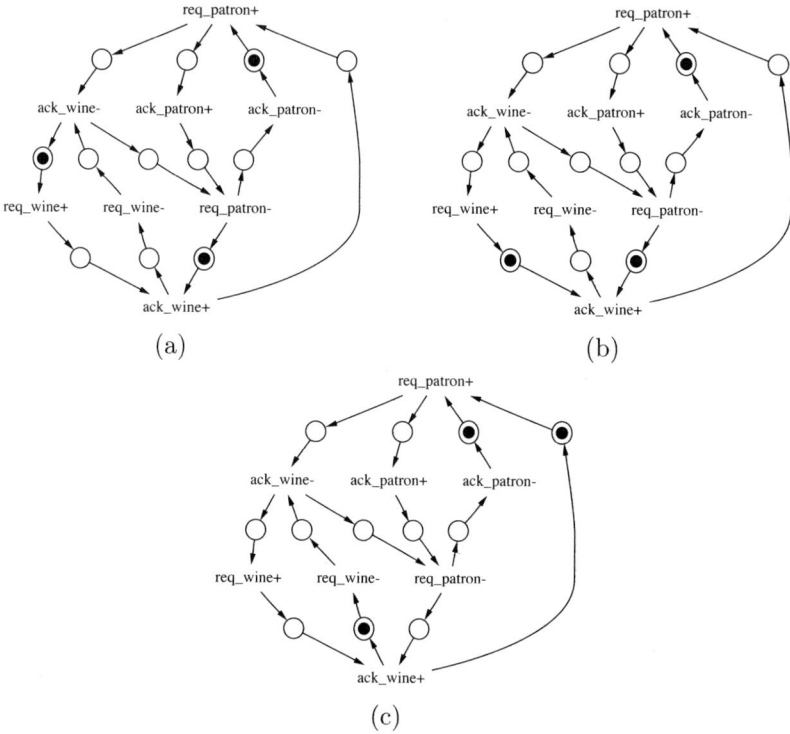

Fig. 1.6 PN for passive/active shop. (a) Initial marking, (b) after *req_wine* goes to '1', and (c) after *ack_wine* goes to '1'.

This TEL structure is composed of three processes that operate concurrently. There is one to describe the behavior of the winery, one to describe the behavior of the patron, and one to describe the behavior of the shop. The difference between a TEL structure and a Petri net is the ability to specify signal *levels* on the arcs. Each level expression corresponds to a `guard`, and each signal transition, or *event*, corresponds to an `assign` statement in the word description. There are four guards and four assign statements in the shop process, which correspond to four levels and four events in the graph for the shop. Note that the dashed arcs represent the initial marking and that in this initial marking all signals are '0'.

1.5 DELAY-INSENSITIVE CIRCUITS

Let's go back now and look at those circuits from before. How do we know they work correctly? Let's look again at our first circuit, redrawn in Figure 1.8

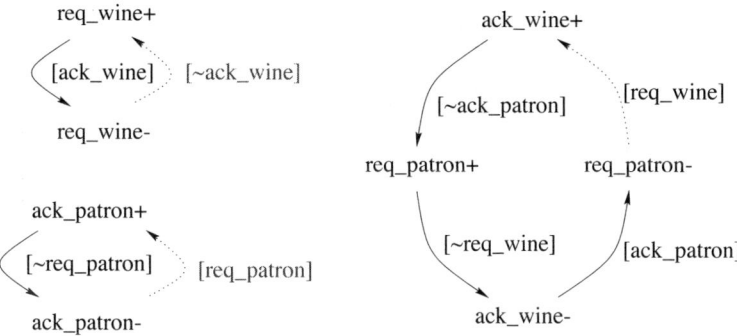

Fig. 1.7 TEL structure for passive/active shop.

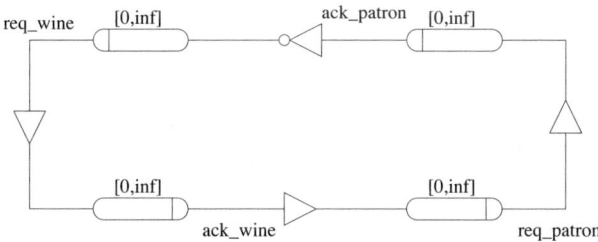

Fig. 1.8 Complete circuit for active/active shop.

as a *complete circuit*, that is, with circuit elements for the environment. Let's also add *delay elements* (cigar-shaped boxes labeled "[0,inf]") to represent an unbounded delay. Now, recall that in the initial state all signal wires are set to '0'. In this state, the only thing that can happen is that *req_wine* can be set to '1'. This is due to the '0' at the input of the inverter, which sets its output to '1'. We assume that this happens instantaneously. The delay element now randomly select a delay d between 0 and infinity. Let's say that it picks five minutes. The signal *req_wine* now changes to '1' after five minutes has elapsed. At that point, the output of the attached buffer also changes to '1', but this change does not affect *ack_wine* until a random delay later due to the attached delay element. If you play with this for awhile, you should be able to convince yourself that regardless of what delay values you choose, the circuit always behaves as you specified originally. We call such a circuit a *delay-insensitive circuit*. That is one whose correctness is independent of the delays of both the gates and the wires, even if these delays are unbounded. This mean that even during a strike of the grape mashers at the winery or when the patron is sleeping one off, the circuit still operates correctly. It is extremely robust.

Let's now look at the second circuit, redrawn with its environment in Figure 1.9. Again, in the initial state, the only transition which can occur is that

12   INTRODUCTION

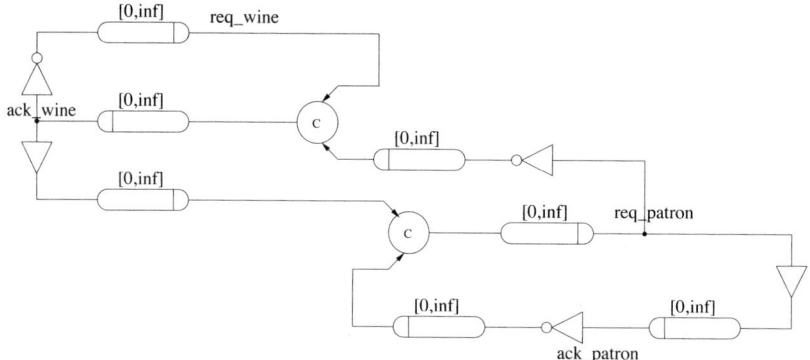

Fig. 1.9 Complete circuit for passive/active shop.

req_wine can be set to '1' after some arbitrary delay. After *req_wine* is '1', the next transition which can occur is that *ack_wine* can be set to '1' after some delay. When *ack_wine* is '1' though, there are two possible next transitions. Either *req_wine* can go '0' or *req_patron* can go to '1', depending on the choices made for the delay values in each delay element along the corresponding path. If you look at this circuit long enough, you can probably convince yourself that the circuit will behave as specified for whichever delay values are chosen each time around. This circuit is also delay-insensitive.

You may at this point begin to believe that you can always build a delay-insensitive circuit. Unfortunately, this is not the case. We are actually pretty fortunate with these two circuit designs. In general, if you use only single-output gates, this class of circuits is severely limited. In particular, you can only use buffers, inverters, and Muller C-elements to build delay-insensitive circuits. As an example, consider the circuit shown in Figure 1.10. It is a circuit implementation of a slightly modified version of our original four-phase protocol, where we have added a *state variable* to get rid of the state coding problem. The protocol is given below.

```
Shop_AA_state_variable:process
begin
    assign(req_wine,'1');    -- call winery
    guard(ack_wine,'1');     -- wine arrives
    assign(x,'1');           -- set state variable
    assign(req_wine,'0');    -- reset req_wine
    guard(ack_wine,'0');     -- ack_wine resets
    assign(req_patron,'1');  -- call patron
    guard(ack_patron,'1');   -- patron buys wine
    assign(x,'0');           -- reset state variable
    assign(req_patron,'0');  -- reset req_patron
    guard(ack_patron,'0');   -- ack_patron resets
end process;
```

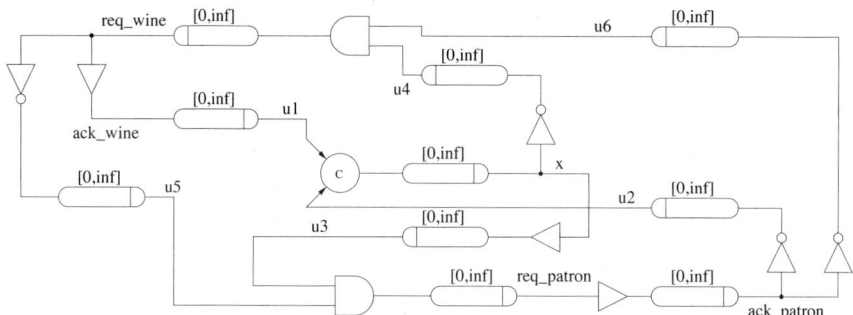

Fig. 1.10 Another complete circuit for active/active shop.

Assume that all signals are initially '0' except $u2$, $u4$, $u5$, and $u6$ which are '1'. Trace the following sequence of actions through the circuit:
req_wine+, ack_wine+, $x+$, req_wine-, ack_wine-, $req\_patron+$, $ack\_patron+$.
At this point, $u2$ and $u6$ are enabled to go to '0', but assume that the delay of $u2-$ is shorter than that of $u6-$. After $u2$ becomes '0', x can go to '0', and assume that it does so before $u6$ goes to '0'. After x becomes '0', $u4$ can go to '1'. If this happens before $u6-$, then req_wine can be enabled to go to '1'. Now, if $u6$ finally changes to '0', then req_wine no longer is enabled to change to '1'. In a real circuit what may happen is that req_wine may experience a small pulse, or glitch. Depending on the duration of the pulse, it may or may not be perceived by the winery as a request for wine. If it is perceived as a request, then when the true request for wine comes, after req_patron and ack_patron have been reset, this will be perceived as a second request. This may lead to a second bottle of wine being delivered before the first has been sold. This potentially catastrophic chain of events would surely spoil the shopkeeper's holiday in Maui. When there exists an assignment of delays in a circuit that allows a glitch to occur, we say the circuit has a *hazard*. Hazards must be avoided in asynchronous circuit design. The engineer discusses the hazard problem with his colleagues Dr. Huffman and Dr. Muller over dinner.

1.6 HUFFMAN CIRCUITS

The first person to arrive at dinner was Dr. Huffman. The first thing he did was redraw the original specification as an AFSM and a Huffman flow table, as shown in Figure 1.11. Note that in the flow table the changing output in each unstable state is made a "don't care," which will help us out later.

The first thing he noticed is that there are more states than necessary. States 0 and 1 are *compatible*. In each entry, either the next states and outputs are the same or a don't care is in one. Similarly, states 2 and 3 are compatible. When states are compatible, they can often be combined to

14 INTRODUCTION

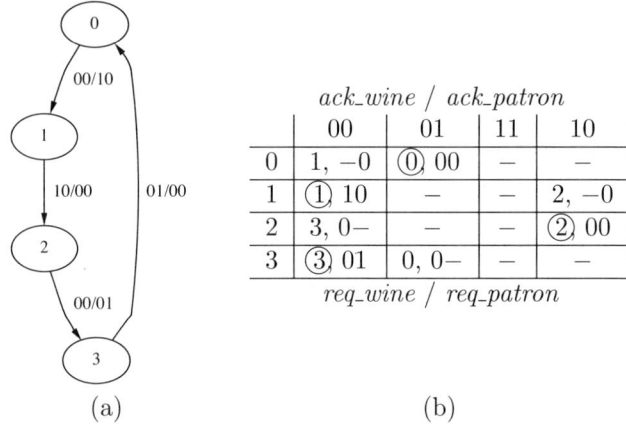

Fig. 1.11 (a) AFSM and (b) Huffman flow table for active/active shop (input/output vector is ⟨ack_wine, ack_patron⟩/⟨req_wine, req_patron⟩).

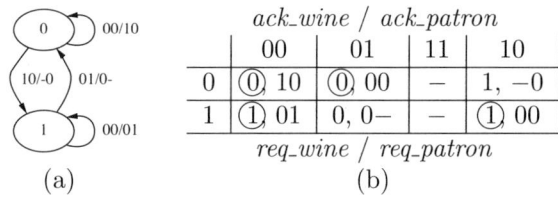

Fig. 1.12 Reduced (a) AFSM and (b) Huffman flow table for active/active shop.

reduce the number of states in the state machine. This process is called *state minimization*. The new AFSM and flow table are shown in Figure 1.12.

Next, we must choose a *state assignment* for our two states. This is a unique binary value for each state. Now, care must be taken when doing this for an asynchronous design, but for this simple design only a single bit is needed. Therefore, encoding state 0 with '0' and state 1 with '1' suffices here.

At this point, it is a simple matter to derive the circuit by creating and solving three *Karnaugh maps* (K-maps). There is one for each output and one for the state signal as shown in Figure 1.13. The circuit implementation is shown in Figure 1.14.

Let's compare this circuit with the one shown in Figure 1.10. The logic for *req_wine* and *req_patron* is identical. The logic for x is actually pretty similar to that of the preceding circuit except that Huffman sticks with simple AND and OR gates. "Muller's C-element is cute, but it is really hard to find in any of the local electronics shops" says Dr. Huffman. Also, all of Huffman's delay elements have an upper bound, U. Huffman assumes a *bounded gate and wire delay model*. Huffman's circuit is also not closed. That is, he does

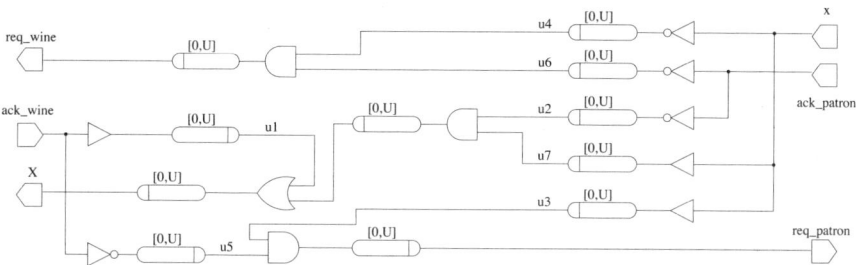

Fig. 1.13 K-maps for active/active shop.

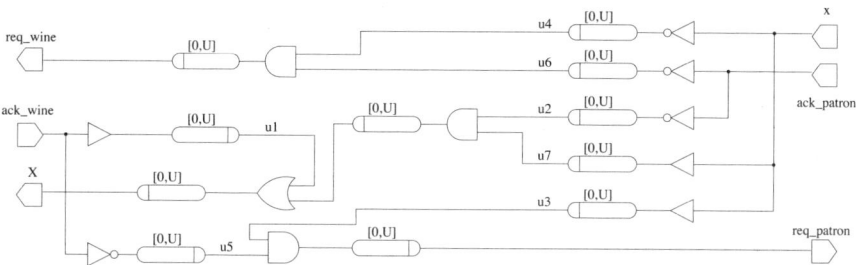

Fig. 1.14 Huffman's circuit for active/active shop.

not explicitly show the relationships between the outputs and inputs, which is not necessary in his method. Huffman also splits the feedback into an output signal, X, and an input signal, x.

These are the obvious differences. There are also some not so obvious ones: namely, Huffman's assumptions on how this circuit will be operated. First, Huffman assumes that the circuit is operated in what is called *single-input-change fundamental mode*. What this means is that the environment will apply only a single input change at a time; it will not apply another until the circuit has stabilized. In other words, the operation of the circuit is as follows: An input changes, outputs and next-state signals change, the next state is fed back to become the current state, and then a new input can arrive. To maintain this order, it may not only be necessary to slow down the environment, but it may also be necessary to delay the state signal change from being fed back too soon by adding a delay between X and x.

Consider again the bad trace. Again, the following sequence can happen:

$$req\_wine+,\ ack\_wine+,\ X+,\ x+,\ req\_wine-,$$
$$ack\_wine-,\ req\_patron+,\ ack\_patron+.$$

At this point, $u2$ and $u6$ are again enabled to go to '0', and let's assume that the delay of $u2-$ is faster than that of $u6-$. After $u2$ becomes '0', X can go to '0' before $u6$ goes to '0'. However, in this case, we have added sufficient delay in the feedback path such that we do not allow x to change to '0' until we have ensured that the circuit has stabilized. In other words, as long as the delay in the feedback path is greater than U, the glitch does not occur.

16   INTRODUCTION

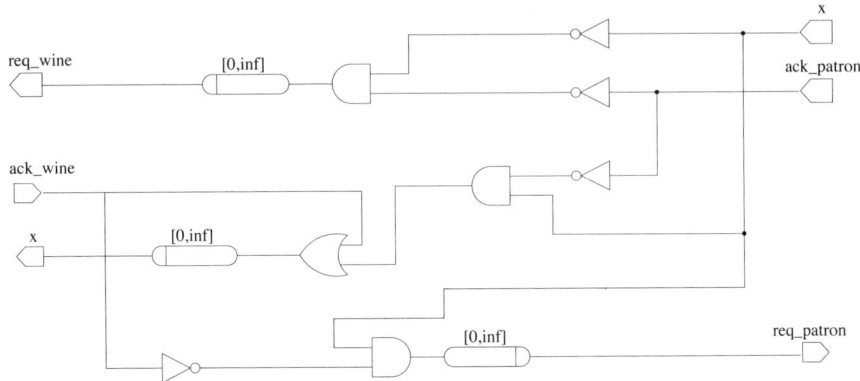

Fig. 1.15   Muller's circuit for active/active shop.

1.7   MULLER CIRCUITS

"What kind of crazy ideas have you been putting in this poor man's head?",

"Well, if it isn't Dr. Muller. You're late as usual," announced Dr. Huffman.

"You know me, I don't believe in bounding time. Let me take a look at what you got there," said Dr. Muller. He stared at the napkin and made some marks on it. "Now, that's much better. Take a look at this" (see Figure 1.15).

Muller's circuit looks similar to Huffman's, but there are some important differences in Muller's model. First, notice that he removed the delay elements from the wires, leaving only a single delay element on each output and next-state signal. He also changed the upper bound of this delay element to infinity. Finally, he changed the signal X to x. In his circuit, Muller has not put any restrictions on the order that inputs, outputs, and state signals change except they must behave according to the protocol. This model is called an *unbounded gate delay* model. Note that in this model, Muller assumes that wire delays are negligible. This means that whenever a signal changes value, all gates it is connected to will see that change immediately. For example, *ack_wine* and *ack_patron* each fork to two other gates while x forks to three. These forks are called *isochronic forks*, meaning that they have no delay. This delay model is called *speed-independent*. A similar model where only certain forks are isochronic is called *quasi-delay insensitive*.

Consider again the bad trace. The following sequence can still happen: *req_wine*+, *ack_wine*+, x+, *req_wine*−, *ack_wine*−, *req_patron*+, *ack_patron*+. At this point, the gates for *req_wine* and x see the change in *ack_patron* at the same time, due to the isochronic fork. Therefore, when x goes to '0', the effect of *ack_patron* being '1' is already felt by *req_wine*, so it does not glitch to '1'. The circuit is hazard-free under Muller's model as well, and he did not need to determine any sort of delay for the state variable feedback path.

"Those isochronic forks can be tricky to design," insisted Dr. Huffman.

This is true. Both models require some special design. By the way, not all the forks actually need to be isochronic. In particular, you can put different delays on each of the branches of the wire fork for x and the circuit still operates correctly.

1.8 TIMED CIRCUITS

The engineer returned from the bar where he had been talking to the wine patron and the head of production from the winery. He learned a few interesting things which can be used to optimize the circuit. First, he learned that with the winery's new wine production machine, they are always well ahead of schedule. In fact, since they are only one block away from the wine shop, they have guaranteed delivery within 2 to 3 minutes after being called. The patron lives about five blocks away and it takes him at least 5 minutes to get to the shop after being called. If he is busy sleeping one off, it may take him even longer. Finally, the circuit delays will all be very small, certainly less than 1 minute. Using this delay information and doing a bit of reshuffling, he came up with this description of the circuit and its environment:

```
Shop_AA_timed: process
begin
    assign(req_wine,'1',0,1);    -- call winery
    assign(req_patron,'1',0,1);  -- call patron
    -- wine arrives and patron arrives
    guard_and(ack_wine,'1',ack_patron,'1');
    assign(req_wine,'0',0,1);
    assign(req_patron,'0',0,1);
    -- wait for ack_wine and ack_patron to reset
    guard_and(ack_wine,'0',ack_patron,'0');
end process;
winery: process
begin
    guard(req_wine,'1');         -- wine requested
    assign(ack_wine,'1',2,3);    -- deliver wine
    guard(req_wine,'0');
    assign(ack_wine,'0',2,3);
end process;
patron: process
begin
    guard(req_patron,'1');       -- shop called
    assign(ack_patron,'1',5,inf); -- buy wine
    guard(req_patron,'0');
    assign(ack_patron,'0',5,7);
end process;
```

The assignment function now takes two additional parameters, which are the lower and upper bounds on the delay the assignment takes to complete.

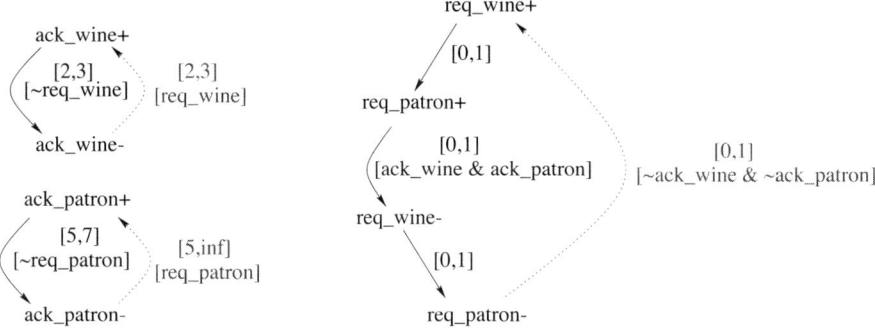

Fig. 1.16 TEL structure for active/active shop.

The **guard_and** function requires multiple signals to attain a given value before the circuit can progress. Notice that in this protocol, the patron is called after the wine is requested but before the wine arrives. This improves the performance of the system by allowing more concurrency. As we will see, timing assures that the patron does not arrive too early.

The TEL structure for this specification is shown in Figure 1.16. Assume that all signals are initially low. In the TEL structure, there are timing annotations on each arc. For example, *req_wine* must go to '1' within 0 to 1 minute from the initial state. Once *req_wine* has gone to '1', *ack_wine* can go to '1' since its level expression is now true. However, it must wait at least 2 minutes and will change within 3 minutes. The signal *req_patron* is also enabled to change and will do so within 0 to 1 minute. Therefore, we know that *req_patron* will change first. After *req_patron* has gone to '1', *ack_patron* is now enabled to change to '1', but it must wait at least 5 minutes. Therefore, we know that *ack_wine* changes next. In other words, we know that the wine will arrive before the patron arrives—a very important property of the system.

From this TEL structure, we derive the state graph in Figure 1.17. In the initial state, all signals are 0, but *req_wine* is labeled with an R to indicate that it is enabled to rise. Once *req_wine* rises, we move to a new state where *req_patron* and *ack_wine* are both enabled to rise. However, as mentioned before, the only possible next-state transition is on *req_patron* rising.

To get a circuit, a K-map is created for each output with columns for each input combination and rows for each output combination. A 1 is placed in each entry, corresponding to a state where the output is either R or 1, a 0 in each entry where the output is F or 0, and a − in the remaining entries. From the maps in Figure 1.18, the simple circuit in Figure 1.19 is derived. Note that the signal *ack_wine* is actually no longer needed.

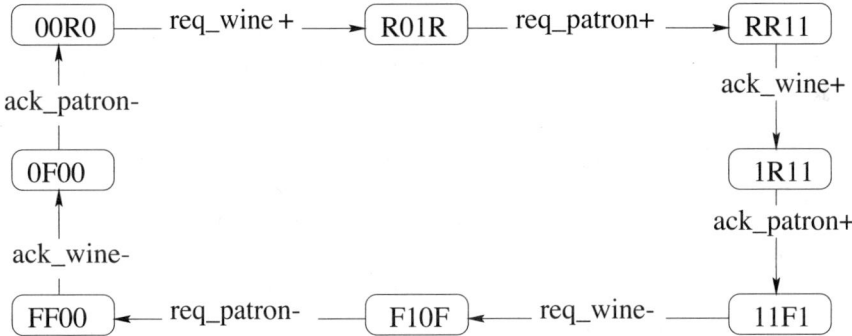

Fig. 1.17 State graph for active/active shop (with state vector $\langle ackwine, ackpatron, reqwine, reqpatron \rangle$).

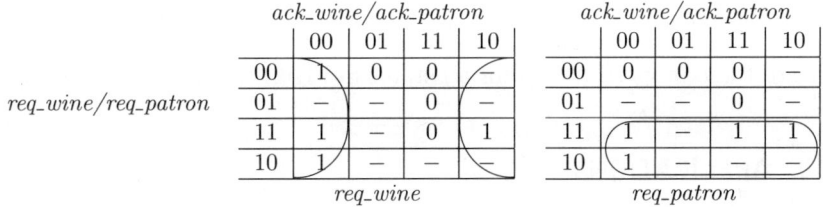

Fig. 1.18 K-maps for active/active shop.

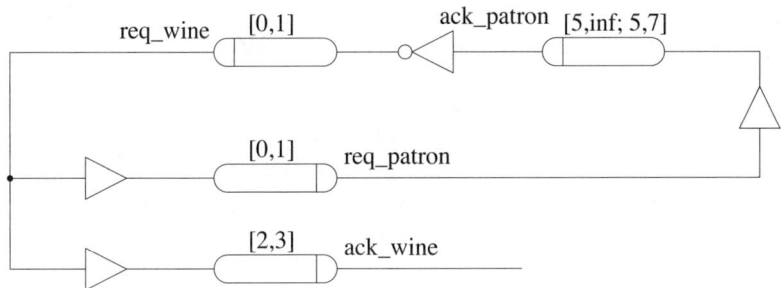

Fig. 1.19 Timed circuit for active/active shop.

1.9 VERIFICATION

The timed circuit made some pretty aggressive assumptions. How does one know if the circuit operates correctly? The first thing to do is to simulate the circuit for some example situations and check that all parties end up happy. This process is *validation*. However, to be really sure, one must exhaustively check all possible situations. This process is *formal verification*. First, one can check that the circuit always does what it is specified to do. In this case, we say that a circuit *conforms* to its specification. In other words, the circuit should follow the protocol. However, this may not be sufficient for the shopkeeper to rest easy on the Maui beaches. For example, in the modified protocol, if the timing assumptions are violated, it is possible for the patron to arrive before the wine. Knowing how irate he gets when this happens, this is clearly not acceptable. Therefore, one should enumerate the properties that the circuit should have and check that the protocol with the timing assumptions satisfies these properties. This process is called *model checking*. Two such properties might be these:

1. The wine arrives before the patron.

2. When the wine is requested, it eventually arrives.

Both of these properties hold for all states in Figure 1.17.

1.10 APPLICATIONS

With all these design alternatives, how does one know which one to choose? You may want to choose the fastest. In a synchronous design, you simply determine the worst-case delay and set your clock cycle to match. Recall that in synchronous wine shopping that the patron received one bottle of wine each hour independent of the speed of production. In an asynchronous circuit, however, it is not that simple to determine performance since you must consider average performance. Let us assume that the delays are those used in the timed circuit and are uniformly distributed over their range, except that the patron's delay in buying the wine actually ranges uniformly between 5 and 10 minutes (it may be more, but that is pretty unlikely). If this is the case, we find that the protocol used by Huffman and Muller has a cycle time of 21.5 minutes on average, while our original one (*ShopAA_reshuffled*) had a cycle time of 20.6 minutes. These means that for these asynchronous designs the patron will receive on average 3 bottles of wine per hour, a substantial improvement over synchronous shopping. The timed circuit's cycle time is only 15.8 minutes. This means that on average the patron will get even one more bottle of wine every hour using the timed circuit!

1.11 LET'S GET STARTED

"Anyway, this is enough talk about work for now. What do you say that we order some dinner?" said the engineer.

"This is just getting interesting. I would like to learn more," said the shopkeeper.

"Why don't you attend our seminar at the local university?" replied the engineer.

Did the shopkeeper attend the seminar? Did he and his wife ever get that trip to Maui? Did they ever get their dinner? To learn more, you must read the next chapter.

1.12 SOURCES

Detailed references on the material in this chapter are given at the end of subsequent chapters when these topics are discussed. The idea of modeling communicating processes through send and receive actions on channels originated with Hoare's *communicating sequential processes* (CSP)[166, 167] and was adapted to the specification of asynchronous circuits by Martin [254]. The Huffman flow table is described in [170] (republished later in [172]). Petri nets were introduced in [312], and they were first adapted to the specification of asynchronous circuits by Seitz [343]. TEL structures were introduced by Belluomini and Myers [34].

In [253], Martin proved that when the gate library is restricted to single-output gates, the class of circuits that can be built using a purely delay-insensitive (DI) model is severely limited. Namely, these circuits can only utilize inverters and *Muller C-elements*. This result and the examples contained in this paper inspired much of the discussion in this chapter. Similar results appeared in [54]. This paper also shows that set-reset latches, C-elements, and toggles cannot be implemented delay insensitively using basic gates (i.e., combinational gates). Leung and Li extend these results and show that when a circuit is not closed (i.e., some input comes from an outside environment and not another gate in the circuit), other gates, such as AND and OR gates, can be used in a limited way [230]. They also show that arbitrary DI behaviors can be implemented using a set of circuit elements that includes an inverter, fork, C-element, toggle, merge, and asynchronous demultiplexer.

Techniques using the fundamental-mode assumption originated with Huffman [170, 172]. The speed-independent model was first proposed by Muller and Bartky [279], and it was used in the design of the ILLIAC and ILLIAC II [46]. In [282], Myers and Meng introduced the first synthesis method which utilized a timed circuit model to optimize the implementation. Earlier, however, most practical asynchronous designs certainly utilized timing assumptions in an ad hoc fashion to improve the implementation.

Problems

1.1 Describe in words the behavior of the channel-level process given below.

```
Shop:process
begin
  receive(WineryShop,bottle1);
  receive(WineryShop,bottle2);
  send(ShopPatron,bottle2);
  send(ShopPatron,bottle1);
end process;
```

1.2 In this problem we design a Huffman circuit for the shop when the winery is passive while the patron is active.

1.2.1. Write a process for a four-phase active/passive shop.

1.2.2. A reshuffled version of an active/passive shop is shown below. Give an AFSM and Huffman flow table that describe its behavior.

```
Shop_AP_reshuffled:process
begin
  guard(req_patron,'1');   -- patron calls
  assign(req_wine,'1');    -- call winery
  guard(ack_wine,'1');     -- wine arrives
  assign(req_wine,'0');    -- reset req_wine
  guard(ack_wine,'0');     -- ack_wine resets
  assign(ack_patron,'1');  -- sells wine
  guard(req_patron,'0');   -- req_patron resets
  assign(ack_patron,'0');  -- reset ack_patron
end process;
```

1.2.3. Combine compatible rows and make a state assignment.

1.2.4. Use K-maps to find a Huffman circuit implementation.

1.3 In this problem we design a Muller circuit for the shop when the winery is passive while the patron is active.

1.3.1. Draw a TEL structure for the reshuffled active/passive shop below.

```
Shop_AP_reshuffled:process
begin
  assign(req_wine,'1');    -- call winery
  guard(ack_wine,'1');     -- wine arrives
  guard(req_patron,'1');   -- patron calls
  assign(ack_patron,'1');  -- sells wine
  assign(req_wine,'0');    -- reset req_wine
  guard(ack_wine,'0');     -- ack_wine resets
  guard(req_patron,'0');   -- req_patron resets
  assign(ack_patron,'0');  -- reset ack_patron
end process;
```

1.3.2. Find the state graph from the TEL structure.

1.3.3. Use K-maps to derive a Muller circuit implementation.

2
Communication Channels

> Inform all the troops that communications have completely broken down.
> —Ashleigh Brilliant

> This "telephone" has too many shortcomings to be seriously considered as a means of communication. The device is inherently of no value to us.
> —Western Union internal memo, 1876

In this chapter we describe a method for the specification of asynchronous systems at a high level through the use of a communication channel model. Any hardware system, synchronous or asynchronous, can be thought of as a set of concurrently operating processes. These processes periodically must communicate data between each other. For example, if one process is a register file and another is an ALU, the register file must communicate operands to the ALU, and the ALU must communicate a result back to the register file. In the *channel model*, this communication takes places when one process attempts to send a message along a *channel* while another is attempting to receive a message along the same channel. We have implemented a package in VHDL to provide a channel abstraction. In this chapter we give a brief overview of VHDL and how to use our package. Since it is beyond the scope of this book to teach VHDL, we primarily provide templates that will allow you to use VHDL to simulate asynchronous systems without needing a thorough understanding of the language.

2.1 BASIC STRUCTURE

Each module that you specify using our channel model will have the same basic structure. A channel is simply a point-to-point means of communication between two concurrently operating processes. One process uses that channel to send data to the other process. As an example, the VHDL model for the wine shop example is given below.

```vhdl
----------------------
-- wine_example.vhd
----------------------
library ieee;
use ieee.std_logic_1164.all;
use ieee.std_logic_arith.all;
use ieee.std_logic_unsigned.all;
use work.nondeterminism.all;
use work.channel.all;
entity wine_example is
end wine_example;
architecture behavior of wine_example is
  type wine_list is (cabernet, merlot, zinfandel,
                     chardonnay, sauvignon_blanc,
                     pinot_noir, riesling, bubbly);
  signal wine_drunk:wine_list;
  signal WineryShop:channel:=init_channel;
  signal ShopPatron:channel:=init_channel;
  signal bottle:std_logic_vector(2 downto 0):="000";
  signal shelf:std_logic_vector(2 downto 0);
  signal bag:std_logic_vector(2 downto 0);
begin
winery:process
begin
  bottle <= selection(8,3);
  wait for delay(5,10);
  send(WineryShop,bottle);
end process winery;
shop:process
begin
  receive(WineryShop,shelf);
  send(ShopPatron,shelf);
end process shop;
patron:process
begin
  receive(ShopPatron,bag);
  wine_drunk <= wine_list'val(conv_integer(bag));
end process patron;
end behavior;
```

Let's consider each part of the VHDL model separately. First, there is a comment indicating the name of the file. All comments in VHDL begin with "−−" and continue until the end of the line. The next section of code indicates what other libraries and packages are used by this file. These are like include statements in C or C++. You should always use the first line, which states that you want access to the ieee library. This a very useful standard library. For example, the next three lines include packages from this library. You probably also always want to use these three lines, as these packages define and implement the std_logic data type and arithmetic operations on them. You will use this data type a lot. The last two lines include the packages nondeterminism and channel, which can be found in Appendix A. The nondeterminism package defines some functions to generate random delays and random selections for simulation. The channel package includes a definition of the channel data type and operations on it, such as *send* and *receive*.

The next part is called the *entity*. The entity describes the interface of the module being designed. In this case, it defines the *wine_example*. In this example we are designing a *closed system* (i.e., no inputs or outputs), so our entity is very simple. We will introduce the syntax for more complex entity descriptions when we get to structural VHDL below.

The next part is called the *architecture*. This is used to define the behavior of the entity being designed. For each entity, there can be multiple architectures describing either different implementations or the same implementation at different levels of abstraction. The name of the architecture is identified in the first line (i.e., *behavior*) and the entity it corresponds to (i.e., *wine_example*).

The architecture has two parts: the *declaration section* and the *concurrent statement section*. In the declaration section, we can define *types*, *signals*, or, as we will see later, *components*. For example, we have defined an *enumerated type* called *wine_list*, which is a list of the different types of wine produced at the winery. We can then declare a signal *wine_drunk* of this type. A signal represents a wire (or possibly a collection of wires) in a design.

For this example, we have also defined two channels for communication. The *WineryShop* channel is used for delivering bottles of wine to the shop and the *ShopPatron* channel is used for selling bottles of wine to the patron. Both channels are initialized in the declaration section by assignment to the return value of the function *init_channel*.

The next three signals are used by the winery, shop, and patron to keep track of what type of wine they have. These signals are of type *std_logic_vector*. The type *std_logic* is a nine-valued enumerated type where the values represent various states of a signal wire (see Table 2.1). In addition to the strong logic values '0' and '1', there is a strong unknown value 'X', which indicates that multiple drivers are forcing the wire to opposite values. There are also weak values which can be overpowered by strong values. The last three indicate a wire that is uninitialized, 'U'; high impedance,'Z'; or don't care, '−'. The table also shows the resolution of the value if two modules drive the same wire

Table 2.1 Std_logic values.

	Meaning	'U'	'X'	'0'	'1'	'Z'	'W'	'L'	'H'	'-'
'U'	Uninitialized	'U'	'U'	'U'	'U'	'U'	'U'	'U'	'U'	'U'
'X'	Forcing unknown	'U'	'X'	'X'	'X'	'X'	'X'	'X'	'X'	'X'
'0'	Forcing 0	'U'	'X'	'0'	'X'	'0'	'0'	'0'	'0'	'X'
'1'	Forcing 1	'U'	'X'	'X'	'1'	'1'	'1'	'1'	'1'	'X'
'Z'	High impedance	'U'	'X'	'0'	'1'	'Z'	'W'	'L'	'H'	'X'
'W'	Weak unknown	'U'	'X'	'0'	'1'	'W'	'W'	'W'	'W'	'X'
'L'	Weak 0	'U'	'X'	'0'	'1'	'L'	'W'	'L'	'W'	'X'
'H'	Weak 1	'U'	'X'	'0'	'1'	'H'	'W'	'W'	'H'	'X'
'-'	Don't care	'U'	'X'	'X'	'X'	'X'	'X'	'X'	'X'	'X'

to different values. A *std_logic_vector* is an array of *std_logic* signals. We are encoding the type of wine using 3-bit-wide *std_logic_vectors*. The bottle going out of the winery is initialized to "000", which corresponds to the 0th type of wine (i.e., a bottle of cabernet).

The concurrent statement section starts with the *begin* statement, and it is where the behavior of the module is defined. The only concurrent statement which we introduce at this time is the *process statement*. In this example, there are three processes which operate concurrently. There is one for the winery, one for the shop, and another for the patron. Each process statement begins with an optional label and the keyword *process*. The statements within the process are executed sequentially.

The behavior of the *winery* begins by randomly selecting a type of wine to produce using the *selection* function. This function is defined in the *nondeterminism* package, and it takes two integer parameters. The first is the number of choices, and the second is the size of the *std_logic_vector* to return. The next step is that the winery waits for some random time between 5 and 10 minutes until it is ready to transmit another bottle of wine. This is accomplished using the *delay* function which is also defined in the *nondeterminism* package. This function takes two integer parameters which set the lower and upper bound for the range of possible delay values. Finally, the winery sends its bottle to the shop with the procedure call *send*. This procedure has two parameters: a channel to communicate on and a *std_logic* or *std_logic_vector* containing the data to be transmitted. The *send* procedure is defined in the channel package. In this example, it will wait until the shop is ready to receive the wine and then delivers it.

The behavior of the *shop* begins by receiving a bottle of wine from the winery with the procedure call *receive*. This procedure also has two parameters: a channel to communicate on and a *std_logic* or *std_logic_vector* where the data is to be copied upon reception. The *receive* procedure is also defined in the channel package. In this example, it waits until the winery has a bottle of wine to deliver; then the shop accepts delivery of it. After receiving the wine, the shop sends it to the patron over the *ShopPatron* channel.

The behavior of the *patron* begins by receiving a bottle of wine which it then identifies (probably with a small sip as the winery does not label its wine). To do this in VHDL, the *bag*, which is a *std_logic_vector*, must first be converted into an integer using the *conv_integer* function. This integer is used as an index to the function *wine_list*'**val**. As mentioned before, *wine_list* is an enumerated type, and **'val** is an *attribute*. When the **'val** attribute is combined with an enumerated type, it produces a function which takes an integer and returns the value of that position in the enumerated type. For example, 0 would return cabernet and 1 would return merlot.

2.2 STRUCTURAL MODELING IN VHDL

Whereas we could model any system using a single entity/architecture pair, it would get quite cumbersome for large designs. This is especially true when many copies of the same process are needed. In this section we introduce a way of specifying separate processes within different entity/architecture pairs and then composing them together using a top-level structural architecture. Let's begin by looking at the entity/architecture pair which specifies just the behavior of the shop.

```
-- shop.vhd
library ieee;
use ieee.std_logic_1164.all;
use work.nondeterminism.all;
use work.channel.all;
entity shop is
   port(wine_delivery:inout channel:=init_channel;
        wine_selling:inout channel:=init_channel);
end shop;
architecture behavior of shop is
   signal shelf:std_logic_vector(2 downto 0);
begin
shop:process
begin
   receive(wine_delivery,shelf);
   send(wine_selling,shelf);
end process shop;
end behavior;
```

The key difference between this model and the previous one is seen in the *port declarations* in the entity. As mentioned before, the entity is used to specify the interface of a module. Considering the shop separately, there are now two channels at the interface which are declared as ports in the entity. Each port declaration is given a unique name. In this example, the ports are *wine_delivery* and *wine_selling*. Each port is also given a *mode* setting the direction of the data flow. The modes can be *in*, *out*, or *inout*. In this example, the channels are given a mode of *inout*. Although it may appear that *wine_delivery* should be *in* and *wine_selling* should be *out*, they are actually *inout* because while data flows only one way, the control for the communication flows in both directions. The next part of the port declaration is the port types. In our example, both ports are of type *channel*. Finally, each port declaration is allowed an optional initialization. In this case, both channels are initialized with calls to the function *init_channel*. The only other differences are that the architecture is linked to the entity *shop* and includes only the signal and process for the *shop*. The entity/architecture pairs for the *winery*, the *patron*, and the composition with the shop are given below.

```
-- winery.vhd
library ieee;
use ieee.std_logic_1164.all;
use work.nondeterminism.all;
use work.channel.all;
entity winery is
  port(wine_shipping:inout channel:=init_channel);
end winery;
architecture behavior of winery is
  signal bottle:std_logic_vector(2 downto 0):="000";
begin
winery:process
begin
  bottle <= selection(8,3);
  wait for delay(5,10);
  send(wine_shipping,bottle);
end process winery;
end behavior;

-- patron.vhd
library ieee;
use ieee.std_logic_1164.all;
use ieee.std_logic_arith.all;
use ieee.std_logic_unsigned.all;
use work.nondeterminism.all;
use work.channel.all;
entity patron is
  port(wine_buying:inout channel:=init_channel);
end patron;
architecture behavior of patron is
  type wine_list is (cabernet, merlot, zinfandel,
                     chardonnay, sauvignon_blanc,
                     pinot_noir, riesling, bubbly);
  signal wine_drunk:wine_list;
  signal bag:std_logic_vector(2 downto 0);
```

```
begin
patron:process
begin
  receive(wine_buying,bag);
  wine_drunk <= wine_list'val(conv_integer(bag));
end process patron;
end behavior;

-- wine_example2.vhd
library ieee;
use ieee.std_logic_1164.all;
use work.nondeterminism.all;
use work.channel.all;
entity wine_example is
end wine_example;
architecture structure of wine_example is
  component winery
    port(wine_shipping:inout channel);
  end component;
  component shop
    port(wine_delivery:inout channel;
         wine_selling:inout channel);
  end component;
  component patron
    port(wine_buying:inout channel);
  end component;
  signal WineryShop:channel:=init_channel;
  signal ShopPatron:channel:=init_channel;
begin
  THE_WINERY:winery
    port map(wine_shipping => WineryShop);
  THE_SHOP:shop
    port map(wine_delivery => WineryShop,
             wine_selling => ShopPatron);
  THE_PATRON:patron
    port map(wine_buying => ShopPatron);
end structure;
```

The descriptions of the *winery* and *patron* are similar to the one for the *shop*. The last entity/architecture pair represents an alternative structural architecture for the *wine_example*. Within the structural architecture, declarations are given for each of the components. Component declarations are forward declarations of existing (but defined elsewhere) entities to define the ports to be used in the instantiations. These can be thought of as being like function prototypes in C or C++. Next, two channels are declared to connect the *winery* to the *shop* and from the *shop* to the *patron*.

The concurrent statement part includes three component instantiations. Each begins with a label for that instance of the component. In this example, the label for the winery is *THE_WINERY*. Following the label is the name

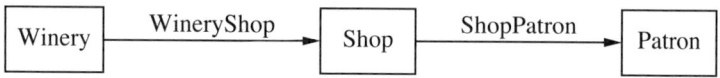

Fig. 2.1 Block diagram for *wine_example*.

of the entity for the component being instantiated. The last part is the *port map*. The port map is used to indicate which wires within a component are connected to which wires at the top level. For example, the *wine_shipping* port within the winery is connected to the *WineryShop* port at the top level. In the instantiation for the *shop*, *wine_delivery* is also connected to the *WineryShop* port, which in effect connects the *winery* to the *shop*. The rest of the instantiations for the *shop* and the *patron* are similar. The block diagram showing the connections between the components is shown in Figure 2.1.

As another example, consider a new shop opening closer to the patron. The patron now buys his wine from the new shop. Due to contracts with the winery, the new shop must buy its wine from the original shop. The new block diagram is shown in Figure 2.2, and the new architecture is shown below.

```
architecture new_structure of wine_example is
  component winery
    port(wine_shipping:inout channel);
  end component;
  component shop
    port(wine_delivery:inout channel;
         wine_selling:inout channel);
  end component;
  component patron
    port(wine_buying:inout channel);
  end component;
  signal WineryShop:channel:=init_channel;
  signal ShopNewShop:channel:=init_channel;
  signal NewShopPatron:channel:=init_channel;
begin
  THE_WINERY:winery
    port map(wine_shipping => WineryShop);
  OLD_SHOP:shop
    port map(wine_delivery => WineryShop,
             wine_selling => ShopNewShop);
  NEW_SHOP:shop
    port map(wine_delivery => ShopNewShop,
             wine_selling => NewShopPatron);
  THE_PATRON:patron
    port map(wine_buying => NewShopPatron);
end new_structure;
```

Fig. 2.2 Block diagram including the new shop.

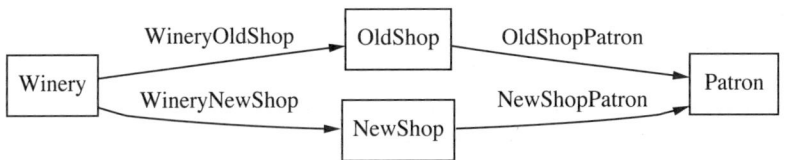

Fig. 2.3 Direct communication between winery and the new shop.

2.3 CONTROL STRUCTURES

So far, the flow of control in all our examples has been very simple in that no choices are being made at any point. In this section we introduce control structures for *selection* and *repetition*. We use as an example the scenario where the winery and patron can deal with either shop. The communication channels are depicted in Figure 2.3.

2.3.1 Selection

There are two ways in VHDL to model selection: **if-then-else** statements and **case** statements. As an example, let's assume that the winery has entered into an agreement to sell its merlot to the new shop. All other types of wine will still go to the old shop. This can be written in two ways:

```
winery2:process
begin
  bottle <= selection(8,3);
  wait for delay(5,10);
  if (wine_list'val(conv_integer(bottle)) = merlot) then
    send(WineryNewShop,bottle);
  else
    send(WineryOldShop,bottle);
  end if;
end process winery2;

winery3:process
begin
  bottle <= selection(8,3);
  wait for delay(5,10);
  case (wine_list'val(conv_integer(bottle))) is
  when merlot =>
    send(WineryNewShop,bottle);
  when others =>
```

```
        send(WineryOldShop,bottle);
      end case;
    end process winery3;
```

Note that in the case statement, **others** is a keyword to indicate any other type of wine.

The previous selection mechanism is *deterministic* in that the action is completely determined by the type of wine being produced. It is often also useful to be able to model a *nondeterministic*, or random, selection. For example, if the winery has decided to choose randomly which shop to sell to each time, it would be written in one of these two ways:

```
winery4:process
variable z:integer;
begin
  bottle <= selection(8,3);
  wait for delay(5,10);
  z:=selection(2);
  if (z = 1) then
    send(WineryNewShop,bottle);
  else
    send(WineryOldShop,bottle);
  end if;
end process winery4;

winery5:process
variable z:integer;
begin
  bottle <= selection(8,3);
  wait for delay(5,10);
  z:=selection(2);
  case z is
  when 1 =>
    send(WineryNewShop,bottle);
  when others =>
    send(WineryOldShop,bottle);
  end case;
end process winery5;
```

In the examples above, the function *selection* is overloaded in the *nondeterminism* package, to take one parameter, a constant indicating the range desired, and returns a random integer result. Since the selection statement is sent the constant 2, it returns either a 1 or a 2. Note that z is an integer variable that is local only to this process.

2.3.2 Repetition

Every process repeats forever, so it implies a loop. If you want to specify a loop internal to a process, there are three different types of looping constructs.

The first type of loop construct that we consider is the **for** loop. Just as in conventional programming languages, the **for** loop allows you to loop a fixed number of times. For example, if the winery decided to send the first four bottles of wine to the old shop and the next three to the new shop, it would be specified like this (note that the loop index variable *i* is declared implicitly):

```
winery6:process
begin
  for i in 1 to 4 loop
    bottle <= selection(8,3);
    wait for delay(5,10);
    send(WineryOldShop,bottle);
  end loop;
  for i in 1 to 3 loop
    bottle <= selection(8,3);
    wait for delay(5,10);
    send(WineryNewShop,bottle);
  end loop;
end process winery6;
```

The next type of loop construct is the **while** loop. Again, its usage should be familiar. The **while** loop allows you to specify behavior that loops until a boolean condition is satisfied. For example, if the winery decided to send out bottles of wine to the old shop until it had a bottle of merlot to send, and then it would send the merlot to the new shop, it could be specified like this:

```
winery7:process
begin
  while (wine_list'val(conv_integer(bottle))) /= merlot)
  loop
    bottle <= selection(8,3);
    wait for delay(5,10);
    send(WineryOldShop,bottle);
  end loop;
  bottle <= selection(8,3);
  wait for delay(5,10);
  send(WineryNewShop,bottle);
end process winery7;
```

The last kind of looping construct is the infinite **loop**. For example, consider the case where the winery has decided to stop doing business with the old shop and deal only with the new shop, but due to existing contract obligations it still owed one final bottle to the old shop. This could be specified as follows:

```
winery8:process
begin
  bottle <= selection(8,3);
  wait for delay(5,10);
  send(WineryOldShop,bottle);
  loop
```

```
      bottle <= selection(8,3);
      wait for delay(5,10);
      send(WineryNewShop,bottle);
    end loop;
  end process winery8;
```

2.4 DEADLOCK

An additional complication that arises with asynchronous design is the potential of introducing a *deadlock* into the system. Deadlock is the state in which a system can no longer make progress toward a required goal. For example, consider the two processes below:

```
producer:process
begin
  send(X,x);
  send(Y,y);
end process producer;

consumer:process
begin
  receive(Y,a);
  receive(X,b);
end process consumer;
```

The producer tries to send x out on channel X while the consumer is trying to receive data on channel Y. They both sit waiting for the other to respond and no progress can be made. In other words, the system is deadlocked. Although this is obvious in this example, such situations are difficult to observe in larger examples.

Let us again consider the wine shop example. In the preceding section we described several different ways the winery can decide to which shop to send the wine. The patron now could go to two different shops to get his wine. The problem that the patron faces now is how to decide which shop has wine. The simplest approach would be to patronize just one shop. This approach, however, can cause the system to deadlock. Consider the case where the winery decides to send its next two bottles of wine to the new shop and the patron patronizes only the old shop. In this case, the winery delivers the first bottle to the new shop, where it is put on their shelf. When they are ready to deliver the second bottle, the new shop is unable to accept it because it still has the first bottle on its shelf. Assuming that the winery is persistent, no more wine will be produced. Another idea might be for the patron to alternate between the two shops. Assuming that the patron begins by going to the old shop, and the winery sends the first two bottles to the new shop, the same problem arises.

2.5 PROBE

To solve this deadlock problem, the patron needs to know who has got wine to sell before he commits to shopping at one particular shop. This is accomplished by adding *probe* to the channel package. The *probe* function takes a channel as a parameter and returns true if there is a pending communication on that channel, and otherwise, returns false. Using this function, the patron can first check if the old shop is trying to sell him wine, and if so, receive it. If not, he can check the new shop. If neither has wine to sell, he will wait for 5 to 10 minutes, and start checking again (he's not very patient). An important subtle note is that the wait statement is essential; otherwise, the simulator will go into an infinite loop. Within every process in VHDL, there must be either a *wait statement* or a *sensitivity list*. Without either one, the process will execute forever and starve all other processes. Note that there is an implicit wait within a *send* or *receive* procedure call. If neither condition is true, a *wait* statement must be inserted to cause the simulation to stop working on this process and give another one a chance to make progress.

```
patron2:process
begin
  if (probe(OldShopPatron)) then
    receive(OldShopPatron,bag);
    wine_drunk <= wine_list'val(conv_integer(bag));
  elsif (probe(NewShopPatron)) then
    receive(NewShopPatron,bag);
    wine_drunk <= wine_list'val(conv_integer(bag));
  end if;
  wait for delay(5,10);
end process patron2;
```

2.6 PARALLEL COMMUNICATION

So far all communication actions have been initiated sequentially. To improve speed, it is often beneficial to launch several communication actions in parallel. Our channel package supports this by allowing you to pass multiple channel/data pairs to the *send* or *receive* procedures. As an example, consider the case where the winery makes two bottles of wine at a time and sends them out concurrently to the two shops. This is described as follows:

```
winery9:process
begin
  bottle1 <= selection(8,3);
  bottle2 <= selection(8,3);
  wait for delay(5,10);
  send(WineryOldShop,bottle1,WineryNewShop,bottle2);
end process winery9;
```

36 COMMUNICATION CHANNELS

We could leave the patron process as described in the preceding section. However, to improve performance, we may want to allow the patron to receive the bottles in parallel (perhaps, he hired someone to run to the other shop for him). This can be described as follows:

```
patron3:process
begin
  receive(OldShopPatron,bag1,NewShopPatron,bag2);
  wine_drunk1 <= wine_list'val(conv_integer(bag1));
  wine_drunk2 <= wine_list'val(conv_integer(bag2));
end process patron3;
```

2.7 EXAMPLE: MINIMIPS

Translating a word description for a design into a good formal specification can be more of an art form than a science. Since word descriptions are, by their very nature, imprecise and unstructured, there can be no set process to a formal specification. Therefore, the best way to learn how to do this is through an example. This section shows the derivation of a specification from a word description for a simple MiniMIPS microprocessor and some initial steps toward its optimization.

The MiniMIPS is a simple microprocessor described in [164]. The MiniMIPS has a simple reduced instruction set computer (RISC)-style architecture and datapath. The design presented here is indeed simple in that it only supports the eight different types of instructions shown in Table 2.2. All instructions are 32 bits wide and come in one of three formats, shown in Figure 2.4.

To specify the MiniMIPS, we have decided to decompose the design into five communicating processes, as shown in Figure 2.5. Two describe the environment: an instruction memory (*imem*) and a data memory (*dmem*). The other three describe the circuit to be designed. Consider arithmetic and logic operations. The first stage fetches instructions from the instruction memory (*fetch*). The second stage takes these instructions, decodes them, and fetches values from the appropriate registers (*decode*). The third stage takes these register values, executes the appropriate function on them, and returns the result to the register file in the decode block (*execute*). If the instruction is a *load*, the *execute* block would generate an address for the data memory which would send the result back to the decode block for storage. If the instruction is a *store*, the *decode* block would send the data to be stored to the data memory while the execute block sends the address. If the instruction is a *branch*, the *execute* block is responsible for sending the result of the comparison back to the *fetch* block, which then uses this to determine the next instruction to fetch. Finally, if the instruction is an unconditional *jump*, the *fetch* block simply adjusts the program counter (PC).

Table 2.2 MiniMIPS instruction set.

Instruction	Opcode	Function	Operation	Example
add	0	32	rd := rs + rt	add r1, r2, r3
sub	0	34	rd := rs − rt	sub r1, r2, r3
and	0	36	rd := rs & rt	and r1, r2, r3
or	0	37	rd := rs | rt	or r1, r2, r3
lw	35	n/a	rt := mem[rs + offset]	lw r1, (32)r2
sw	43	n/a	mem[rs + offset] := rt	sw r1, (32)r2
beq	4	n/a	if (rs==rt) then PC := PC + offset	beq r1, r2, Loop
j	6	n/a	PC := address	j Loop

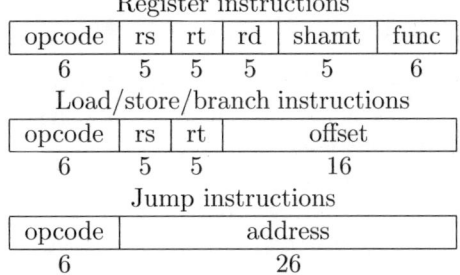

Fig. 2.4 Instruction formats for MiniMIPS.

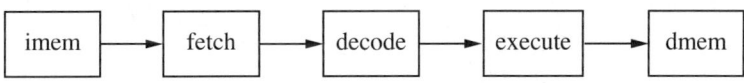

Fig. 2.5 Block diagram for MiniMIPS.

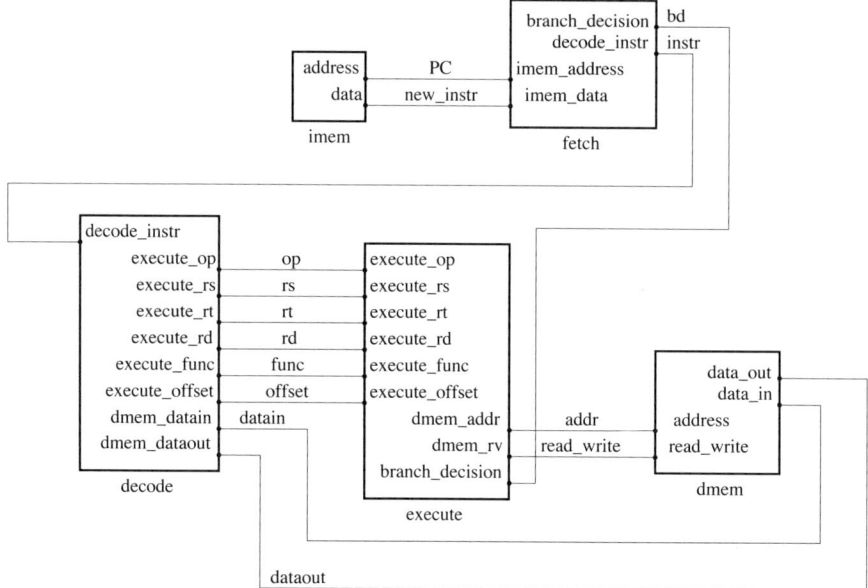

Fig. 2.6 Channel-level diagram for MiniMIPS.

2.7.1 VHDL Specification

The first step in the design process is to determine the needed communication channels between the blocks and draw a block diagram. The result for the MiniMIPS is shown in Figure 2.6. Beginning with the communications between the *imem* and *fetch* blocks, the *fetch* block needs the channel *PC* to send the PC to the *imem* block, and the *imem* block needs the channel *new_instr* to send the new instruction back to the *fetch* block. The *fetch* block also needs the channel *instr* to send the new instruction to the *decode* block, and the channel *bd* from the *execute* block to notify it whether or not a branch is to be taken. The *decode* block parses the instruction and needs many channels to send the opcode (*op* channel), ALU function code (*func* channel), source operands (*rs* and *rt* channels), and offset (*offset* channel) to the *execute* block. Note that not all channels are used for every instruction. The *decode* block also needs the channel *rd* from the *execute* block to get the result of arithmetic/logic operations, and the channels *datain* and *dataout* from and to the *dmem* block for receiving and sending data for loads and stores. Finally, the *execute* block needs the channels *addr* and *read_write* to the *dmem* block to send the address and read/write signals. The structural VHDL code connecting the blocks together is shown below.

```
--------------------
-- minimips.vhd
--------------------
library IEEE;
use IEEE.std_logic_1164.all;
use work.channel.all;
entity minimips is
end minimips;
architecture structure of minimips is
  component imem
    port(data:inout channel;
         address:inout channel);
  end component;
  component fetch
    port(branch_decision:inout channel;
         decode_instr:inout channel;
         imem_data:inout channel;
         imem_address:inout channel);
  end component;
  component decode
    port(dmem_dataout:inout channel;
         dmem_datain:inout channel;
         execute_offset:inout channel;
         execute_func:inout channel;
         execute_rd:inout channel;
         execute_rt:inout channel;
         execute_rs:inout channel;
         execute_op:inout channel;
         decode_instr:inout channel);
  end component;
  component execute
    port(branch_decision:inout channel;
         dmem_rw:inout channel;
         dmem_addr:inout channel;
         execute_offset:inout channel;
         execute_func:inout channel;
         execute_rd:inout channel;
         execute_rt:inout channel;
         execute_rs:inout channel;
         execute_op:inout channel);
  end component;
  component dmem
    port(read_write:inout channel;
         data_out:inout channel;
         data_in:inout channel;
         address:inout channel);
  end component;
  signal addr:channel;
  signal bd:channel;
```

```
    signal datain:channel;
    signal dataout:channel;
    signal func:channel;
    signal instr:channel;
    signal new_instr:channel;
    signal offset:channel;
    signal op:channel;
    signal PC:channel;
    signal rd:channel;
    signal read_write:channel;
    signal rs:channel;
    signal rt:channel;
begin
  imem1:imem
     port map(data => new_instr,
              address => PC);
  fetch1:fetch
     port map(branch_decision => bd,
              decode_instr => instr,
              imem_data => new_instr,
              imem_address => PC);
  decode1:decode
     port map(dmem_dataout => dataout,
              dmem_datain => datain,
              execute_offset => offset,
              execute_func => func,
              execute_rd => rd,
              execute_rt => rt,
              execute_rs => rs,
              execute_op => op,
              decode_instr => instr);
  execute1:execute
     port map(branch_decision => bd,
              dmem_rw => read_write,
              dmem_addr => addr,
              execute_offset => offset,
              execute_func => func,
              execute_rd => rd,
              execute_rt => rt,
              execute_rs => rs,
              execute_op => op);
  dmem1:dmem
     port map(read_write => read_write,
              data_out => dataout,
              data_in => datain,
              address => addr);
end structure;
```

Now let's consider specification of each of the individual blocks using our channel model. First, the *imem* block models a small (eight-word) memory using an array to store the instructions. The array is hardcoded to store a particular program. Therefore, to simulate a different program, it is necessary to edit the code. The *imem* block receives an address from the *fetch* block. This address is a *std_logic_vector*, so it is converted to an integer and then used to index into the memory array to find the instruction requested. The memory than waits for the specified amount of time to simulate the delay of the memory (this is also necessary to make sure that *instr* has taken its new value). Then it sends the fetched instruction to the *fetch* block. The VHDL code is shown below.

```vhdl
--------------------
-- imem.vhd
--------------------
library ieee;
use ieee.std_logic_1164.all;
use ieee.std_logic_arith.all;
use ieee.std_logic_unsigned.all;
use work.nondeterminism.all;
use work.channel.all;
entity imem is
  port(address:inout channel:=init_channel;
       data:inout channel:=init_channel);
end imem;
architecture behavior of imem is
  type memory is array (0 to 7) of
    std_logic_vector(31 downto 0);
  signal addr:std_logic_vector(31 downto 0);
  signal instr:std_logic_vector(31 downto 0);
begin
process
  variable imem:memory:=(
    X"8c220000", -- L: lw r2,0(r1)
    X"8c640000", --    lw r4,0(r3)
    X"00a42020", --    add r4,r5,r4
    X"00642824", --    and r5,r3,r4
    X"ac640000", --    sw r4,0(r3)
    X"00822022", -- M: sub r4,r4,r2
    X"1080fff9", --    beq r4,r0,L
    X"18000005");--    j M
begin
  receive(address,addr);
  instr <= imem(conv_integer(addr(2 downto 0)));
  wait for delay(5,10);
  send(data,instr);
end process;
end behavior;
```

The *fetch* block fetches instructions and determines the PC for the next instruction. The behavior of the *fetch* block is that it sends the current PC to the *imem* block, it waits to receive a new instruction, and it increments the PC. It then partially decodes the instruction to determine if it is a branch or jump. If it is a branch, it forwards the instruction to the *decode* block and waits to hear from the *execute* block whether or not the branch is to be taken. If it is taken, it determines the new PC by adding a sign-extended version of the *offset* to the current PC. Otherwise, no action is necessary since the PC has already been incremented. If the new instruction is a jump, the *fetch* block can simply insert the address into the lower 26 bits of the current PC. All other types of instructions are simply forwarded to the decode block. The VHDL code is shown below. Note the use of *aliases* for the opcode, offset, and address. These allow you to associate names for partial ranges of an array.

```
--------------------
-- fetch.vhd
--------------------
library ieee;
use ieee.std_logic_1164.all;
use ieee.std_logic_arith.all;
use ieee.std_logic_unsigned.all;
use work.nondeterminism.all;
use work.channel.all;
entity fetch is
  port(imem_address:inout channel:=init_channel;
       imem_data:inout channel:=init_channel;
       decode_instr:inout channel:=init_channel;
       branch_decision:inout channel:=init_channel);
end fetch;
architecture behavior of fetch is
  signal PC:std_logic_vector(31 downto 0):=(others=>'0');
  signal instr:std_logic_vector(31 downto 0);
  signal bd:std_logic;
  alias opcode:std_logic_vector(5 downto 0) is
    instr(31 downto 26);
  alias offset:std_logic_vector(15 downto 0) is
    instr(15 downto 0);
  alias address:std_logic_vector(25 downto 0) is
    instr(25 downto 0);
begin
process
  variable branch_offset:std_logic_vector(31 downto 0);
begin
  send(imem_address,PC);
  receive(imem_data,instr);
  PC <= PC + 1;
  wait for delay(5,10);
  case opcode is
```

```
      when "000100" => -- beq
        send(decode_instr,instr);
        receive(branch_decision,bd);
        if (bd = '1') then
          branch_offset(31 downto 16):=(others=>instr(15));
          branch_offset(15 downto 0):=offset;
          PC <= PC + branch_offset;
          wait for delay(5,10);
        end if;
      when "000110" => -- j
        PC <= (PC(31 downto 26) & address);
        wait for delay(5,10);
      when others =>
        send(decode_instr,instr);
      end case;
    end process;
  end behavior;
```

The major component of the *decode* block is the register file. For simplicity, our register file has only eight registers, whose values are stored in an array of *std_logic_vectors* in this block. The *decode* block receives an instruction from the *fetch* block, from which it extracts the *rs* and *rt* fields to index into the register array to find these registers' current values. It also extracts the opcode and sends it to the execute block. If the instruction is an ALU operation (i.e., add, sub, and, or), it sends the function field and the register values to the *execute* block. It then waits to receive the result from the *execute* block, and it stores this result into the register pointed to by the *rd* field in the instruction. If it had been a branch instruction, it sends the *rs* and *rt* registers to the *execute* block. If it is a load word, it sends the *rs* register and the offset to the *execute* block. It then waits to receive the result of the load from the *dmem* block, and it stores this value in the *rt* register. Finally, if it is a store, it sends the *rs* register and the offset to the *execute* block, and it sends the *rt* register to the *dmem* block. The VHDL code is shown below.

```
  --------------------
  -- decode.vhd
  --------------------
  library ieee;
  use ieee.std_logic_1164.all;
  use ieee.std_logic_arith.all;
  use ieee.std_logic_unsigned.all;
  use work.nondeterminism.all;
  use work.channel.all;
  entity decode is
    port(decode_instr:inout channel:=init_channel;
         execute_op:inout channel:=init_channel;
         execute_rs:inout channel:=init_channel;
         execute_rt:inout channel:=init_channel;
         execute_rd:inout channel:=init_channel;
```

```
            execute_func:inout channel:=init_channel;
            execute_offset:inout channel:=init_channel;
            dmem_datain:inout channel:=init_channel;
            dmem_dataout:inout channel:=init_channel);
end decode;
architecture behavior of decode is
  type registers is array (0 to 7) of
    std_logic_vector(31 downto 0);
  signal instr:std_logic_vector(31 downto 0);
  alias op:std_logic_vector(5 downto 0) is
    instr(31 downto 26);
  alias rs:std_logic_vector(2 downto 0) is
    instr(23 downto 21);
  alias rt:std_logic_vector(2 downto 0) is
    instr(18 downto 16);
  alias rd:std_logic_vector(2 downto 0) is
    instr(13 downto 11);
  alias func:std_logic_vector(5 downto 0) is
    instr(5 downto 0);
  alias offset:std_logic_vector(15 downto 0) is
    instr(15 downto 0);
  signal reg:registers:=(X"00000000",
                         X"11111111",
                         X"22222222",
                         X"33333333",
                         X"44444444",
                         X"55555555",
                         X"66666666",
                         X"77777777");
  signal reg_rs:std_logic_vector(31 downto 0);
  signal reg_rt:std_logic_vector(31 downto 0);
  signal reg_rd:std_logic_vector(31 downto 0);
begin
process
begin
  receive(decode_instr,instr);
  reg_rs <= reg(conv_integer(rs));
  reg_rt <= reg(conv_integer(rt));
  wait for delay(5,10);
  send(execute_op,op);
  case op is
  when "000000" => -- ALU op
    send(execute_func,func,execute_rs,reg_rs,
         execute_rt,reg_rt);
    receive(execute_rd,reg_rd);
    reg(conv_integer(rd)) <= reg_rd;
    wait for delay(5,10);
  when "000100" => -- beq
    send(execute_rs,reg_rs,execute_rt,reg_rt);
```

EXAMPLE: MINIMIPS   45

```vhdl
      when "100011" => -- lw
        send(execute_rs,reg_rs,execute_offset,offset);
        receive(dmem_dataout,reg_rt);
        reg(conv_integer(rt)) <= reg_rt;
        wait for delay(5,10);
      when "101011" => -- sw
        send(execute_rs,reg_rs,execute_offset,offset,
             dmem_datain,reg_rt);
      when others => -- undefined
        assert false
          report "Illegal instruction"
          severity error;
    end case;
  end process;
end behavior;
```

The *execute* block begins by waiting to receive an opcode from the *decode* block. If this opcode is an ALU-type instruction, it then waits to receive the *func* field and its two operands from the *decode* block. Using the *func* field, it determines which type of ALU operation to perform, and it performs it on the two operands. After waiting for a short time to model delay through the ALU, it sends the result back to the *decode* block. If the opcode had been a branch, it would then wait for two operands, which it needs to compare. If they are equal, it will send back a '1' to the *fetch* block to indicate that a branch has been taken. If they are not equal, it will send a '0' back to the *fetch* block to indicate that the branch is not taken. If it is a load, it waits to receive the register containing the base address and the offset from the *decode* block. It sign extends the offset and adds it to the base to determine the address. It then sends the address and a read indication to the *dmem* block. Finally, if it is a store, it again waits to receive the base address and offset from the *decode* block and computes the address. It then sends the address and a write indication to the *dmem* block. The VHDL code is shown below.

```vhdl
----------------------
-- execute.vhd
----------------------
library ieee;
use ieee.std_logic_1164.all;
use ieee.std_logic_arith.all;
use ieee.std_logic_unsigned.all;
use work.nondeterminism.all;
use work.channel.all;
entity execute is
  port(execute_op:inout channel:=init_channel;
       execute_rs:inout channel:=init_channel;
       execute_rt:inout channel:=init_channel;
       execute_rd:inout channel:=init_channel;
       execute_func:inout channel:=init_channel;
       execute_offset:inout channel:=init_channel;
```

46 COMMUNICATION CHANNELS

```vhdl
           dmem_addr:inout channel:=init_channel;
           dmem_rw:inout channel:=init_channel;
           branch_decision:inout channel:=init_channel);
end execute;
architecture behavior of execute is
  signal rs:std_logic_vector(31 downto 0);
  signal rt:std_logic_vector(31 downto 0);
  signal rd:std_logic_vector(31 downto 0);
  signal op:std_logic_vector(5 downto 0);
  signal func:std_logic_vector(5 downto 0);
  signal offset:std_logic_vector(15 downto 0);
  signal rw:std_logic;
  signal bd:std_logic;
begin
process
  variable addr_offset:std_logic_vector(31 downto 0);
begin
  receive(execute_op,op);
  case op is
  when "000000" => -- ALU op
    receive(execute_func,func,execute_rs,rs,
            execute_rt,rt);
    case func is
    when "100000" => -- add
      rd <= rs + rt;
    when "100010" => -- sub
      rd <= rs - rt;
    when "100100" => -- and
      rd <= rs and rt;
    when "100101" => -- or
      rd <= rs or rt;
    when others =>
      rd <= (others => 'X'); -- undefined
    end case;
    wait for delay(5,10);
    send(execute_rd,rd);
  when "000100" => -- beq
    receive(execute_rs,rs,execute_rt,rt);
    if (rs = rt) then bd <= '1';
    else bd <= '0';
    end if;
    wait for delay(5,10);
    send(branch_decision,bd);
  when "100011" => -- lw
    receive(execute_rs,rs,execute_offset,offset);
    addr_offset(31 downto 16):=(others => offset(15));
    addr_offset(15 downto 0):=offset;
    rd <= rs + addr_offset;
    rw <= '1';
```

```
      wait for delay(5,10);
      send(dmem_addr,rd);
      send(dmem_rw,rw);
   when "101011" => -- sw
      receive(execute_rs,rs,execute_offset,offset);
      addr_offset(31 downto 16):=(others => offset(15));
      addr_offset(15 downto 0):=offset;
      rd <= rs + addr_offset;
      rw <= '0';
      wait for delay(5,10);
      send(dmem_addr,rd);
      send(dmem_rw,rw);
   when others => -- undefined
      assert false
         report "Illegal instruction"
         severity error;
   end case;
   end process;
end behavior;
```

The last block is the *dmem* block, which models a small memory (eight words), which is again modeled as an array. This block first waits to receive an address and a read/write indication from the *execute* block. If the $read/\overline{write}$ bit, *rw*, is '1', it is a read. A read uses the address it receives to index into the memory array. It then sends this value to the *decode* block. If it is a write, it waits to receive the data to write from the *decode* block. It then writes this value into the memory array. The VHDL code is shown below.

```
-------------------
-- dmem.vhd
-------------------
library ieee;
use ieee.std_logic_1164.all;
use ieee.std_logic_arith.all;
use ieee.std_logic_unsigned.all;
use work.nondeterminism.all;
use work.channel.all;
entity dmem is
   port(address:inout channel:=init_channel;
        data_in:inout channel:=init_channel;
        data_out:inout channel:=init_channel;
        read_write:inout channel:=init_channel);
end dmem;
architecture behavior of dmem is
   type memory is array (0 to 7) of
      std_logic_vector(31 downto 0);
   signal addr:std_logic_vector(31 downto 0);
   signal d:std_logic_vector(31 downto 0);
   signal rw:std_logic;
```

```
               signal dmem:memory:=(X"00000000",
                                    X"11111111",
                                    X"22222222",
                                    X"33333333",
                                    X"44444444",
                                    X"55555555",
                                    X"66666666",
                                    X"77777777");
    begin
    process
    begin
      receive(address,addr);
      receive(read_write,rw);
      case rw is
      when '1' =>
        d <= dmem(conv_integer(addr(2 downto 0)));
        wait for delay(5,10);
        send(data_out,d);
      when '0' =>
        receive(data_in,d);
        dmem(conv_integer(addr(2 downto 0))) <= d;
        wait for delay(5,10);
      when others =>
        wait for delay(5,10);
      end case;
    end process;
    end behavior;
```

2.7.2 Optimized MiniMIPS

In synchronous design, the performance of a processor is improved through *pipelining*. The idea of pipelining is whenever possible to keep all parts of the circuit doing useful work concurrently. This is analogous to an assembly line in an auto factory. One worker puts the tires on a car and passes the car to the next person, who puts on the doors. However, while that worker is putting the doors on that car, the previous worker is putting the tires on the next car. In a microprocessor, a simple assembly line could have one stage fetching an instruction, another decoding an instruction, while a third executes an instruction. Ideally, all three stages could be working on different instructions at the same time.

The question, therefore, is whether our asynchronous design exhibits behavior akin to pipelining. The answer is "yes and no." If you simulate our design and watch the flow of instructions through our three stages (*fetch, decode,* and *execute*), what you see is that the *fetch* block can fetch the next instruction (assuming that the current instruction is not a branch) while the *decode* block is decoding the current instruction. However, the *decode* block and *execute* block operate sequentially. The reason for this is that even though the *decode* block passes an instruction to the *execute* block, it must wait for that block to complete before it can decode the next instruction. For example, if the instruction is an ALU operation, the *decode* block is stuck waiting for the result of the operation, which is to be stored in the register file.

To eliminate this problem, we must split the *decode* block into two processes. The first process is responsible for decoding the instruction and forwarding it to the *execute* block. The second process is responsible for collecting the results of the execution and writing them back to the register file. There is a problem, though. If we naively split the process into two parts, we can introduce the possibility of *data hazards*. For example, a read after write (RAW) hazard is one where an instruction that writes a register is followed by one that reads that same register. If we allow the *decode* block to work concurrently with the *execute* block in this case, it is possible that the old value of the register will be read instead of the new value. This would obviously cause the program to generate wrong results. For example, let's assume that initially, $r1$ contains 1 and $r2$ contains 2, and we execute the following two instructions:

```
add r1,r2,r2
add r4,r1,r1
```

If the second instruction is decoded and has its operands fetched before the first completes its execution and has written back its result, $r4$ will end up containing 2 when it should contain 8.

Perhaps the simplest way to eliminate data hazards is through *register locking*. In register locking, while an instruction is being decoded and before it is dispatched to the *execute* block, the destination register for that instruction is locked. When the next instruction arrives, if it needs to read that register, it will find that it is locked and stall until the lock is released. Thus, this mechanism prevents stale data from ever being read. Our *decode* block, rewritten to allow concurrent decoding and execution without data hazards, is shown below.

```vhdl
----------------------
-- decode.vhd
----------------------
library ieee;
use ieee.std_logic_1164.all;
use ieee.std_logic_arith.all;
use ieee.std_logic_unsigned.all;
use work.nondeterminism.all;
use work.channel.all;
entity decode is
  port(decode_instr:inout channel:=init_channel;
       execute_op:inout channel:=init_channel;
       execute_rs:inout channel:=init_channel;
       execute_rt:inout channel:=init_channel;
       execute_rd:inout channel:=init_channel;
```

```vhdl
               execute_func:inout channel:=init_channel;
               execute_offset:inout channel:=init_channel;
               dmem_datain:inout channel:=init_channel;
               dmem_dataout:inout channel:=init_channel);
  end decode;
  architecture behavior of decode is
    type registers is array (0 to 7) of
        std_logic_vector(31 downto 0);
    type booleans is array (natural range <>) of boolean;
    signal instr :  std_logic_vector( 31 downto 0);
    alias op:std_logic_vector(5 downto 0) is
      instr(31 downto 26);
    alias rs:std_logic_vector(2 downto 0) is
      instr(23 downto 21);
    alias rt:std_logic_vector(2 downto 0) is
      instr(18 downto 16);
    alias rd:std_logic_vector(2 downto 0) is
      instr(13 downto 11);
    alias func:std_logic_vector(5 downto 0) is
      instr(5 downto 0);
    alias offset:std_logic_vector(15 downto 0) is
      instr(15 downto 0);
    signal reg:registers:=(X"00000000",
                           X"11111111",
                           X"22222222",
                           X"33333333",
                           X"44444444",
                           X"55555555",
                           X"66666666",
                           X"77777777");
    signal reg_rs:std_logic_vector(31 downto 0);
    signal reg_rt:std_logic_vector(31 downto 0);
    signal reg_rd:std_logic_vector(31 downto 0);
    signal reg_locks:booleans(0 to 7):=(others => false);
    signal decode_to_wb:channel:=init_channel;
    signal wb_instr:std_logic_vector(31 downto 0);
    alias wb_op:std_logic_vector(5 downto 0) is
      wb_instr(31 downto 26);
    alias wb_rt:std_logic_vector(2 downto 0) is
      wb_instr(18 downto 16);
    alias wb_rd:std_logic_vector(2 downto 0) is
      wb_instr(13 downto 11);
    signal lock:channel:=init_channel;
  begin
  decode:process
  begin
    receive(decode_instr,instr);
    if ((reg_locks(conv_integer(rs))) or
        (reg_locks(conv_integer(rt)))) then
```

```
          wait until ((not reg_locks(conv_integer(rs))) and
                      (not reg_locks(conv_integer(rt))));
        end if;
        reg_rs <= reg(conv_integer(rs));
        reg_rt <= reg(conv_integer(rt));
        send(execute_op,op);
        wait for delay(5,10);
        case op is
        when "000000" => -- ALU op
          send(execute_func,func,execute_rs,reg_rs,
               execute_rt,reg_rt);
          send(decode_to_wb,instr);
          receive(lock);
        when "000100" => -- beq
          send(execute_rs,reg_rs,execute_rt,reg_rt);
        when "100011" => -- lw
          send(execute_rs,reg_rs,execute_offset,offset);
          send(decode_to_wb,instr);
          receive(lock);
        when "101011" => -- sw
          send(execute_rs,reg_rs,execute_offset,offset,
               dmem_datain,reg_rt);
        when others => -- undefined
          assert false
            report "Illegal instruction"
            severity error;
        end case;
      end process;
      writeback:process
      begin
        receive(decode_to_wb,wb_instr);
        case wb_op is
        when "000000" => -- ALU op
          reg_locks(conv_integer(wb_rd)) <= true;
          wait for 1 ns;
          send(lock);
          receive(execute_rd,reg_rd);
          reg(conv_integer(wb_rd)) <= reg_rd;
          wait for delay(5,10);
          reg_locks(conv_integer(wb_rd)) <= false;
          wait for delay(5,10);
        when "100011" => -- lw
          reg_locks(conv_integer(wb_rt)) <= true;
          wait for 1 ns;
          send(lock);
          receive(dmem_dataout,reg_rd);
          reg(conv_integer(wb_rt)) <= reg_rd;
          wait for delay(5,10);
          reg_locks(conv_integer(wb_rt)) <= false;
```

```
          wait for delay(5,10);
      when others => -- undefined
          wait for delay(5,10);
        end case;
      end process;
    end behavior;
```

Let's first consider the *decode* process. After this process receives an instruction, it checks if either of its source operands, *rs* or *rt*, is locked, and if either is locked, it stalls until they are both released. After checking that all the necessary registers are unlocked, it can then dispatch the instruction to the *execute* block. If this instruction results in data being written into the register file (i.e., it is an ALU operation or a load word), it then also forwards the instruction to the *write_back* process and waits to receive from this process a lock on the destination register. Note that the *receive* procedure call here takes a channel but no data, since it is used only for synchronization. Once it has received a lock, it can then go ahead and begin decoding the next instruction.

The *write_back* process waits to receive an instruction. Once it does, it locks the appropriate destination register and sends the lock back to the *decode* process. It then waits to receive the result from the *execute* block or the *dmem* block. Once it does, it writes the result to the register file and removes the lock on that register.

2.8 SOURCES

The channel model described in this chapter is inspired by Hoare's *communicating sequential processes* (CSP) [166, 167], with Martin's addition of the probe [248]. Martin first adapted CSP to model asynchronous circuits and systems [249, 250, 252, 254, 255]. A similar channel based model is used in the *Tangram* language proposed by van Berkel et al.[37, 39]. *Occam*, a parallel programming language based on CSP, is used by Brunvand to model asynchronous circuits at a communication level [52, 53]. Gopalakrishnan and Akella use a language called *hopCP* [150].

When designing systems using the channel model, it is particularly important to avoid deadlock. Friedman and Menon investigated interconnections of modules operating using asynchronous codes and determined conditions under which they could deadlock [130]. They also showed that by adding buffers the system performance could actually be improved. Bruno and Altman developed conditions under which an interconnected control structure composed of JUNCTION, WYE, SEQUENCE, ITERATE, and SELECT would deadlock [51]. Jump and Thiagarajan developed a methodology for checking for deadlock in an interconnection of asynchronous control structures which translates the network into a marked graph and then performs a simple analysis [184].

Problems

2.1 In this problem you will specify a 4-bit adder using the channel model. The adder has three input ports and two output ports. The input ports are X and Y, which are used to pass in the 4-bit integer operands, and Cin, which is a 1-bit carry-in. The output ports are Sum, which is a 4-bit integer sum, and $Cout$, which is a 1-bit carry-out.

2.1.1. Specify the 4-bit adder in VHDL using the channel model. The *adder4* process should accept the two operands a and b and the carry-in c from the corresponding input ports (X, Y, Cin). It should then compute the sum, s, and carry-out, d, and output the results on the corresponding output ports $(Sum, Cout)$. Create environment processes to generate random data using the selection procedure and consume the data. Simulate until you are convinced that it works.

2.1.2. Specify a 1-bit full adder using VHDL. The *adder1* process should accept two 1-bit operands a and b and the carry-in c from the corresponding ports (X, Y, Cin). It should then compute the 1-bit sum, s, and carry-out, d, and output the results on the corresponding output ports $(Sum, Cout)$. You may use only logic functions (no arithmetic functions). Create an environment process which communicates with the 1-bit adder and simulate until you are convinced that it works.

2.1.3. Use structural VHDL to build a 4-bit adder using your 1-bit full adders from Problem 2.1.2. Create an environment process which communicates with the 4 bits of the adder. Convince yourself through simulation that it performs the same function as your 4-bit adder from Problem 2.1.1.

2.1.4. Optimize your 1-bit full adder from Problem 2.1.2 so that it takes advantage of the fact that for most computations, the longest carry chain is significantly shorter than n, where n is the number of bits in your adder.

2.2 Specify and simulate a four-element stack using a channel-level model. The stack should be constructed using a number of identical modules which will each hold a single 8-bit data value. Each module has two passive channels to its left, *Push* and *Pop*, and two active channels to its right, *Put* and *Get*. The module should wait to get either a communication on its *Push* or its *Pop* channel. If it detects a communication on its *Push* channel and it is empty, it should receive the data and store the value received internally. If it is full, it should communicate its data to the right using its *Put* channel, then complete the receive from the *Push* channel. If it gets a communication on its *Pop* channel and it is full, it should send its data over the *Pop* channel. If it is empty, it should request data over its *Get* channel and then forward the received data out the *Pop* channel. For simplicity, assume that the environment will never push data into a full stack or pop data from an empty stack.

2.3 Specify and simulate a 4-bit shifter. It should be constructed using three identical modules which each hold a single bit of data, and a special module at the end of the shifter. Each module has a *Load* channel, *Shift_in* channel, *Shift_out* channel, *Done_in* channel, *Done_out* channel, and an *Output* channel. Each module waits to receive a communication on its *Load* channel, at which time it accepts a bit of data. Next, the most significant bit waits to receive a communication from the environment on either its *Shift_in* channel or its *Done_in* channel. If it is on the *Shift_in* channel, it sends its bit to the next most significant bit over the *Shift_out* channel and it accepts a new bit from the *Shift_in* channel. If it receives a communication on its *Done_in* channel, it sends its bit out the *Output* channel and sends a communication on its *Done_out* channel. Note that the *Done_in* and *Done_out* channels do not need to carry any data; they are used only for synchronization. The block at the end does not have a *shift_out* or *Done_out* channel.

2.4 In this problem you are going to specify a simple entropy decoder in VHDL using the channel model. An entropy decoder relies on a standard entropy code that represents fixed-length symbols from a source alphabet as variable-length code symbols. The idea is that common symbols are represented using short codes, and uncommon symbols are represented using longer codes. The result is that the average code length is significantly smaller than the symbol length. Such entropy codes are used in numerous audio, video, and data compression schemes (e.g., JPEG, MPEG, etc.). As an example, consider a set of 2-bit symbols (i.e., 00, 01, 10, 11) where 00 has a probability of 90 percent, 01 has a probability of 9 percent, 10 has a probability of 0.9 percent, and 11 has a probability of 0.1 percent. If we encode the symbols as follows:

- 00 - 0
- 01 - 10
- 10 - 110
- 11 - 1110

the average code length would be

$$0.9 \times 1 + 0.09 \times 2 + 0.009 \times 3 + 0.001 \times 4 = 1.11$$

This is nearly half the size of the fixed-length size of the symbols.

2.4.1. Specify in VHDL using the channel model an entropy decoder for 2-bit symbols which are encoded as described above. Assume that you have a 1-bit input port, *In*, and a 2-bit output port, *Out*. The module should receive 1 bit of data at a time and output the 2-bit symbol once it has recognized the code word. Specify environment processes to generate and consume the data. Simulate the design until you are convinced that it works.

2.4.2. We have decided to break up the entropy decoder from the first problem into four identical blocks, and we also would like to make it programmable. Each block has a 1-bit input port, *L*, which receives bits from its left neighbor, a 1-bit output port, *R*, which transmits bits to its right neighbor, a 2-bit input port, *Load*, which receives the 2-bit symbol this block transmits when it recognizes the code, and a 2-bit output port, *Out*, which transmits its stored symbol when it recognizes the code. The behavior of each block is that it can either receive a 2-bit symbol from the *Load* port or 1 bit from its left neighbor. If it receives a 2-bit symbol from the *Load* port, that symbol is stored in an internal register and becomes the symbol for which this stage is responsible. If it receives a bit from its left neighbor, it checks if it is a 0 or 1. If it is a 0, it will output its stored symbol. If it is a 1, it strips this bit and passes the remaining bits that it receives from its left neighbor one by one to its right neighbor until it sees a 0. At that point it loops back to the beginning of the cycle and waits for the next code to be transmitted. Specify one block in VHDL. Create an environment and simulate until you are convinced that it works.

2.4.3. Use structural VHDL to model an entropy decoder which can recognize codes of length 4.

2.4.4. The results being transmitted from the *Out* ports from each of the four blocks from the last problem need to be multiplexed together to produce the symbol that has just been recognized. There is a potential race problem with this design as described above. One stage may come to the end of forwarding bits (i.e., it sees the trailing 0) and immediately sees another 0, causing it to transmit its symbol on its *Out* port. This transmission may occur before the transmission by the later stage of the previous symbol. In other words, the symbols may get recognized out of order. Add another channel to forward acknowledgments of transmissions of symbols back to the previous blocks. Try to do this in a way which will prevent transmissions from getting out of order and does not hold up the recognition step.

2.5 Add a jump and link (*jal*) instruction to the MiniMIPS example from Section 2.7. This instruction is of the jump format, and the opcode is 000011. This instruction should send the old PC+1 to register 31 (effectively register 7 in our scaled-down design), and set the PC to the address.

2.6 Extend the MiniMIPS to include the set less than (*slt*) instruction. This instruction is of the register format, and the opcode is 000000 with function code 101010. It checks if register *rs* is less than *rt*. If so, it sets *rd* to 1. Otherwise, it sets *rd* to 0. This instruction typically is used before a branch.

2.7 Extend the MiniMIPS to include immediate instructions: add immediate (*addi*, op = 001000), and immediate (*andi*, op = 001100), or immediate (*ori*, op = 001101), and set less than immediate (*slti*, op = 001010). These instructions are of the load/store/branch format. The *rs* field points to one operand and the offset is the other operand. The offset is sign extended for *addi* and *slti*, while it is zero extended for *andi* and *ori*. The result is stored in the register pointed to by the *rt* field.

2.8 Extend the MiniMIPS to include shift instructions: shift left logical (*sll*), shift right logical (*srl*), and shift right arithmetic (*sra*). These instructions are of the register format with opcode 000000. The function codes are 000000 for *sll*, 000010 for *srl*, and 000011 for *sra*. They shift *rs* by the amount in *rt* and store the result in *rd*. In an arithmetic shift, the high-order bit is shifted in. In logical shifts, a 0 is shifted in.

2.9 Extend the MiniMIPS to include a *trap* and a return from exception (*rfe*) instruction.

2.10 Extend the MiniMIPS to support exceptions. In particular, *add* and *sub* instructions can cause an arithmetic overflow exception, and illegal instructions should also cause an exception.

3
Communication Protocols

(The Chief Programmer) needs great talent, ten years experience and considerable systems and applications knowledge, whether in applied mathematics, business data handling, or whatever.
—Fred P. Brooks, *The Mythical Man Month*

Never put off until run time what you can do at compile time.
—David Gries

In the beginning was the word. But by the time the second word was added to it, there was trouble. For with it came syntax ....
—John Simon

In this chapter we describe the *handshake protocols* that are used to implement channel communication and methods of translation from a channel-level specification to a protocol-level description.

3.1 BASIC STRUCTURE

As we saw in Chapter 1, a channel communication can be implemented through a handshake protocol on two or more signal wires. One or more wires are used to *request* the communication; the others are used to *acknowledge* completion of the communication. In this section we describe how handshake protocols can be described in VHDL. Throughout the rest of the chapter, we use similar language constructs to describe alternative protocols. One of the alternative protocols from this chapter is given below. This protocol is for a passive/active shop which communicates one bit of data (i.e., one of two different types of wine) using a *dual-rail protocol*. In this section we describe the basic structure and syntax. The behavior and rationale of this and alternative protocols are described in future sections.

```
library ieee;
use ieee.std_logic_1164.all;
use work.nondeterminism.all;
use work.handshake.all;
entity shopPA_dualrail is
  port(bottle1:in std_logic;
       bottle0:in std_logic;
       ack_wine:buffer std_logic:='0';
       shelf1:buffer std_logic:='0';
       shelf0:buffer std_logic:='0';
       ack_patron:in std_logic);
end shopPA_dualrail;
architecture hse of shopPA_dualrail is
begin
shopPA_dualrail:process
begin
  guard(ack_patron,'0');
  guard_or(bottle0,'1',bottle1,'1');
  if bottle0 = '1' then assign(shelf0,'1',1,3);
  elsif bottle1 = '1' then assign(shelf1,'1',1,3);
  end if;
  assign(ack_wine,'1',1,3);
  guard(ack_patron,'1');
  vassign(shelf0,'0',1,3,shelf1,'0',1,3);
  guard_and(bottle0,'0',bottle1,'0');
  assign(ack_wine,'0',1,3);
end process;
end hse;
```

The first difference from the channel-level model is inclusion of the *handshake* package instead of the *channel* package. This package can be found in Appendix A, and it includes the definitions of the procedures: *guard*, *guard_or*, *guard_and*, *assign*, and *vassign*. The next change is the replacement of the *WineryShop* port of type channel with ports of type *std_logic*. The first of these new ports are *bottle1* and *bottle0*, which are used to tell the shop to accept a new bottle of wine of type 1 or type 0, respectively. The next port is *ack_wine*, which is used to indicate acknowledgment of the wine delivery to the shop. Although this signal is an *output*, it must be of mode *buffer* because it is also tested within the architecture. This port is initialized to '0'. The *ShopPatron* channel is implemented with the next three ports. The first two ports, *shelf1* and *shelf0*, are output ports used to communicate a single bit of data to the patron (i.e., the type of wine). The last port is *ack_patron*, which is used by the *patron* to acknowledge receipt of the wine.

In this protocol, the first thing the shop does is wait until *ack_patron* is '0'. This is accomplished using the *guard* procedure. The procedure *guard(s,v)* takes a signal, *s*, and a value, *v*, and replaces the following code:

```
if (s /= v) then
    wait until s = v;
end if;
```

The reason for the *if* statement in the *guard* procedure is that when a wait statement is encountered, if the expression is *already* true, the process stalls until the expression goes false and becomes true again. In this case, *ack_patron* starts low, so if we did not first test the signal, the process above would stall waiting for an event on *ack_patron*. However, *ack_patron* is already '0', and it does not change again unless this process sets *shelf1* or *shelf0*. This process, however, will not change these signals until it is woken up with an event on *ack_patron*. Therefore, this process is deadlocked. In order to address this problem, each wait must be predicated with a test to make sure that the expression is false before the wait statement is executed.

The next step in the protocol is to wait until either *bottle0* or *bottle1* goes high. This is accomplished using the *guard_or* procedure. The procedure *guard_or(s1,v1,s2,v2,...)* takes a set of signals and values and stalls a process until some signal s_i has taken value v_i, and is defined as follows:

```
if ((s1 /= v1) and (s2 /= v2) ... ) then
    wait until (s1 = v1) or (s2 = v2) ...;
end if;
```

After *bottle0* or *bottle1* goes high, the protocol next sets *shelf0* or *shelf1* high, depending on which of the bottle wires goes high. It sets the appropriate shelf signal using the *assign* procedure. The procedure *assign(s,v,l,u)* takes a signal, *s*, a value, *v*, a lower bound of delay, *l*, and an upper bound of delay, *u*, and it replaces the following code:

```
assert (s /= v)
    report ''Vacuous assignment!''
    severity failure;
s <= v after delay(l,u);
wait until s = v;
```

The *assert statement* causes the assignment to fail if the signal already has the value being assigned. It is often the case that this indicates a error in the specification. The *delay* function in the actual signal assignment is from the *nondeterminism* package (see Appendix A). This function is used to simulate assignments that happen after a random delay. The function delay takes a lower and upper bound and returns a delay between the two. The use of delay makes it easier to debug, as it separates the transitions in time. The wait statement at the end is necessary because in VHDL when signal assignment statements are executed, they schedule events but do not actually change the value of the signal. If we removed the wait statement from the *assign* procedure, the shelf signal could be scheduled to change in two time units. Next, *ack_wine* could be scheduled to change in one time unit. The result would be that *ack_wine* would go high before the shelf signal changes value. Since the desired behavior is to order the two signal assignments, it is necessary to add the wait statement in the *assign* procedure.

The *assign* procedure also allows parallel assignments. For example, the procedure call $assign(s1,v1,l1,u1,s2,v2,l2,u2)$ replaces the following code:

```
assert ((s1 /= v1) or (s2 /= v2))
  report ''Vacuous assignment!''
  severity failure;
s1 <= v1 after delay(l1,u1);
s2 <= v2 after delay(l2,u2);
wait until (s1 = v1) and (s2 = v2);
```

After the appropriate shelf signal goes high, *ack_wine* is set high and the shop waits for *ack_patron* to go high. Next, the shop resets the shelf signal. At this point, only one of the two shelf signals is high. Therefore, the assignment will only result in a change in one of the two signals. Therefore, we use the *vacuous assign (vassign)* procedure, as defined below.

```
if (s /= v) then
  s <= v after delay(l,u);
  wait until s = v;
end if;
```

The *vassign* procedure also allows parallel assignments as defined below.

```
if (s1 /= v1) then
  s1 <= v1 after delay(l1,u1);
end if;
if (s2 /= v2) then
  s2 <= v2 after delay(l2,u2);
end if;
if (s1 /= v1) or (s2 /= v2) then
  wait until s1 = v1 and s2 = v2;
end if;
```

After the shelf signals are reset, the shop waits until both bottle signals are low. This is done using the *guard_and* procedure. The procedure $guard\_and(s1,v1,s2,v2,\ldots)$ takes a set of signals (i.e., $s1, s2, \ldots$) and a set of values (i.e., $v1, v2, \ldots$), and it stalls a process until each signal s_i has taken value v_i. In other words, it replaces the following code:

```
if ((s1 /= v1) or (s2 /= v2) ... ) then
  wait until s1 = v1 and s2 = v2 ...;
end if;
```

Finally, the shop protocol assigns *ack_wine* to '0' and loops back to the beginning.

3.2 ACTIVE AND PASSIVE PORTS

A channel connects two processes for communication. The point of connection to each process is called a *port*. For each channel, one port must be *active* while the other is *passive*. The process connected to the *active* port initiates each communication on the channel. The process connected to the *passive* port must patiently wait for communication to be initiated. The first step in translation from a channel-level specification to a handshaking-level specification is to assign the *active* and *passive* side of each channel. The decision can be annotated by replacing the function call *init_channel* with the call *active* or *passive* within the entity. Although this does not change the simulation behavior, it documents the decision. If we have decided to make the connection with the winery passive and the connection with the patron active, we would change the entity for the *shop* as follows:

```
entity shopPA is
  port(wine_delivery:inout channel:=passive;
       wine_selling:inout channel:=active);
end shopPA;
```

One must always assign exactly one side to be passive and the other side to be active. If the *probe* function is not used by either process that communicates on the channel, the choice of which port to make passive is arbitrary. If, on the other hand, the *probe* function is used in one of the processes, the process using the *probe* function must be connected to the passive port. It is illegal for the *probe* function to be used on both sides of a channel, as this would imply that both sides of the channel wait passively for communication.

3.3 HANDSHAKING EXPANSION

The next step of the compilation process is to introduce signal wires that are to be used to implement the channel communication. We first consider simple *bundled data* communication and later introduce methods of encoding the data. Synchronization of a bundled data communication is typically achieved using two wires: one for *requests* and another for *acknowledgments*. The *request* wire is controlled by the *active* side of the communication, and the *acknowledge* wire is controlled by the *passive* side.

For the passive/active *shop*, we would introduce the wires *req_wine* and *ack_wine* between the winery and the shop. We would also introduce the wires *req_patron* and *ack_patron* between the shop and the patron. There is also data that must be transmitted between the processes, namely the bottle of wine which is transmitted on the *bottle* and *shelf* wires. The bottle of wine can be one of eight different types, which means that it would require a minimum of 3 bits to encode the type. The bundled data method can use this minimum encoding. The channels after handshaking expansion are shown

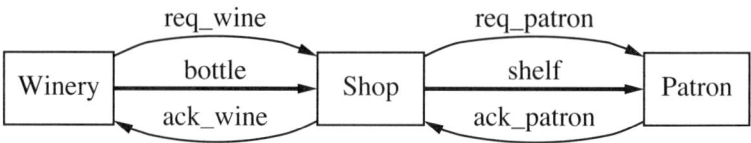

Fig. 3.1 Handshaking expansion for passive/active *shop* using bundled-data.

in Figure 3.1. The *bottle* is transmitted using a 3-bit *std_logic_vector* in the bundled data protocol. One possible description of the active *winery* process after handshaking expansion is shown below.

```
library ieee;
use ieee.std_logic_1164.all;
use work.nondeterminism.all;
use work.handshake.all;
entity winery_bundled is
   port(req_wine:buffer std_logic:='0';
        ack_wine:in std_logic;
        bottle:buffer std_logic_vector(2 downto 0):="000");
end winery_bundled;
architecture two_phase of winery_bundled is
begin
winery_bundled_2phase:process
begin
   bottle <= selection(8,3);
   wait for delay(5,10);
   assign(req_wine,not req_wine,1,3);  -- call shop
   guard(ack_wine,req_wine);           -- wine delivered
end process;
end two_phase;
```

The channel in the entity has been replaced by a request, *req_wine*, and acknowledge, *ack_wine*, wire of type *std_logic*, and a 3-bit *std_logic_vector*, *bottle*, to carry the type of wine. As in the channel-level specification, the first two lines of the process randomly select a type of wine and wait a random delay of 5 to 10 minutes before delivering it to the shop. With bundled data, it is necessary that the data be applied to the data lines before the request signal is asserted. In other words, the *shop* needs to be able to safely assume that when it is requested to take that bottle of wine, there is valid data sitting at its inputs when it sees *req_wine* change. This approach requires a timing assumption called a *bundling constraint*. This assumption basically says that the data must be applied to the data lines before the request is asserted. The bundling constraint is simulated in the model using a wait statement.

The next two lines of the process implement the *send* procedure call from the channel-level model. We initially assume that we have a new bottle of wine to transmit. To indicate this to the shop, the winery changes *req_wine* from '0' to '1'. This is accomplished by the call to *assign*, which sets *req_wine* to its complement. The winery then waits for acknowledgment that the wine has been delivered. This is indicated by observing *ack_wine* change from '0' to '1'. This is accomplished by the call to *guard*, which stalls until *ack_wine* equals *req_wine*. These two signal changes complete the transaction of delivery of wine from the winery to the shop.

The model now loops back around and selects a new type of wine. The second time around, the process changes *req_wine* from '1' to '0', and it then waits for *ack_wine* to go to '0'. These two transitions complete another transaction. Since two transitions are required per transaction, this protocol is known as *two-phase handshaking* or *two-cycle signaling*. This protocol is also called *transition signaling*, since communication transactions happen on each transition of the control signals. Next, let us consider the two-phase handshaking expansion of the passive *patron* shown below.

```
patronP_bundled_2phase: process
begin
    guard(req_patron,not ack_patron);      -- shop calls
    bag <= shelf after delay(2,4);
    wait for delay(5,10);
    assign(ack_patron,not ack_patron,1,3); -- buys wine
end process;
```

The handshaking expansion above implements the *receive* procedure call from the channel-level model. It begins by waiting for a request to receive a new bottle of wine. This is indicated when *req_patron* differs from *ack_patron*. After *req_patron* changes to '1', the patron knows there is stable data on the *shelf* data lines which it can copy into the *bag*. It then must wait for a delay equal to the bundling constraint. At this point, it can acknowledge acceptance of the wine. This is accomplished by toggling the *ack_patron* line. The shop does a *receive* followed by a *send*, and it is expanded in a similar fashion, as shown below.

```
shop_bundled_2phase: process
begin
    guard(req_wine,not ack_wine);          -- winery calls
    shelf <= bottle after delay(2,4);
    wait for delay(5,10);
    assign(ack_wine,not ack_wine,1,3);     -- wine arrives
    assign(req_patron,not req_patron,1,3); -- call patron
    guard(ack_patron,req_patron);          -- patron buys wine
end process;
```

64 COMMUNICATION PROTOCOLS

Alternatively, we could have decided to use *four-phase handshaking*, also known as *four-cycle signaling*. The entity remains the same, only the protocol in the architecture changes, as shown below.

```
winery_bundled_4phase: process
begin
  bottle <= selection(8,3);
  wait for delay(5,10);
  assign(req_wine,'1',1,3);    -- call shop
  guard(ack_wine,'1');          -- wine delivered
  assign(req_wine,'0',1,3);    -- reset req_wine
  guard(ack_wine,'0');          -- ack_wine resets
end process;
```

Again, the protocol starts with *req_wine* being changed from '0' to '1', and it waits for *ack_wine* to change from '0' to '1'. These two signal changes again complete the transaction of delivery of wine from the winery to the shop. However, at this point, the *req_wine* and *ack_wine* signals are reset before there can be a new transaction. This protocol is also called *level signaling*, since communication happens only when the request signal is '1'. The four-phase bundled data specification for the passive *patron* is below.

```
patronP_bundled_4phase: process
begin
  guard(req_patron,'1');        -- shop calls
  bag <= shelf after delay(2,4);
  wait for delay(5,10);
  assign(ack_patron,'1',1,3);  -- patron buys wine
  guard(req_patron,'0');        -- req_patron resets
  assign(ack_patron,'0',1,3);  -- reset ack_patron
end process;
```

The *patron* again waits for *req_patron* to go to '1'. This indicates that it is safe to copy the data from the *shelf* into the *bag*. After waiting 5 ns to satisfy the bundling constraint, it then sets *ack_patron* to '1' to acknowledge that it has latched the wine. Then it waits for *req_patron* to reset to '0', and it resets *ack_patron* to '0'. The *shop* process is similar, and it is shown below.

```
shop_bundled_4phase: process
begin
  guard(req_wine,'1');          -- winery calls
  shelf <= bottle after delay(2,4);
  wait for delay(5,10);
  assign(ack_wine,'1',1,3);    -- shop receives wine
  guard(req_wine,'0');          -- req_wine resets
  assign(ack_wine,'0',1,3);    -- reset ack_wine
  assign(req_patron,'1',1,3);  -- call patron
  guard(ack_patron,'1');        -- patron buys wine
  assign(req_patron,'0',1,3);  -- reset req_patron
  guard(ack_patron,'0');        -- ack_patron resets
end process;
```

Although this protocol may appear to be more complex in that it requires twice as many transitions of signal wires, it often leads to simpler circuitry. This means that when operator delays dominate communication costs, four-phase is better. However, if transmission delays dominate communication costs, two-phase is better.

3.4 RESHUFFLING

Consider the *shop_bundled_4phase* process above. It is not possible to implement this specification directly as a circuit. Initially, all the signal wires are '0', and the circuit is supposed to wait for a call from the winery (i.e, wait for *req_wine* to change to '1'). After the wine has arrived and the signal *req_wine* and *ack_wine* have been reset, the value of the signal wires are again all '0'. At this point, however, the circuit for the shop must call the patron (i.e., *req_patron* is enabled to change to '1'). In other words, when all signal wires are '0', the circuit is in a state of confusion. At this point, should the shop wait for a call from the winery or call the patron?

One way to fix this problem is to *reshuffle* the order of the assignments and guards. Since in the four-phase protocol the setting of the request and acknowledgment signals to '0' are used only to reset the variables, they can be asserted anywhere as long as cyclic order is maintained. In other words, we could move up calling the patron to immediately after the wine arrives. This would change the specification of the shop protocol as follows:

```
Shop_PA_reshuffled: process
begin
    guard(req_wine,'1');        -- winery calls
    shelf <= bottle after delay(2,4);
    wait for delay(5,10);
    assign(ack_wine,'1',1,3);   -- shop receives wine
    assign(req_patron,'1',1,3); -- call patron
    guard(req_wine,'0');        -- req_wine resets
    assign(ack_wine,'0',1,3);   -- reset ack_wine
    guard(ack_patron,'1');      -- patron buys wine
    assign(req_patron,'0',1,3); -- reset req_patron
    guard(ack_patron,'0');      -- ack_patron resets
end process;
```

Another interesting reshuffling is to delay the wait on the reset of the acknowledgment for an active communication. Consider the guard on the signal *ack_patron* going to '0'. It is not actually necessary to wait for it to reset until it is necessary to set *req_patron* to '1' again. Therefore, it can be delayed until just before this point, as shown below. The first time the guard on *ack_patron* is encountered, it is already '0', so the guard does nothing. For this reason, this guard is called *vacuous*. This protocol is termed *lazy-active*.

```
Shop_PA_lazy_active: process
begin
  guard(req_wine,'1');            -- winery calls
  shelf <= bottle after delay(2,4);
  wait for delay(5,10);
  assign(ack_wine,'1',1,3);       -- shop receives wine
  guard(ack_patron,'0');          -- ack_patron resets
  assign(req_patron,'1',1,3);     -- call patron
  guard(req_wine,'0');            -- req_wine resets
  assign(ack_wine,'0',1,3);       -- reset ack_wine
  guard(ack_patron,'1');          -- patron buys wine
  assign(req_patron,'0',1,3);     -- reset req_patron
end process;
```

Care must, however, be taken when reshuffling as it can introduce *deadlock*. Consider what would happen if the patron moved in at the winery and the handshaking is reshuffled as shown below.

```
Winery_Patron: process
begin
  bottle <= selection(8,3);
  wait for delay(5,10);
  assign(req_wine,'1',1,3);       -- call shop
  guard(ack_wine,'1');            -- wine delivered
  guard(req_patron,'1');          -- shop calls patron
  bag <= shelf after delay(2,4);
  wait for delay(5,10);
  assign(ack_patron,'1',1,3);     -- patron buys wine
  guard(req_patron,'0');          -- req_patron resets
  assign(ack_patron,'0',1,3);     -- reset ack_patron
  assign(req_wine,'0',1,3);       -- reset req_wine
  guard(ack_wine,'0');            -- ack_wine resets
end process;
```

The following sequence of events would cause the system to deadlock:

req_wine+, *ack_wine+*, *req_patron+*, *ack_patron+*

The *shop* is waiting for *req_wine* to go to '0' while the combined *winery/patron* would be waiting for *req_patron* to go to '0'.

3.5 STATE VARIABLE INSERTION

As we saw in Chapter 1, another alternative to solving the state coding problem is to add a *state variable*. Consider again the passive/active shop before reshuffling. Another way to solve the state coding problem is to add a state variable as shown below.

```
Shop_PA_SV:process
begin
  guard(req_wine,'1');           -- winery calls
  shelf <= bottle after delay(2,4);
  wait for delay(5,10);
  assign(ack_wine,'1',1,3);      -- shop receives wine
  assign(x,'1',1,3);             -- set x
  guard(req_wine,'0');           -- req_wine resets
  assign(ack_wine,'0',1,3);      -- reset ack_wine
  assign(req_patron,'1',1,3);    -- call patron
  guard(ack_patron,'1');         -- patron buys wine
  assign(x,'0',1,3);             -- reset x
  assign(req_patron,'0',1,3);    -- reset req_patron
  guard(ack_patron,'0');         -- ack_patron resets
end process;
```

We will see later how to find a good insertion point for state variables.

3.6 DATA ENCODING

The advantages of the bundled data approach described above are that it is simple, it allows the use of standard combinational datapath components, and it still allows adaptivity to changes in operating conditions. The disadvantage is that it does not allow any speedup due to data dependencies. A second approach encodes each data line using two wires, and it is thus called *dual-rail*. One wire is set to '1' to indicate that the data line is '0' while another is set to '1' to indicate that the data line is '1'. If both wires are '0', this indicates there is no valid data on the data lines. Both wires are not allowed to be '1' at the same time. This scheme can require up to twice as many wires (8 wires per channel for the wine shop example as opposed to 5 for bundled data). However, it does allow for data-dependent delay variations.

To simplify the discussion, we first consider encoding a single-bit of data, and later we show how to expand it to multibit channels. Figure 3.2 shows the wine shop example using dual-rail encoded data. Observe that there is no longer any need for explicit request wires. For example, the data wires, *bottle0* and *bottle1*, replace the *req_wine* signal. The *winery* process using a dual-rail data encoding is shown below.

```
winery_dual_rail:process
  variable z:integer;
begin
  z:=selection(2);
  case z is
    when 1 =>
      assign(bottle1,'1',1,3);
    when others =>
      assign(bottle0,'1',1,3);
```

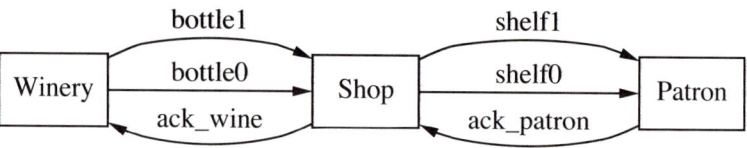

Fig. 3.2 Handshaking expansion for passive/active *wine_shop* example using dual-rail data for a single bit.

```
      end case;
      guard(ack_wine,'1');
      vassign(bottle1,'0',1,3,bottle0,'0',1,3);
      guard(ack_wine,'0');
    end process;
```

Initially, both *bottle0* and *bottle1* are low. This indicates that there is no bottle of wine on the channel. The *selection* procedure is used to randomly select between two different types of wine. Depending on the result, either *bottle0* or *bottle1* is set high. The winery then waits for *ack_wine* to go to '1' to indicate that the bottle of wine has been accepted by the shop. It then resets its data wires and waits for *ack_wine* to go to '0'. The processes describing the dual rail protocols for the *shop* and the *patron* are shown below.

```
    shopPA_dual_rail:process
    begin
      guard(ack_patron,'0');
      guard_or(bottle0,'1',bottle1,'1');
      if bottle0 = '1' then assign(shelf0,'1',1,3);
      elsif bottle1 = '1' then assign(shelf1,'1',1,3);
      end if;
      assign(ack_wine,'1',1,3);
      guard(ack_patron,'1');
      vassign(shelf0,'0',1,3,shelf1,'0',1,3);
      guard_and(bottle0,'0',bottle1,'0');
      assign(ack_wine,'0',1,3);
    end process;

    patronP_dualrail:process
    begin
      guard_or(shelf1,'1',shelf0,'1');
      assign(ack_patron,'1',1,3);
      guard_and(shelf1,'0',shelf0,'0');
      assign(ack_patron,'0',1,3);
    end process;
```

DATA ENCODING 69

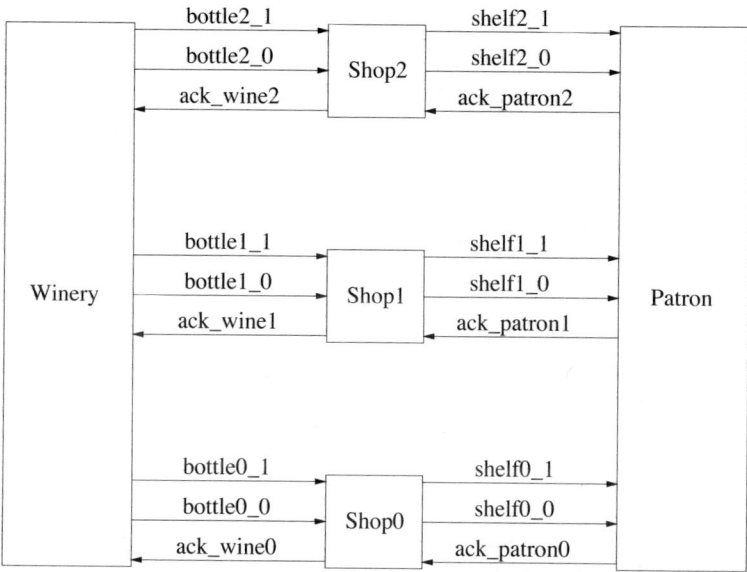

Fig. 3.3 Handshaking expansion for passive/active *wine_shop* example using dual-rail data for 3 bits.

The protocol used for the shop has been reshuffled to produce a better circuit. It starts by making sure that the *ack_patron* signal is '0'. This indicates that the *patron* is currently ready to accept a new bottle of wine. It then waits until the winery sends a bottle of wine (i.e., *bottle0* or *bottle1* is '1'). Once there is a new bottle of wine, it determines which of the two data wires went to '1' and sets the corresponding output data wire to '1' (i.e., if *bottle0* is '1', then *shelf0* should be set to '1'). It then acknowledges that it received the wine from the winery by setting *ack_wine* to '1'. It then waits to receive an acknowledgment that the patron received the wine (i.e., *ack_patron* has gone to '1'). It then resets its data wires. Finally, it waits for the winery to reset its data wires, and it resets the *ack_wine* signal. The patron waits for a *shelf* wire to go high, sets *ack_patron* high, and resets the handshake.

In order to transmit 3 bits of data, you can simply instantiate three copies of the *shop* process and connect them as shown in Figure 3.3. The winery needs to be changed as shown below to send out the three pairs of dual-rail encoded bits. It also must collect three acknowledgments before continuing.

```
winery_dual_rail:process
  variable z:integer;
begin
  z:=selection(8);
  case z is
    when 1 =>
      assign(bottle2_0,'1',1,3,bottle1_0,'1',1,3,
```

```
              bottle0_0,'1',1,3);
  when 2 =>
    assign(bottle2_0,'1',1,3,bottle1_0,'1',1,3,
           bottle0_1,'1',1,3);
  when 3 =>
    assign(bottle2_0,'1',1,3,bottle1_1,'1',1,3,
           bottle0_0,'1',1,3);
  when 4 =>
    assign(bottle2_0,'1',1,3,bottle1_1,'1',1,3,
           bottle0_1,'1',1,3);
  when 5 =>
    assign(bottle2_1,'1',1,3,bottle1_0,'1',1,3,
           bottle0_0,'1',1,3);
  when 6 =>
    assign(bottle2_1,'1',1,3,bottle1_0,'1',1,3,
           bottle0_1,'1',1,3);
  when 7 =>
    assign(bottle2_1,'1',1,3,bottle1_1,'1',1,3,
           bottle0_0,'1',1,3);
  when others =>
    assign(bottle2_1,'1',1,3,bottle1_1,'1',1,3,
           bottle0_1,'1',1,3);
end case;
guard_and(ack_wine2,'1',ack_wine1,'1',ack_wine0,'1');
vassign(bottle2_0,'0',1,3,bottle1_0,'0',1,3,
        bottle0_0,'0',1,3);
vassign(bottle2_1,'0',1,3,bottle1_1,'0',1,3,
        bottle0_1,'0',1,3);
guard_and(ack_wine2,'0',ack_wine1,'0',ack_wine0,'0');
end process;
```

The patron must collect data from the three *shop* processes. It also must send out three separate acknowledgments as shown below.

```
patronP_dualrail:process
begin
  guard_or(shelf2_1,'1',shelf2_0,'1');
  guard_or(shelf1_1,'1',shelf1_0,'1');
  guard_or(shelf0_1,'1',shelf0_0,'1');
  assign(ack_patron2,'1',1,3,ack_patron1,'1',1,3,
         ack_patron0,'1',1,3);
  guard_and(shelf2_1,'0',shelf2_0,'0');
  guard_and(shelf1_1,'0',shelf1_0,'0');
  guard_and(shelf0_1,'0',shelf0_0,'0');
  assign(ack_patron2,'0',1,3,ack_patron1,'0',1,3,
         ack_patron0,'0',1,3);
end process;
```

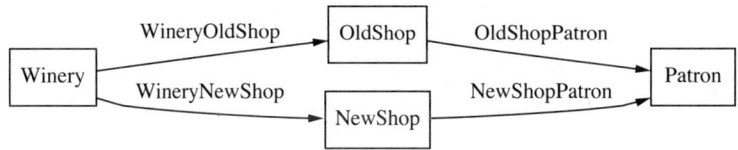

Fig. 3.4 Example with two wine shops.

3.7 EXAMPLE: TWO WINE SHOPS

As another example, let us consider the handshaking expansion of the example from Chapter 2 with two wine shops depicted in Figure 3.4. In this example, the winery randomly decides which shop to deliver its wine to, and the patron probes each shop before deciding which shop to buy wine from. The processes for the *winery*, *shop*, and *patron* are shown below.

```
winery5:process
variable z:integer;
begin
  bottle <= selection(8,3);
  wait for delay(5,10);
  z:=selection(2);
  case z is
  when 1 =>
    send(WineryNewShop,bottle);
  when others =>
    send(WineryOldShop,bottle);
  end case;
end process winery5;

shop:process
begin
  receive(WineryShop,shelf);
  send(ShopPatron,shelf);
end process shop;

patron2:process
begin
  if (probe(OldShopPatron)) then
    receive(OldShopPatron,bag);
    wine_drunk <= wine_list'val(conv_integer(bag));
  elsif (probe(NewShopPatron)) then
    receive(NewShopPatron,bag);
    wine_drunk <= wine_list'val(conv_integer(bag));
  end if;
  wait for delay(5,10);
end process patron2;
```

72 COMMUNICATION PROTOCOLS

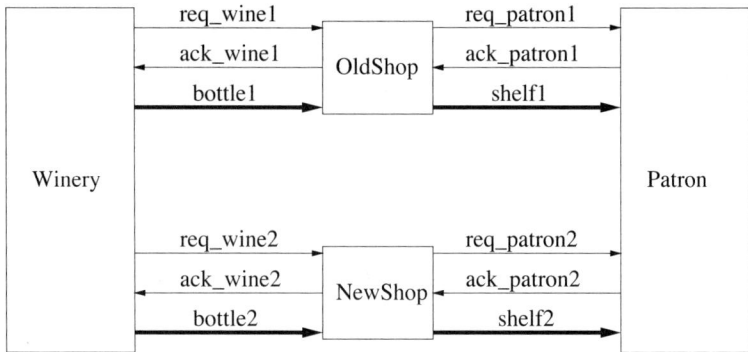

Fig. 3.5 Handshaking expansion with two wine shops.

The first thing to notice is that the patron probes its two channels. Therefore, the patron must be connected to the passive ports of the *OldShopPatron* and *NewShopPatron* channels. This implies that the *shop* must be connected to the active ports. For the *WineryOldShop* and *WineryNewShop* channels, the choice of active and passive ports is arbitrary since neither process probes these channels. Let us connect the winery to the active ports of the *WineryOldShop* port and the *WineryNewShop* port. Assuming bundled data, the block diagram after handshaking expansion is shown in Figure 3.5. The *winery* and *patron* processes after handshaking expansion are shown below. The *shop* process is identical to the one we saw earlier. One key thing to notice is in the *patron* process, the *probe* function calls turn into tests on *req_patron1* and *req_patron2*.

```
winery:process
variable z : integer;
begin
  z := selection(2);
  bottle <= selection(8,3);
  wait for delay(5,10);
  case z is
    when 1 =>
      bottle1 <= bottle after delay(2,4);
      wait for 5 ns;
      assign(req_wine1,'1',1,3);    -- call winery
      guard(ack_wine1,'1');         -- wine delivered
      assign(req_wine1,'0',1,3);    -- reset req_wine
      guard(ack_wine1,'0');         -- ack_wine resets
    when others =>
      bottle2 <= bottle after delay(2,4);
      wait for 5 ns;
      assign(req_wine2,'1',1,3);    -- call winery
      guard(ack_wine2,'1');         -- wine delivered
```

```
              assign(req_wine2,'0',1,3);   -- reset req_wine
              guard(ack_wine2,'0');        -- ack_wine resets
          end case;
        end process;

        patronP:process
        begin
          if (req_patron1 = '1') then
            bag <= shelf1 after delay(2,4);
            wait for delay(5,10);
            assign(ack_patron1,'1',1,3); -- patron buys wine
            guard(req_patron1,'0');      -- req_patron resets
            assign(ack_patron1,'0',1,3); -- reset ack_patron
            wine_drunk <= wine_list'val(conv_integer(bag));
          elsif (req_patron2 = '1') then
            bag <= shelf2 after delay(2,4);
            wait for delay(5,10);
            assign(ack_patron2,'1',1,3); -- patron buys wine
            guard(req_patron2,'0');      -- req_patron resets
            assign(ack_patron2,'0',1,3); -- reset ack_patron
            wine_drunk <= wine_list'val(conv_integer(bag));
          end if;
          wait for delay(1,2);
        end process;
```

3.8 SYNTAX-DIRECTED TRANSLATION

We conclude this chapter by describing an asynchronous design procedure where the handshaking expansion step directly determines the circuit implementation. These procedures are based upon the software compiler idea of *syntax-directed translation*. In these procedures, each language construct in the high-level channel-based specification corresponds to a particular circuit structure. These circuit structures are then composed together as dictated by the structure of the program.

Consider the *shop* process, repeated for convenience below.

```
shop:process
begin
  receive(WineryShop,shelf);
  send(ShopPatron,shelf);
end process shop;
```

First, the process statement implies a loop. A loop of the form *while(cond) loop S; end loop* can be translated into the circuit shown in Figure 3.6(a). This circuit operates under the two-phase model. In this circuit, the *req* wire feeds a *merge gate*, M. Typically, the merge gate is implemented with an XOR. Assuming that the input coming from the process is initially 0, a change on the *req* wire to 1 causes the input to the *selector module*, SEL, to go to 1. The

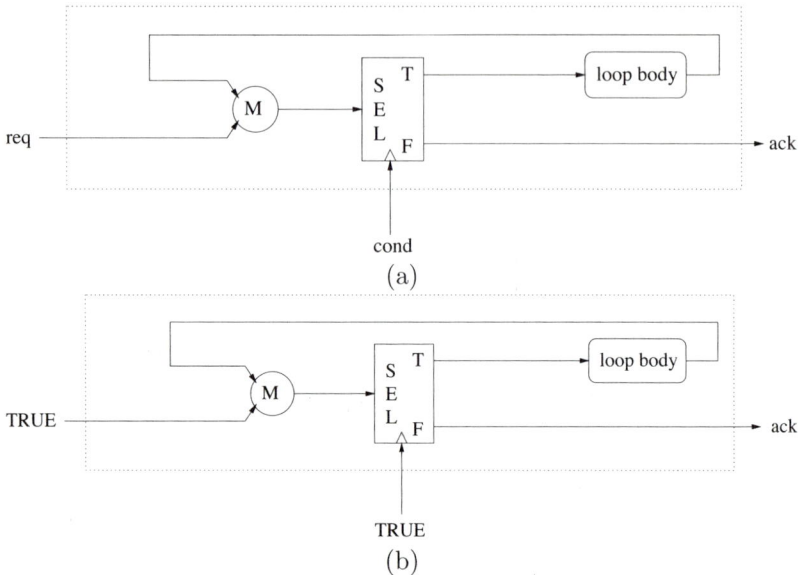

Fig. 3.6 (a) Circuit for looping constructs. (b) Circuit for the *shop* process statement.

SEL module then samples the *cond* wire. If the wire is *true*, it requests the loop to execute. When the loop is done, it acknowledges to the merge gate, which causes another request to the selector. If the condition is false, the entire block acknowledges to the environment that it is done looping. For the process statement, the *req* is *true*, which starts the *shop* operating. A process never terminates, so the condition, *cond*, is set to *TRUE*. The result is shown in Figure 3.6(b).

The next part to consider is the signal *shelf*. The *receive* statement assigns data from the channel *WineryShop* to the signal *shelf*. The circuit to implement the *signal assignment* is shown in Figure 3.7. In this figure, thin lines represent handshake signals and thick lines represent data. When this circuit gets a request, the *enable module* (EN) allows the data on the input lines, *In*, to drive the input bus to the register, *datain*. The EN module then issues a request to the *CALL module*, which makes a request to the register to latch its input data, which will appear on the signal *shelf* when the register latches. The purpose of the CALL module is to coordinate multiple assignments to a single signal. For example, if a signal has two assignments to it in the specification, it can be implemented as shown in Figure 3.8.

To complete implementation of the *receive* statement, we must now construct a circuit to handle synchronization on the channel. Such a circuit is shown in Figure 3.9. This circuit starts with a request to the CALL module. The CALL module forwards the request to the C-element. When the *winery*

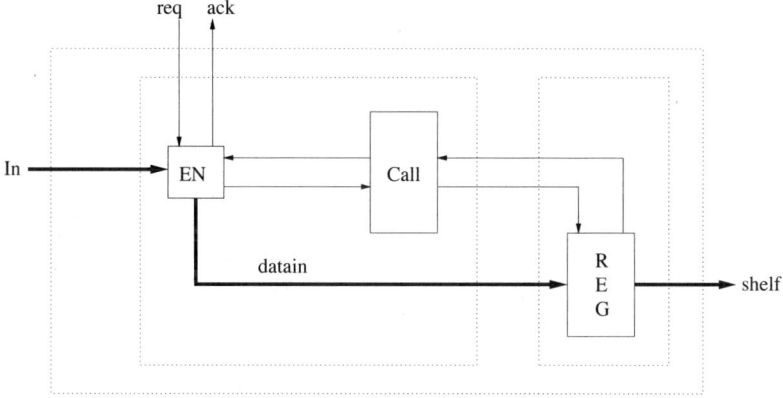

Fig. 3.7 Circuit for assignment to *shelf*.

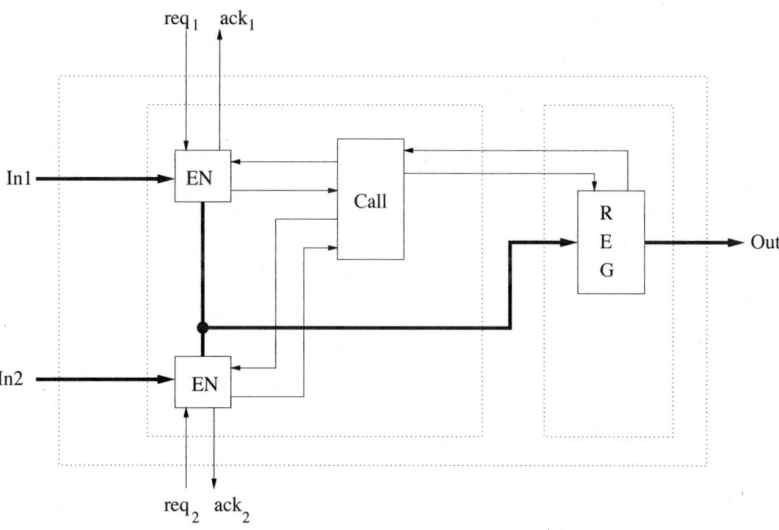

Fig. 3.8 Circuit to implement assignment to a signal from two locations.

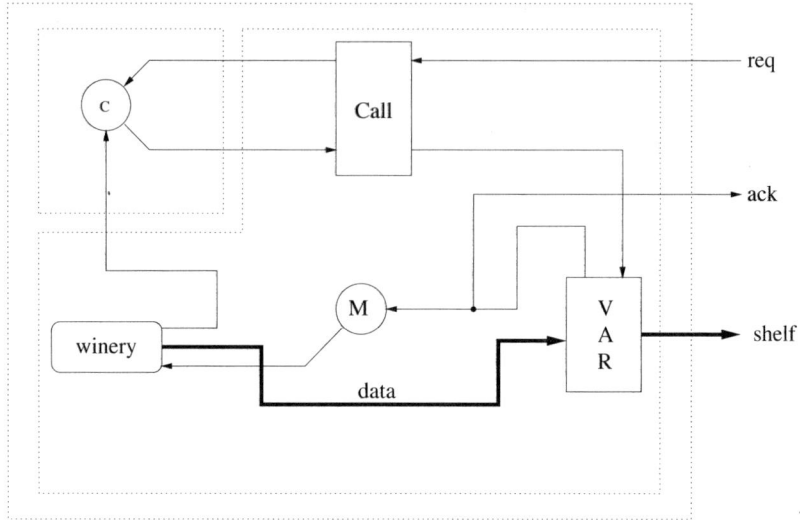

Fig. 3.9 Circuit for a receive statement.

is ready to send data, the other input to the C-element is also asserted. This causes the C-element to generate a transition to the CALL module, which then issues a request for the data sitting on the channel to be latched into the variable (VAR). Note that VAR is a symbol for the circuit shown in Figure 3.7. Once the data has been latched, it is output on the signal *shelf* and the VAR circuit sends an acknowledgment to the unit that requested the communication and also to the *winery* to tell it that the data sent has been latched. Note that the CALL module and the Merge (M) are included because there may be multiple places in the specification where a receive on a given channel may get executed.

The last circuit module needed is one to implement the *send* statement, which is shown in Figure 3.10. This circuit begins with a request to the EN module. This causes the data sitting on the signal *shelf* to be copied to the data wires connected to the *patron*. The request is also forwarded to the CALL module, which forwards it the patron. When the patron acknowledges that the data has been received, the CALL module forwards this acknowledgment back to the EN module. This completes the send, so the block acknowledges its requester.

Although not needed for this circuit, another interesting circuit construct is for selections of the form

      if (cond1) then
        S1;
      elsif (cond2) then
        S2;

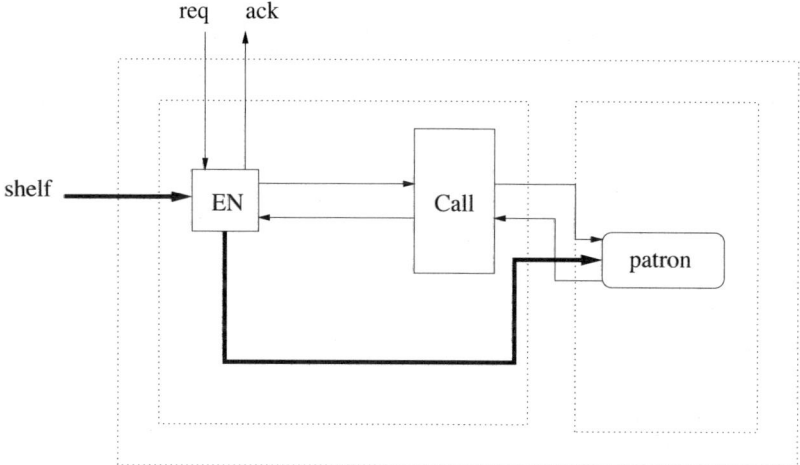

Fig. 3.10 Circuit for a send statement.

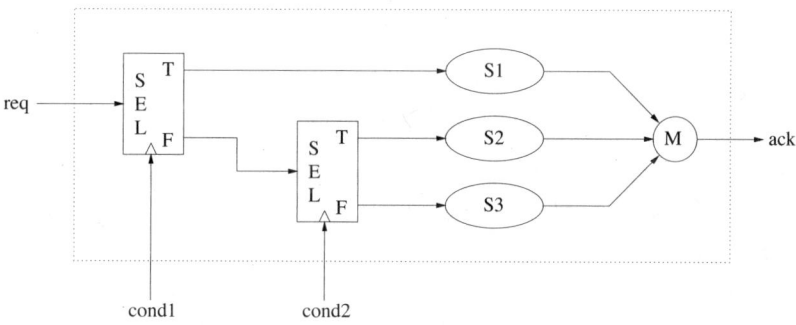

Fig. 3.11 Circuit for a selection statement.

```
    else
        S3;
    end if;
```

A circuit implementing this type of construct is shown in Figure 3.11. This circuit is started by a transition on the *req* signal. It then samples the *cond1* signal. If this signal is *true*, it starts the operations labeled $S1$. Otherwise, it requests the next *SEL* module to check *cond2*. If this signal is *true*, it starts $S2$. Otherwise, it starts $S3$. As soon as either $S1$, $S2$, or $S3$ is complete, this circuit acknowledges that the selection has completed.

Using these circuit modules, we can now stitch together a complete circuit for the *shop*. If two statements are executed in sequence, they can be connected by taking the acknowledgment from one and using it as the request

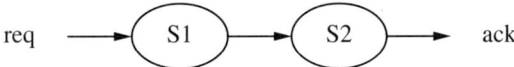

Fig. 3.12 Circuit for sequential composition of two statements.

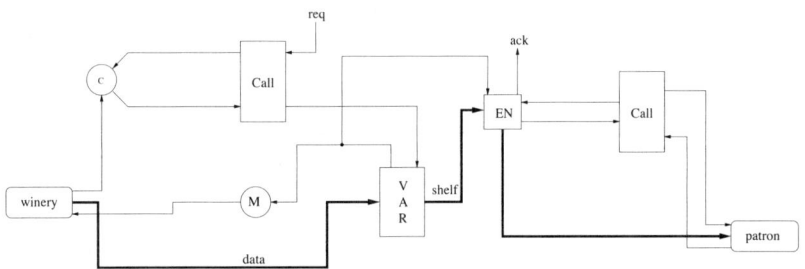

Fig. 3.13 Circuit for a receive followed by a send.

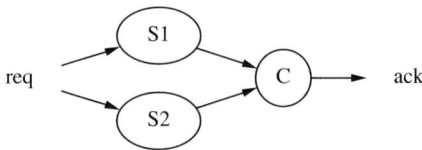

Fig. 3.14 Circuit for parallel composition of two statements.

to the next, as shown in Figure 3.12. For example, the composition of the *receive* and *send* statements from the *shop* yield the circuit shown in Figure 3.13. Although again not needed for this circuit, another useful construct is for parallel composition, and it is shown in Figure 3.14.

Composing the circuit shown in Figure 3.13 with the loop construct and expanding the VAR block yields the complete circuit shown in Figure 3.15. This circuit is clearly more complex than it needs to be. There are several *peephole optimizations* that can be performed to improve this circuit. For example, any *CALL* module that has only a single input request can be removed. After this optimization, the circuit becomes the one shown in Figure 3.16. Another optimization is to eliminate any select modules where the condition is a constant true or false. Also, if a merge only has a single input, it can be removed. The result is shown in Figure 3.17. When a data bus has only a single driver, the EN module can be removed. Finally, since one input of the merge gate is always true, we can replace the merge gate with an inverter. The final simple circuit is shown in Figure 3.18.

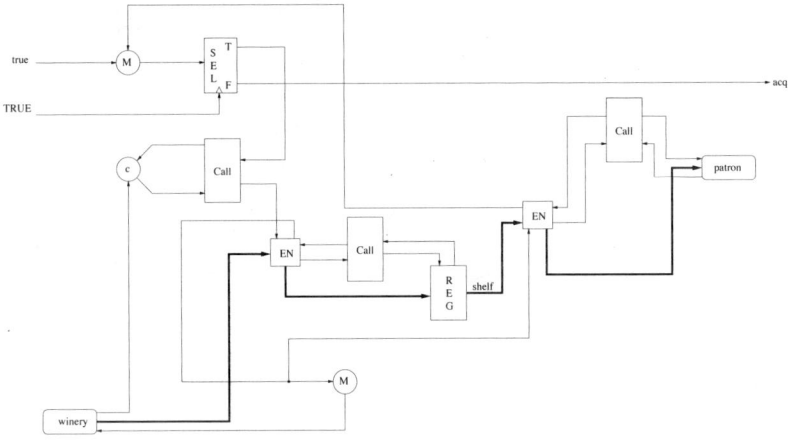

Fig. 3.15 Unoptimized circuit for the *shop*.

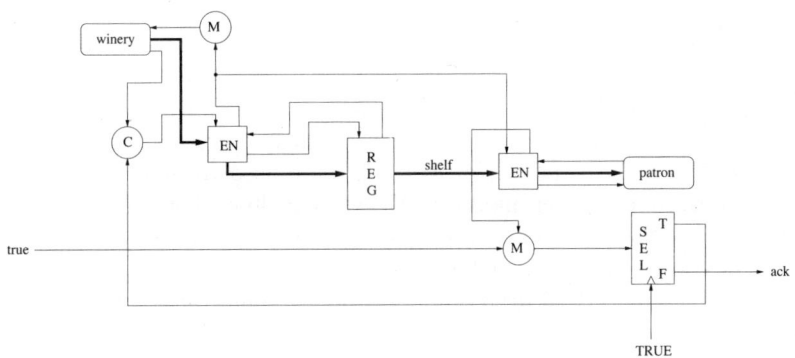

Fig. 3.16 *Shop* circuit after CALL module optimization.

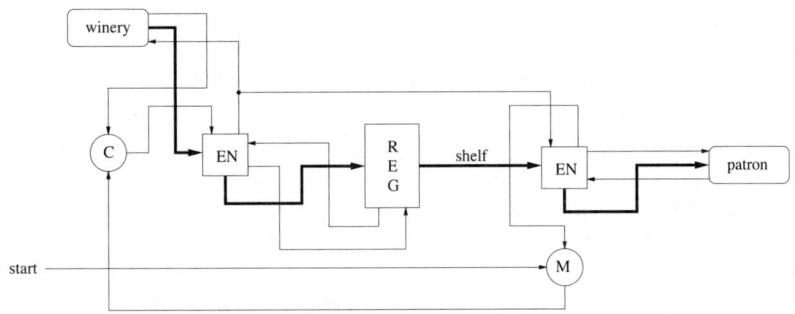

Fig. 3.17 *Shop* circuit after SEL and merge module optimizations.

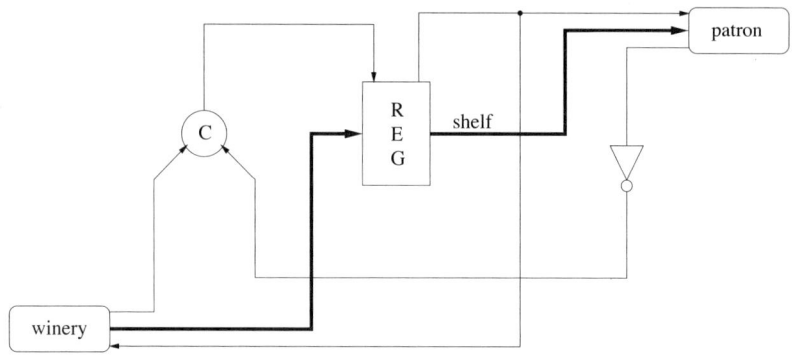

Fig. 3.18 Final circuit for the *shop*.

3.9 SOURCES

The translation process described in this chapter from communication channels to handshaking expansions follows Martin's method [254].

Some of the earliest bundled data asynchronous designs were constructed in the macromodule project at the Washington University in St. Louis. In this project, asynchronous design methods were used to build large computation systems out of macromodules [87, 88, 302, 369]. These macromodules operated using the bundled data design approach, where data wires carried with them control signals to indicate data validity. One of the most famous more recent bundled data design methods is *micropipelines*, which was proposed by Sutherland in his Turing Award paper [370]. Improvements in the latch controller for micropipelines have been proposed by several researchers to reduce the control overhead [104, 133, 275, 375]. A semicustom controllable delay element for micropipeline design is proposed by De Gloria and Olivieri [147]. In the STRiP microprocessor architecture, Dean et al. demonstrated that by using different bundling delays for each type of instruction a significant performance gain could be achieved [106]. An extension of this idea to a finer level of granularity was proposed by Nowick et al. in which based upon a quick analysis of the data, a bundled data constraint is chosen [297, 300]. This technique allows for some advantage due to data dependency even in a bundled data design.

The idea of dual-rail encoding dates back to Gilchrist et al.'s "fast" adder design (40-digit addition in $0.21\mu s$ using vacuum tube logic) [139]. Waite generalized these results to arbitrary iterative networks [396]. Armstrong et al. extended the coding scheme to m/n *codes*, where m is the *weight* of the code word (number of 1's in it) and n is the length of the code word [12]. Blaum and Bruck introduced *skew-tolerent codes*, which permit bits to be transmitted without waiting for acknowledgments, allowing for pipelined communication [43, 44]. They also allow skew in which parts of a second message can be received before the entire first message has arrived. One of the most exciting asynchronous designs which made use of dual-rail encoding is Williams and Horowitz's self-timed floating-point divider [401, 404]. It achieved faster operation than that of a comparable synchronous design since intermediate results did not need to pass through clocked registers. This divider was later used in a commercial SPARC microprocessor [400]. Recently, Singh and Nowick developed a scheme capable of doubling the throughput over Williams pipeline designs by developing a better protocol and achieving faster completion detection [354].

Dean et al. proposed an alternative data encoding scheme called *level encoded two-phase dual rail* (LEDR), which like two-phase does not return data inputs to zero, but like four-phase it uses the signal levels to encode the data values [105].

The TRIMOSBUS proposed by Sutherland et al. detected that data had been transferred to the bus simply by driving the data onto it, checking that the bus had the value it transmitted, and releasing its drive, allowing the staticizer to hold the state [371]. The TRIMOSBUS also used a three-wire protocol which allowed for a one-to-many communication to be completed without risky timing assumptions.

A completely different approach to detecting completion is to monitor current draw from the power supply [107]. When a circuit ceases to draw current, it has completed its operation. This technique has been explored further by other researchers [137, 153, 154, 218, 355].

The syntax-directed translation methods described in Section 3.8 follows the methods proposed by [52, 53]. A similarity can be seen with this work and the macromodule work described earlier in the types of modules used and the way the systems are interconnected. A peephole optimizer for macromodule-based syntax-directed translation appears in [152]. In [273] and [237], translation methods from Petri nets are developed. Nordmann proposed a method that translates from a type of flowchart, and this method was used in the design of the ILLIAC III [296]. Hirayama developed a silicon compiler for asynchronous architectures [165]. Burns and Martin proposed a translation method from CSP [64, 67, 68]. Another technique was proposed by van Berkel and Rem to automatically compile a circuit described using a language called *Tangram* to a *handshake circuit* implementation [36, 37, 40]. Akella and Gopalakrishnan developed a syntax-directed method using a new channel language called *hopCP* [3, 4, 5]. Snepscheut developed a technique to derive circuits from a trace-based language [360]. Ebergen also translated a trace-based language to DI circuits [119, 121]. A similar approach was proposed by Leung and Li [229]. Keller showed in [190] that a library composed of a merge, select, fork, and an arbitrating test-and-set module are sufficient to design any circuit.

Problems

3.1 The patron has decided that he is going to choose which of the two shops to patronize, but first he will give that shop a call to tell them that he is coming. The chosen shop must then call the winery to request a bottle of wine to be delivered. The configuration is again as shown in Figure 3.4 using bundled data. The channel-level models of the *winery*, *shop*, and *patron* are shown below. Derive the handshaking expansion, and simulate until you are convinced that it is correct.

```
winery:process
begin
  if (probe(wine_to_new_shop)) then
     bottle <= selection(8,3);
     wait for delay(5,10);
     send(wine_to_new_shop,bottle);
  elsif (probe(wine_to_old_shop)) then
     bottle <= selection(8,3);
     wait for delay(5,10);
     send(wine_to_old_shop,bottle);
  else
     wait for delay(5,10);
  end if;
end process winery;

shop:process
begin
  if (probe(wine_selling)) then
     receive(wine_delivery,shelf);
     send(wine_selling,shelf);
  end if;
  wait for delay(1,2);
end process shop;

patron:process
variable z:integer;
begin
  z := selection(2);
  if (z = 1) then
     receive(old_shop_to_patron,bag);
     wine_drunk<=wine_type'val(conv_integer(bag));
  else
     receive(new_shop_to_patron,bag);
     wine_drunk<=wine_type'val(conv_integer(bag));
  end if;
  wait for delay(1,2);
end process patron;
```

3.2 Find two legal reshufflings for the handshaking expansion from Problem 3.1. Simulate until you are convinced that they are correct.

3.3 Perform handshaking expansion on the simple fetch process shown below using bundled data.

```
process
begin
  if (probe(increment)) then
    send(pcadd,'1');
    receive(pcadd_result,pc);
    send(imem_address,pc);
    receive(imem_data,instr);
    send(instr_decode,instr);
    receive(increment);
  elsif (probe(jump)) then
    send(pcadd,offset);
    receive(pcadd_result,pc);
    send(imem_address,pc);
    receive(imem_data,instr);
    send(instr_decode,instr);
    receive(jump);
  end if;
end process;
```

3.4 Find four legal reshufflings of the handshaking expansion from Problem 3.3.

3.5 Perform handshaking expansion on your 4-bit adder design from Problem 2.1.1 using bundled data. Also, perform handshaking expansion on your environment processes. Simulate your design with its environment.

3.6 Perform handshaking expansion on your 1-bit adder design from Problem 2.1.2. Use dual-rail for data encoding. Also, perform handshaking expansion on your environment processes. Simulate your design. Using these single-bit blocks, build a 4-bit adder with its environment and simulate.

3.7 Perform handshaking expansion on your stack design from Problem 2.2 using bundled data. Create an environment and simulate a four-stage stack.

3.8 Perform handshaking expansion on your stack design from Problem 2.2 assuming only 1 bit of data encoded using dual-rail. Create an environment and simulate a four-stage stack.

3.9 Perform handshaking expansion on your shifter design from Problem 2.3 using bundled data. Create an environment and simulate a four-element shifter.

3.10 Perform handshaking expansion on your shifter design from Problem 2.3 using dual-rail data encoding. Create an environment and simulate a four-element shifter.

3.11 Perform handshaking expansion on your entropy decoder design from Problem 2.4.1 using bundled data. Create an environment and simulate.

3.12 Perform handshaking expansion on your entropy decoder design from Problem 2.4.2 using dual-rail data encoding. Create an environment and simulate a 4-bit decoder.

3.13 Using syntax-directed translation, find a circuit to implement Euclid's greatest common divisor (gcd) algorithm given below.

```
receive(A,a,B,b);
while (a != b) loop
  if (a > b) then
    a <= a - b;
    wait for 5 ns;
  else
    b <= b - a;
    wait for 5 ns;
  end if;
end loop;
end if;
send(C,a);
```

You may assume that the user is not allowed to send in operands a or b equal to 0. You can also assume the existence of a comparator which has two input buses, and after a request sets a condition which is true when operand 1 is greater than operand 2. After the condition is stable, an acknowledgment is provided. You may also assume the existence of a subtractor which has two input buses and one output bus. After getting a request, it returns an acknowledgment to indicate a stable result on the output bus.

4
Graphical Representations

One ought, every day at least, to hear a little song, read a good poem, see a fine picture, and if it were possible, to speak a few reasonable words.
—Johann Wolfgang von Goethe

Take nothing but pictures. Leave nothing but footprints. Kill nothing but time.
—Motto of the Baltimore Grotto (caving society)

One picture is worth a thousand words.
—Fred R. Barnard

In this chapter we present several methods for representing asynchronous circuits using graphs. While for large designs hardware description languages allow a clearer specification of behavior, graphs are a useful pictorial tool for small examples. They are also the underlying data structure used by virtually all automated analysis, synthesis, and verification tools. Most graphical representation methods can be loosely categorized as either *state machine*-based or *Petri net*-based. We present various different types of each and conclude with a description of *timed event/level* (TEL) *structures*, which unify some of the key properties of both.

4.1 GRAPH BASICS

In this section we present a brief introduction to basic graph terminology used in this chapter. A *graph*, G, is composed of a finite nonempty set of *vertices*,

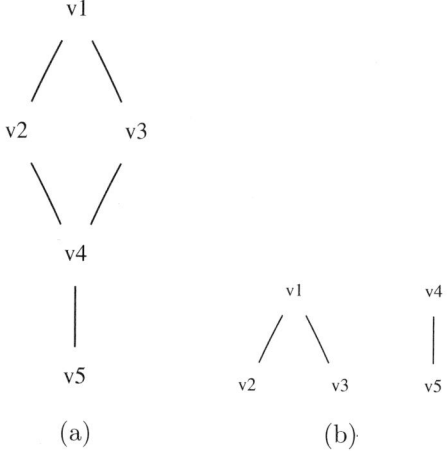

Fig. 4.1 (a) Simple undirected graph. (b) Graph which is not connected.

V, and a binary relation R on V (i.e., $R \subseteq V \times V$). A graph can be either undirected or directed.

In an *undirected graph*, R is an irreflexive and symmetric relation. By making it irreflexive, self-loops are not allowed. Since R is symmetric, for each ordered pair $(u, v) \in R$, the pair (v, u) is also in R. The set of *edges*, E, are the set of symmetric pairs in R. Each pair $\{(u, v), (v, u)\}$ is denoted (u, v) by convention. This set of edges can be empty.

Example 4.1.1 Consider the undirected graph shown in Figure 4.1(a). This graph has the following vertex set, relation, and edge set:

$$
\begin{aligned}
V &= \{v_1, v_2, v_3, v_4, v_5\} \\
R &= \{(v_1, v_2), (v_2, v_1), (v_1, v_3), (v_3, v_1), (v_2, v_4), \\
  &\quad (v_4, v_2), (v_3, v_4), (v_4, v_3), (v_4, v_5), (v_5, v_4)\} \\
E &= \{(v_1, v_2), (v_1, v_3), (v_2, v_4), (v_3, v_4), (v_4, v_5)\}
\end{aligned}
$$

In a *directed graph* or *digraph*, each ordered pair of R is called a *directed edge* or *arc*. Note that as opposed to undirected graphs, R does not need to be irreflexive or symmetric. This means that if (u, v) is an arc of a digraph that (v, u) need not also be an arc. Also, self-loops, edges from a vertex to itself, are allowed.

Example 4.1.2 Consider the *digraph* shown in Figure 4.2(a). The graph has the following set of vertices and arcs:

$$
\begin{aligned}
V &= \{v_1, v_2, v_3, v_4, v_5\} \\
E &= \{(v_1, v_2), (v_1, v_3), (v_2, v_4), (v_3, v_4), (v_4, v_5)\}
\end{aligned}
$$

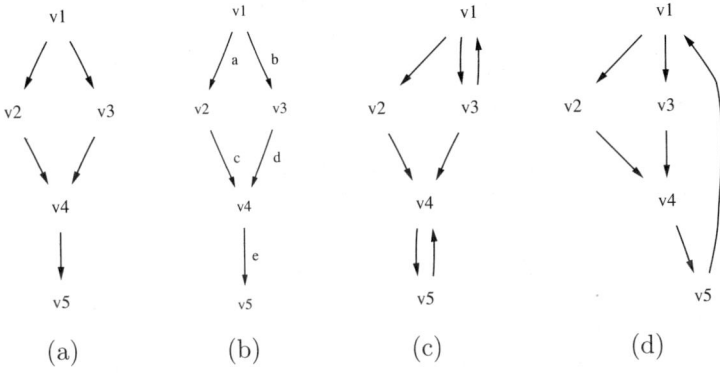

Fig. 4.2 (a) Simple digraph. (b) Labeled digraph. (c) Cyclic digraph. (d) Strongly connected digraph.

The number of elements in V (denoted $|V|$) is called the *order* of G. The number of elements in E (denoted $|E|$) is called the *size* of G. For a graph G, it is sometimes convenient to denote the vertex and edge sets as $V(G)$ and $E(G)$, respectively. If $e = (u,v) \in E(G)$, then e *joins* the vertices u and v. In an undirected graph, the edge (u,v) is *incident on* u and v. In a digraph, the arc (u,v) is *incident from* or *leaves* vertex u and is *incident to* or *enters* v. The vertex v is *adjacent* to u in a graph G if $(u,v) \in E(G)$. If $(u,v) \notin E(G)$, then u and v are *nonadjacent* vertices.

Example 4.1.3 The undirected graph in Figure 4.1(a) has order 5 and size 5. The edge (v_1, v_2) joins v_1 and v_2. The vertex v_1 is adjacent to v_2, but v_1 and v_4 are nonadjacent. In the digraph in Figure 4.2(a), the vertex v_1 is not adjacent to v_2, but v_2 is adjacent to v_1.

Let u and v be vertices of a graph G. A *u-v path* in G is an alternating sequence of vertices and edges of G, beginning with u and ending with v, such that every edge joins the vertices immediately preceding it and following it. The length of a *u-v* path is the number of edges in the path. If there exists a *u-v* path in a graph G, then v is *reachable* from u. A *u-v* path is *simple* if it does not repeat any vertex. If for every pair of vertices u and v there exists a *u-v* path, the graph is *connected*.

Example 4.1.4 v_1, (v_1, v_2), v_2, (v_2, v_4), v_4, (v_4, v_5), v_5, (v_5, v_4), v_4, (v_4, v_3), v_3 is a v_1-v_3 path in the graph in Figure 4.1(a). We actually only need to give the vertices, since the edges are obvious. Therefore, the path can be described more concisely by v_1, v_2, v_4, v_5, v_4, v_3. This path is not simple, since it repeats vertex v_4. The graph in Figure 4.1(a) is connected while the one in Figure 4.1(b) is not since there is no path between v_4 or v_5 and the other vertices.

In a digraph, a *u-v* path forms a *cycle* if $u = v$. If the *u-v* path excluding u is simple, then the cycle is also *simple*. A cycle of length 1 is a self-loop. A digraph with no self-loops is *simple*. In an undirected graph, a u-v path can be a cycle only if it is simple. A graph which contains no cycles is *acyclic*. An acyclic digraph is often called a *directed acyclic graph* or DAG.

Example 4.1.5 The undirected graph shown in Figure 4.1(a) contains the simple cycle $(v_1, v_2, v_4, v_3, v_1)$. The undirected graph shown in Figure 4.1(b) is acyclic. The digraph shown in Figure 4.2(a) is a DAG while the digraphs in Figure 4.2(c) and (d) are not. For example, the digraph in Figure 4.2(d) has the simple cycle $(v_1, v_3, v_4, v_5, v_1)$.

A digraph G is *strongly connected* if for every two distinct vertices u and v in G, there exists a u-v path and a v-u path. A graph G is *bipartite* if it is possible to partition the set of vertices into two subsets V_1 and V_2 such that every edge of G joins a vertex of V_1 with V_2.

Example 4.1.6 The digraph in Figure 4.2(d) is strongly connected while all the rest are not. The graph in Figure 4.1(a) is bipartite with $V_1 = \{v_1, v_4\}$ and $V_2 = \{v_2, v_3, v_5\}$.

A labeled graph is a triple $\langle V, R, L \rangle$ in which L is a labeling function associated either to the set of vertices (i.e., $L : V \rightarrow label$) or edges (i.e., $L : V \times V \rightarrow label$).

Example 4.1.7 The digraph shown in Figure 4.2(b) has a labeling function L defined as follows:

$$L = \{((v_1, v_2), a), ((v_1, v_3), b), ((v_2, v_4), c), ((v_3, v_4), d), ((v_4, v_5), e)\}$$

4.2 ASYNCHRONOUS FINITE STATE MACHINES

One of the most common graphical models for sequential logic is the *finite state machine* (FSM). The major difference between a synchronous FSM and an asynchronous FSM is when state changes are allowed. In the synchronous FSM, a clocked register is used as shown in Figure 4.3(a). In the asynchronous FSM, delay is inserted in the feedback path as shown in Figure 4.3(b). Since FSMs can be used in a similar fashion to describe both synchronous and asynchronous designs, they are a good framework for discussing sequential design somewhat independent of timing methodology.

4.2.1 Finite State Machines and Flow Tables

A FSM can be described using a 6-tuple $\langle I, O, S, S_0, \delta, \lambda \rangle$, where:

- I is the input alphabet (a finite, nonempty set of input values).
- O is the output alphabet.
- S is the finite, nonempty set of states.
- $S_0 \subseteq S$ is the set of initial (reset) states.

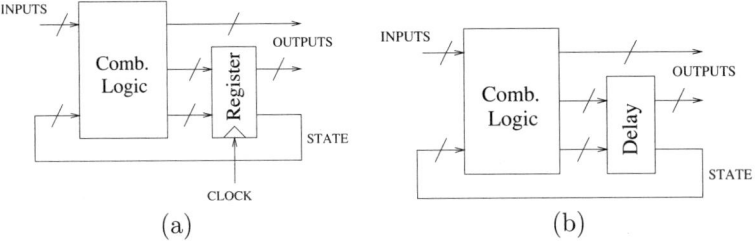

Fig. 4.3 (a) Synchronous FSM. (b) Asynchronous FSM.

- $\delta : S \times I \to S$ is the next-state function.
- $\lambda : S \times I \to O$ is the output function for a *Mealy* machine (or $\lambda : S \to O$ for a *Moore* machine).

In a synchronous FSM, state changes can occur only on a clock edge. In an asynchronous FSM (AFSM), state changes can occur immediately after the inputs change.

FSMs are often represented using a labeled digraph called a *state diagram*. The vertex set contains the states (i.e., $V = S$). The edge set contains the set of state transitions [i.e., $(u, v) \in E$ iff $\exists i \in I$ s.t. $((u, i), v) \in \delta$]. The labeling function, L, is defined by the next-state and output functions. In other words, each edge (u, v) is labeled with i/o where $i \in I$ and $o \in O$ and $((u, i), v) \in \delta$ and $((u, i), o) \in \lambda$.

Example 4.2.1 Consider the AFSM shown in Figure 4.4(a), which models the passive/active handshaking expansion for the wine shop given below.

```
shop_PA_1:process
begin
    guard(req_wine,'1');            -- winery calls
    assign(ack_wine,'1',1,3);       -- receives wine
    guard(req_wine,'0');            -- req_wine reset
    assign(req_patron,'1',1,3);     -- call patron
    guard(ack_patron,'1');          -- wine purchased
    assign(req_patron,'0',1,3);     -- reset req_patron
    guard(ack_patron,'0');          -- ack_patron reset
    assign(ack_wine,'0',1,3);       -- reset ack_wine
end process;
```

This AFSM has the following FSM characterization:

$$I = \{00, 01, 10\}$$
$$O = \{00, 10, 11\}$$
$$S = \{s0, s1, s2, s3\}$$
$$S_0 = \{s0\}$$

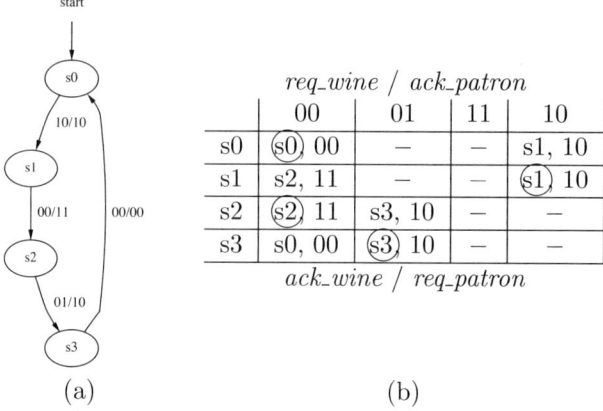

Fig. 4.4 (a) AFSM and (b) Huffman flow table for a passive/active wine shop (state vector *req_wine ack_patron/ ack_wine req_patron*).

$$\delta = \{((s0, 10), s1), ((s1, 00), s2), ((s2, 01), s3), ((s3, 00), s0)\}$$
$$\lambda = \{((s0, 10), 10), ((s1, 00), 11), ((s2, 01), 10), ((s3, 00), 00)\}$$

This defines a graph of the following form:

$$V = \{s0, s1, s2, s3\}$$
$$E = \{(s0, s1), (s1, s2), (s2, s3), (s3, s0)\}$$
$$L = \{((s0, s1), 10/10), ((s1, s2), 00/11), ((s2, s3), 01/10),$$
$$((s3, s0), 00/00)\}$$

Each input gives the value of *req_wine* and *ack_patron*, respectively. Each output gives the value of *ack_wine* and *req_patron*, respectively. An example state transition in the AFSM would be when in state *s0*, if *req_wine* changes to '1', then *ack_wine* would be set to '1' and *req_patron* would be held at '0' as the machine makes a transition to state *s1*.

An AFSM can also be represented in a tabular form called a *Huffman flow table*. A Huffman flow table has one row for each state and one column for each input value. Each entry in the table consists of an ordered pair representing the next state and output value. When the next state is equal to the current state (i.e., the row label), it is circled to indicate that it is a *stable state*.

Example 4.2.2 Consider the Huffman flow table in Figure 4.4(b). Row *s2* and column 00 has entry *s2*, 11 which states that the next state is also *s2* with output 11, and it is a stable state. For row *s2* and column 01, the entry is *s3*, 10, which indicates that the next state is *s3* and the output should be changed to 10.

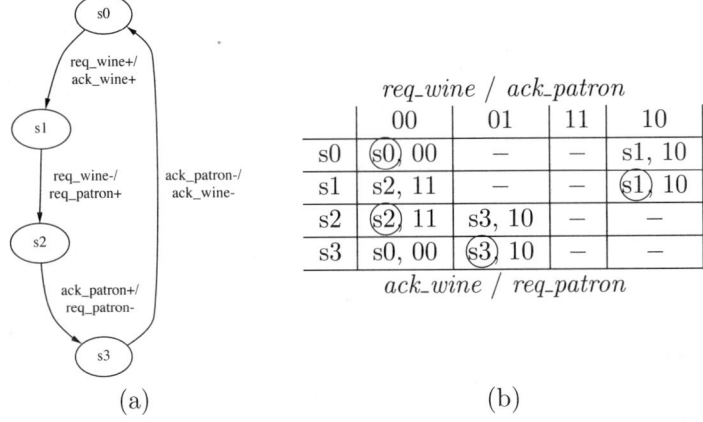

Fig. 4.5 (a) BM machine and (b) Huffman flow table for a passive/active wine shop.

4.2.2 Burst-Mode State Machines

An alternative method of specifying an AFSM is using a *burst-mode* (BM) *state machine*. An example BM machine which describes the same behavior as the AFSM in Figure 4.4 is shown in Figure 4.5(a). The key difference is that in the BM machine the arcs between the states are labeled with the input and output signal transitions rather than their values. The set of input transitions is called the *input burst* and the set of output transitions is called the *output burst*. Note that the set of transitions in the input burst cannot be empty but the set of transitions in the output burst can be empty. When the set of transitions in the input burst change value, the system generates the transitions in the corresponding output burst and moves to a new state. Any BM machine can be translated into a Huffman flow table. The Huffman flow table for the BM machine in Figure 4.5(a) is shown in Figure 4.5(b). Note that, as expected, it is the same as the one from Figure 4.4(b).

> **Example 4.2.3** For the BM machine shown in Figure 4.5, in order to move from state $s0$ to $s1$, it is necessary for *req_wine* to change from '0' to '1' (indicated by *req_wine+*). Note that since *ack_patron* does not need to change value (nor is it allowed to), it is omitted from the input burst. At this point, *ack_wine* is enabled to change from '0' to '1'. Since *req_patron* does not need to change value (nor is it allowed to), it is omitted from the output burst.

Formally, a burst-mode specification is a labeled digraph which is described by a 7-tuple $\langle V, E, I, O, v_0, in, out \rangle$, where:

- V is a finite set of vertices (or states).
- $E \subseteq V \times V$ is the set of edges (or transitions).

- $I = \{x_1, \ldots, x_m\}$ is the set of inputs.
- $O = \{z_1, \ldots, z_n\}$ is the set of outputs.
- $v_0 \in V$ is the start state.
- $in : V \to \{0,1\}^m$ defines the values of the m inputs upon entry to each state.
- $out : V \to \{0,1\}^n$ defines the values of the n outputs upon entry to each state.

The value of input x_i on entering state v is denoted by $in_i(v)$, and the value of output z_j on entering state v is denoted by $out_j(v)$. Intuitively, the formalism requires every state in the (unminimized) BM specification to be entered at a single *unique entry point*.

For a BM machine, two edge-labeling functions $trans_i : E \to 2^I$ and $trans_o : E \to 2^O$ can be derived to specify the transitions in the input burst and output burst, respectively. For any edge $e = (u,v) \in E$, an input x_i is in the input burst of e if it changes value between u and v [i.e., $x_i \in trans_i(e)$ iff $in_i(u) \neq in_i(v)$]. The function $trans_o$ can be defined in a similar manner.

A BM machine must satisfy the *maximal set property*. This property states that no input burst leaving a given state can be a subset of another leaving the same state, since the behavior in that state would then be ambiguous [i.e., $\forall (u,v), (u,w) \in E : trans_i(u,v) \subseteq trans_i(u,w) \Rightarrow v = w$].

Example 4.2.4 The BM machine in Figure 4.6(a) violates the maximal set property since in state s0 the input burst a+ is a subset of the input burst a+, b+. This means that if a changes from '0' to '1' in state s0 before b changes, the circuit would not know whether to move to state s1 or wait for b to change.

Not every *BM state diagram* represents a legal BM machine. If a BM state diagram is mislabeled with transitions that are not possible, it is impossible to define the *in* and *out* functions in a consistent fashion. These are cases where the labeling of arcs is obviously incorrect in that there must be a strict alternation of rising and falling transitions on every input and output signal, across all paths of the specification.

Example 4.2.5 Consider the BM machine in Figure 4.6(b), which begins with all signals low in state s0. In state s0, the transition b− does not make sense since b is already low. After a changes to '1', the machine can set x to '1' and change to state s1. At this point, b can change to '1', causing y to be set to '1' and return to state s0. However, now the machine is in state s0 with all signals high. In this case, the transition b− is okay, but the transition a+ does not make sense. Furthermore, the value on entry of all input and output signals is not unique. Therefore, the functions *in* and *out* cannot be defined. The BM state diagram shown in Figure 4.6(c) has the same behavior as the one in Figure 4.6(b), but it is now a legal BM machine.

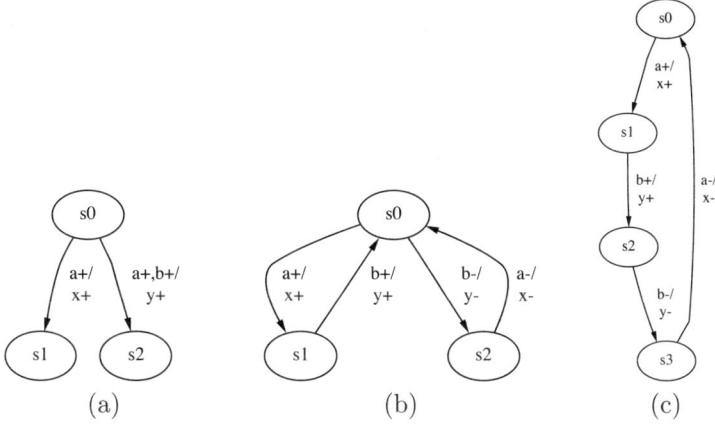

Fig. 4.6 (a) BM machine which violates maximal set property. (b) BM state diagram which is not a BM machine. (c) BM state diagram equivalent to that in (b) which is a BM machine.

4.2.3 Extended Burst-Mode State Machines

One important limitation of AFSM and BM machines is the strict regulation that changes in signal values must follow a prescribed order: inputs change, outputs and state signals change, then state signals are fed back. In *extended burst-mode* (XBM) *state machines*, this limitation is loosened a bit by the introduction of *directed don't cares*. These allow one to specify that an input change may or may not happen in a given input burst. The idea is that some inputs in a burst may be allowed to change once monotonically along a sequence of bursts rather than having to change in a particular burst.

Example 4.2.6 The handshaking expansion shown below cannot be specified with an ordinary BM machine.

```
Shop_PA_2: process
begin
    guard(req_wine,'1');           -- winery calls
    assign(ack_wine,'1',1,3);      -- receives wine
    guard(req_wine,'0');           -- req_wine reset
    assign(ack_wine = '0',1,3,req_patron,'1',1,3);
    guard(ack_patron,'1');         -- wine purchased
    assign(req_patron,'0',1,3);    -- reset req_patron
    guard(ack_patron,'0');         -- ack_patron reset
end process;
```

The problem is *req_wine* can change from '0' to '1' at any point after *ack_wine* goes to '0'. It can, however, be specified as an XBM machine using directed don't cares as shown in Figure 4.7(a). Initially, this machine waits in state $s0$ for *req_wine* to change to '1', and it then sets

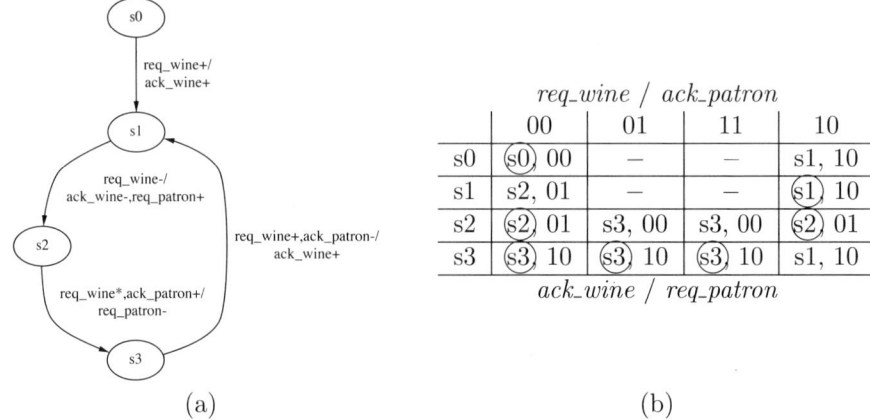

Fig. 4.7 (a) XBM machine and (b) Huffman flow table for a passive/active wine shop.

ack_wine to '1' and moves to state $s1$. In state $s1$, the machine waits for req_wine to change to '0', and it then sets ack_wine to '0' and req_patron to '1' and moves to state $s2$. Once ack_wine goes to '0', it is possible for req_wine to be set to '1' again at any time. This is indicated by putting req_wine∗ in the input burst leaving $s2$. The "∗" means that req_wine is allowed to change but is not required to change in state $s2$. In other words, the machine moves from state $s2$ to state $s3$ when ack_patron changes to '1' regardless of whether or not req_wine has been set to '1'. In state $s3$, in addition to ack_patron changing to '0', req_wine must complete (if it has not already) its transition to '1' before the machine moves to state $s1$. The Huffman flow table for the XBM machine is shown in Figure 4.7(b). Note that the effect of the directed don't care in state $s2$ is that req_wine can change without creating a state change. The signal req_wine also may not change before moving to state $s3$.

The use of directed don't cares is restricted. Before describing their restriction, it is useful to define a few terms. First, a transition is *terminating* when it is of the form $t+$ or $t-$. A directed don't care transition is of the form $t*$. A *compulsory* transition is a terminating transition which is not preceded by a directed don't care transition in any state directly preceding the current one. Each input burst must include at least one compulsory transition.

Example 4.2.7 The XBM machine in Figure 4.8 is illegal. The transition leaving state $s0$ has no compulsory transition, since req_wine has a directed don't care transition from the preceding state, $s3$.

The presence of directed don't cares requires a modification of the maximal set property. In addition to the subset violation from Figure 4.6(a), one must also restrict out specifications of the form shown in Figure 4.9. In Figure 4.9(a), if b changes to '1' before a changes, it is not possible to distinguish whether the machine should move to state $s2$ or wait for a to change and

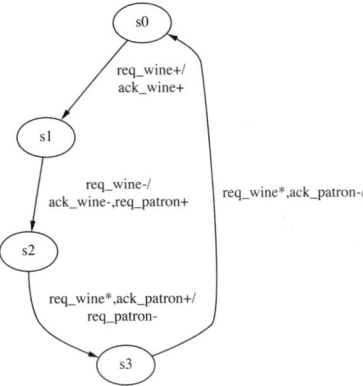

Fig. 4.8 Illegal XBM machine for a passive/active wine shop.

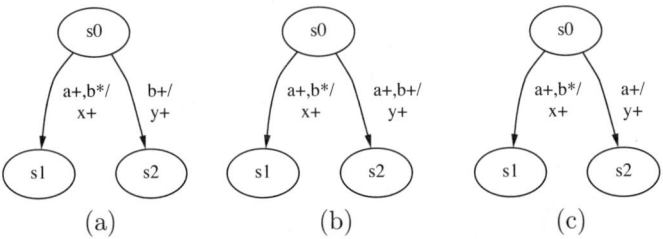

Fig. 4.9 Violations of the maximal set property in three XBM machines.

move to state s1. In Figure 4.9(b), after a and b have both changed to '1', the machine does not know whether to go to state s1 or state s2. In Figure 4.9(c), after a changes to '1' again, both transitions are possible, leaving the machine in a state of confusion. In general, if the compulsory transitions of one input burst leaving a state is a subset of all possible transitions in another input burst leaving the same state, then the maximal set property is violated.

Another important limitation of BM machines is their inability to express conditional behavior. It is often useful to make a decision of future behavior based on the level of a particular signal. To support this type of behavior, XBM machines allow *conditional input bursts*. A conditional input burst includes a regular input burst as defined earlier and a *conditional clause* that restricts the validity of the input burst. A clause of the form $<s->$ indicates that the transition is taken only if s is low. A clause of the form $<s+>$ indicates that the transition is taken only if s is high. The signal in the conditional clause must be stable before every compulsory transition in the input burst. This is a form of setup requirement, analogous to the one in a synchronous system. A specified conditional signal must be stable and valid at some setup time before any compulsory signal arrives. The compulsory signals therefore act like a clock, to sample the conditional signal's value. The conditionals also have a hold-time requirement before they can change again. When a conditional signal is not specified, it is free to change in an arbitrary fashion.

Example 4.2.8 Let us consider the situation where a second patron moves to town. The shop has decided to sell wine of type '0' to *patron1* and type '1' to *patron2*. The VHDL model is given below. In this model, after *req_wine* goes to '0', the value of *shelf* is sampled. If *shelf* is low, the machine sets *req_patron1* to '1'. If *shelf* is high, the machine sets *req_patron2* to '1'.

```
Shop_PA_2:process
begin
  guard(req_wine,'1');
  shelf <= bottle after delay(2,4);
  wait for delay(5,10);
  assign(ack_wine,'1',1,3);
  guard(req_wine,'0');
  if (shelf = '0') then
    assign(ack_wine,'0',1,3,req_patron1,'1',1,3);
    guard(ack_patron1,'1');
    assign(req_patron1,'0',1,3);
    guard(ack_patron1,'0');
  elsif (shelf = '1') then
    assign(ack_wine,'0',1,3,req_patron2,'1',1,3);
    guard(ack_patron2,'1');
    assign(req_patron2,'0',1,3);
    guard(ack_patron2,'0');
  end if;
end process;
```

The XBM machine for the two-patron wine shop is shown in Figure 4.10. The two transitions leaving state $s1$ are annotated with conditional clauses on the signal *shelf*. The signal shelf must be stable a setup time before the compulsory transition which is *req_wine−* in both cases. The Huffman flow table for this XBM machine is shown in Figure 4.11. Since there are three input signals and one level signal, it gets quite complicated. Note that in all states except $s1$, the level signal, *shelf*, is free to change back and forth without changing the behavior. In state $s1$, *shelf* can change up to a setup time before *req_wine* goes low. At that point, the value of *shelf* is sampled and the next state depends on its value. A hold time after reaching the appropriate next state, *shelf* is again free to change value.

The use of conditional input bursts again changes the maximal set property. For example, in Figure 4.12(a), if we ignore the conditional, this machine violates the maximal set property. However, if s is stable before a changes to 1, the transitions are distinguishable, so there is no violation of the maximal set property. If, on the other hand, s is stable before b changes to 1, but

ASYNCHRONOUS FINITE STATE MACHINES   97

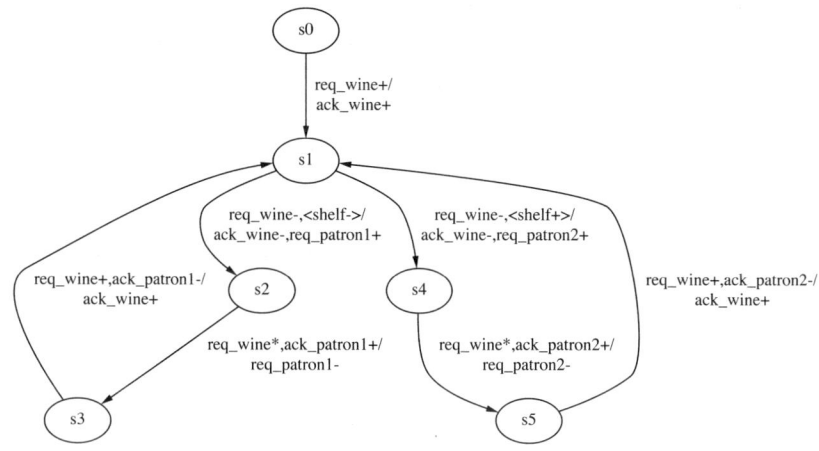

Fig. 4.10 XBM machine for wine shop with two patrons.

req_wine / ack_patron1 / ack_patron2 / shelf

	0000	0001	0011	0010	0110	0111	0101	0100
s0	(s0) 000	(s0) 000	—	—	—	—	—	—
s1	s2, 010	s4, 001	—	—	—	—	—	—
s2	(s2) 010	(s2) 010	—	—	—	—	s3, 000	s3, 000
s3	(s3) 010	(s3) 010	—	—	—	—	(s3) 000	(s3) 000
s4	(s4) 010	(s4) 010	—	—	—	—	s5, 000	s5, 000
s5	(s5) 010	(s5) 010	(s5) 000	(s5) 000	—	—	—	—

ack_wine / req_patron1 / req_patron2

req_wine / ack_patron1 / ack_patron2 / shelf

	1100	1101	1111	1110	1010	1011	1001	1000
s0	—	—	—	—	—	—	s1, 100	s1, 100
s1	—	—	—	—	—	—	(s1) 100	(s1) 100
s2	s3, 000	s3, 000	—	—	—	—	(s2) 010	(s2) 010
s3	(s3) 000	(s3) 000	—	—	—	—	s1, 100	s1, 100
s4	s5, 000	s5, 000	—	—	—	—	(s4) 010	(s4) 010
s5	—	—	—	—	(s5) 000	(s5) 000	s1, 100	s1, 100

ack_wine / req_patron1 / req_patron2

Fig. 4.11 Huffman flow table for wine shop with two patrons.

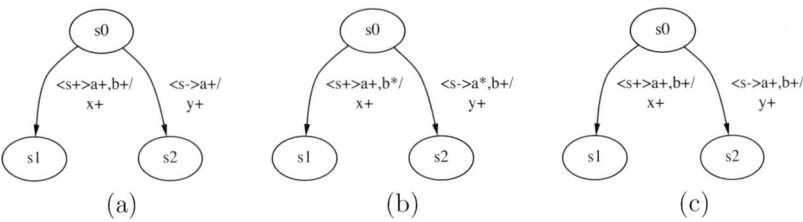

Fig. 4.12 Maximal set property and conditional input bursts in three XBM machines.

not before a changes, we would have a maximal set property violation. For example, for the sequence of transitions $a+$, $s+$, $b+$, it is not clear whether the machine ends up in state $s2$ or state $s1$. Therefore, as a conservative set up requirement, the conditional signal must be setup before any compulsory transition in any input burst leaving the state. This means that for this example we would require that s is set up before either a or b is allowed to change. In Figure 4.12(b), $a+$ is a compulsory transition for one transition and $b+$ is a compulsory transition for the other, so we again require that s is set up before either is allowed to change. In Figure 4.12(c), assume that b is a directed don't care in the state preceding state $s0$. This means that we don't know b's value when we enter state $s0$. It also means that b is not a compulsory transition for either input burst. Therefore, in this case, s must stabilize only before a can change, but b is allowed to change before s has stabilized.

XBM machines can also be described by a labeled digraph which is given formally by a 9-tuple $\langle V, E, I, O, C, v_0, \mathit{in}, \mathit{out}, \mathit{cond} \rangle$, where:

- V is a finite set of vertices (or states).
- $E \subseteq V \times V$ is the set of edges (or transitions).
- $I = \{x_1, \ldots, x_m\}$ is the set of inputs.
- $O = \{z_1, \ldots, z_n\}$ is the set of outputs.
- $C = \{c_1, \ldots, c_l\}$ is the set of conditional signals.
- $v_0 \in V$ is the start state.
- $\mathit{in} : V \to \{0, 1, *\}^m$ defines the values of the m inputs upon entry to each state.
- $\mathit{out} : V \to \{0, 1\}^n$ defines the values of the n outputs upon entry to each state.
- $\mathit{cond} : E \to \{0, 1, *\}^l$ defines the values of the conditional inputs needed to take a state transition.

ASYNCHRONOUS FINITE STATE MACHINES

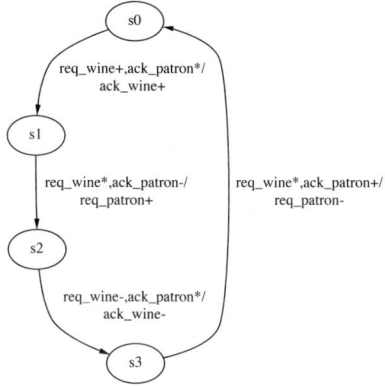

Fig. 4.13 Illegal XBM machine for reshuffled passive/lazy-active wine shop.

Again for XBM machines, the two edge-labeling functions $trans_i : E \rightarrow 2^I$ and $trans_o : E \rightarrow 2^O$ can be derived to specify the transitions in the input burst and output burst, respectively. For any edge $e = (u, v) \in E$, an input x_i is in the input burst of e if it changes or can change value between u and v [i.e., $x_i \in trans_i(e)$ iff $in_i(u) \neq in_i(v) \vee in_i(v) = *$]. The transition x_i+ is in the input burst iff $in_i(v) = 1$ and $in_i(u) \neq 1$, x_i- is in the burst iff $in_i(v) = 0$ and $in_i(u) \neq 0$, and x_i* is in the burst iff $in_i(v) = *$.

Although XBM machines are capable of specifying more general behavior than either AFSMs or BM machines, they are still incapable of specifying arbitrarily concurrent systems.

Example 4.2.9 Consider the following handshaking expansion and the XBM machine shown in Figure 4.13. This XBM machine appears to model the handshaking expansion. However, it is not legal since none of the input bursts have a compulsory transition.

```
Shop_PA_lazy_active:process
begin
    guard(req_wine,'1');          -- winery calls
    assign(ack_wine,'1',1,3);     -- receives wine
    guard(ack_patron,'0');        -- ack_patron reset
    assign(req_patron,'1',1,3);   -- call patron
    guard(req_wine,'0');          -- req_wine reset
    assign(ack_wine,'0',1,3);     -- reset ack_wine
    guard(ack_patron,'1');        -- wine purchased
    assign(req_patron,'0',1,3);   -- reset req_patron
end process;
```

4.3 PETRI NETS

In order to specify highly concurrent systems, variants of another graphical representation method, the *Petri net*, are often used. Rather than specifying system states, these methods describe the allowed interface behavior. The interface behavior is described by means of allowed sequences of transitions, or *traces*. In this section we describe Petri nets and one of their variants, the *signal transition graph* (STG), which has been used to model asynchronous circuits and systems.

4.3.1 Ordinary Petri Nets

An *ordinary Petri net* is a bipartite digraph. The vertex set is partitioned into two disjoint subsets P, the set of *places*, and T, the set of *transitions*. Recall that in a bipartite graph, nodes from one subset must only connect with nodes from the other. In other words, the set of arcs given by the *flow relation*, F, is composed of pairs where one element is from P and the other is from T [i.e., $F \subseteq (P \times T) \cup (T \times P)$]. A *marking*, M, for a Petri net is a function that maps places to natural numbers (i.e., $M : P \to N$). A Petri net is defined by a quadruple $\langle P, T, F, M_0 \rangle$, where M_0 is the *initial marking*.

Example 4.3.1 As an example, a model of a wine shop with infinite shelf space is shown in Figure 4.14(a). This Petri net is characterized as follows:

$$P = \{p_1, p_2, p_3, p_4, p_5\}$$
$$T = \{t_1, t_2, t_3, t_4\}$$
$$F = \{(t_1, p_1), (p_1, t_2), (t_2, p_2), (p_2, t_1), (t_2, p_3), (p_3, t_3), (t_3, p_4),$$
$$(p_4, t_4), (t_4, p_5), (p_5, t_3)\}$$
$$M_0 = \{(p_1, 0), (p_2, 1), (p_3, 2), (p_4, 0), (p_5, 2)\}$$

The place and transition numbering is from top to bottom and left to right. A *labeled Petri net* is one which is extended with a labeling function from the places and/or the transitions to a set of labels. The Petri nets shown in Figure 4.14 have labeled transitions, and the labeling function $L : T \to label$ is given below.

$$L = \{(t_1, produce), (t_2, receive), (t_3, send), (t_4, consume)\}$$

For a transition $t \in T$, we define the *preset* of t (denoted $\bullet t$) as the set of places connected to t [i.e., $\bullet t = \{p \in P \mid (p,t) \in F\}$]. The *postset* of t (denoted $t\bullet$) is the set of places t is connected to [i.e., $t\bullet = \{p \in P \mid (t,p) \in F\}$]. The presets and postets of a place $p \in P$ can be similarly defined [i.e., $\bullet p = \{t \in T \mid (t,p) \in F\}$ and $p\bullet = \{t \in T \mid (p,t) \in F\}$].

Markings can be added or subtracted. They can also be compared:

$$M \geq M' \quad \text{iff} \quad \forall p \in P \,.\, M(p) \geq M'(p)$$

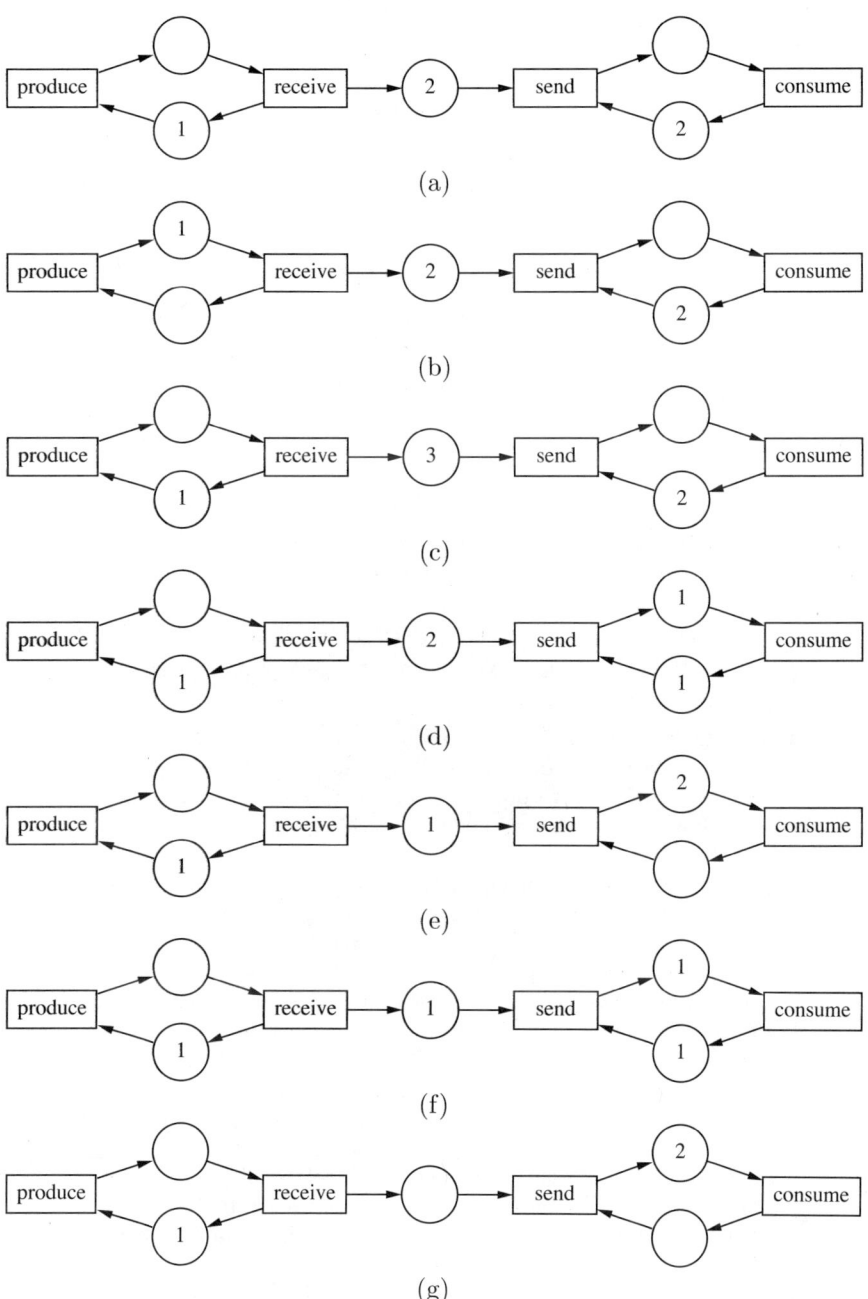

Fig. 4.14 Simple Petri net model of a shop with infinite shelf space.

For a set of places, $A \subseteq P$, C_A denotes the *characteristic marking* of A:

$$C_A(p) \quad = \quad \text{if } p \in A \text{ then } 1 \text{ else } 0.$$

The basic operation on Petri nets is the firing of a transition $t \in T$. A transition t is *enabled* under the marking M if $M \geq C_{\bullet t}$. In other words, $M(p) \geq 1$ for each $p \in \bullet t$. The firing of an enabled transition transforms the marking as follows:

$$M' \quad = \quad M - C_{\bullet t} + C_{t \bullet}$$

(denoted $M[t\rangle M'$). This means that when a transition t fires, a token is removed from each place in its preset (i.e., $p \in \bullet t$) and a token is added to each place in its postset (i.e., $p \in t\bullet$). The condition that a transition must be enabled before it can be fired ensures that this operation can complete without making a marking negative.

In order to understand the behavior of a system described using a Petri net, it is typically necessary to find the set of *reachable markings*. As described above, the firing of a transition transforms the marking of the Petri net into a new marking. A sequence of transition firings, or *firing sequence* (denoted $\sigma = t_1, t_2, \ldots, t_n$) produces a sequence of markings (i.e., M_0, M_1, ..., M_n). If such a firing sequence exists, we say that the marking M_n is *reachable* from M_0 by σ (denoted $M_0[\sigma\rangle M_n$). We denote the set of all markings reachable from a given marking by $[M\rangle$.

Example 4.3.2 A sample firing sequence for our Petri-net example is shown in Figure 4.14. In the initial marking shown in Figure 4.14(a), the winery is working to produce a new bottle of wine while the shop has two bottles on its shelf. For this marking, the enabled transitions are *produce* and *send*, since all places in the preset of these transitions have a marking greater or equal to 1. If the winery produces a new bottle of wine first, the marking shown in Figure 4.14(b) results. In this marking, the enabled transitions are *receive* and *send*. If the next transition is that the shop receives the bottle of wine from the winery, the new marking is shown in Figure 4.14(c). Now the shop has three bottles of wine on its shelf. In this marking, the enabled transitions are *produce* and *send*. At this point, if the shop sells a bottle of wine to the patron, the marking shown in Figure 4.14(d) results. Notice that the shop now has two bottles of wine on its shelf, the patron has one in his hand, and the patron is still capable of accepting another bottle of wine (after all, he does have two hands). Therefore, in this marking, the enabled transitions are *produce*, *send*, and *consume*. If the shop sends the patron another bottle of wine, the marking shown in Figure 4.14(e) results. In this marking, the shop's shelf still has one bottle of wine, but the patron is unable to accept it (both hands are now full). The enabled transitions are now *produce* and *consume*. The patron decides to consume one of the bottles of wine while still in the shop, leading to the marking in Figure 4.14(f). Now, the patron can go ahead and accept the last bottle of wine in the shop.

This example sequence obviously demonstrates only one of the infinite possible behaviors of this model. The number of behaviors is infinite because we could simply have a sequence of *produce* and *receive* transitions, incrementing the value of the marking of place p_3 (the shelf) each time.

A Petri net is *k-bounded* if there does not exist a reachable marking which has a place with more than k tokens. A 1-bounded Petri net is also called a *safe* Petri net [i.e., $\forall p \in P, \forall M \in [M_0).M(p) \leq 1$]. When working with safe Petri nets, a marking can be denoted as simply a subset of places. If $M(p) = 1$, we say that $p \in M$, and if $M(p) = 0$, we say that $p \notin M$. $M(p)$ is not allowed to take on any other values in a safe Petri net. Since a marking can only take on the values 1 and 0, the place can be annotated with a token when it is 1 and without when it is 0.

Example 4.3.3 The Petri nets in Figure 4.14 are not k-bounded for any value of k. However, by adding a single place as shown in Figure 4.15(a), the net is now 2-bounded. This is accomplished by restricting the winery not to deliver a new bottle of wine unless the shop has a place to put it on its shelf. Changing the initial marking as shown in Figure 4.15(b) results in a safe Petri net. An equivalent Petri net to the one in Figure 4.15(b) using tokens is shown in Figure 4.15(c).

Another important property of Petri nets is *liveness*. A Petri net is *live* if from every reachable marking, there exists a sequence of transitions such that any transition can fire. This property can be expressed formally as follows:

$$\forall M \in [M_0), \forall t \in T, \exists M' \in [M).M' \geq C_{\bullet t}$$

Example 4.3.4 The Petri net shown in Figure 4.15(c) is live, but the one shown in Figure 4.15(d) is not. In this case, this shop can receive a bottle of wine, but it will not be able to send it anywhere. For this marking, the transitions *send* and *consume* are dead.

To determine if a Petri net is live, it is typically necessary to find all the reachable markings. This can be a very expensive operation for a complex Petri net. There are, however, different categories of liveness that can be determined more easily. In particular, a transition t for a given Petri net is said to be:

1. *dead (L0-live)* if there does not exist a firing sequence in which t can be fired.

2. *L1-live (potentially firable)* if there exists at least one firing sequence in which t can be fired.

3. *L2-live* if t can be fired at least k times.

4. *L3-live* if t can be fired infinitely often in some firing sequence.

5. *L4-live* or *live* if t is L1-live in every marking reachable from the initial marking.

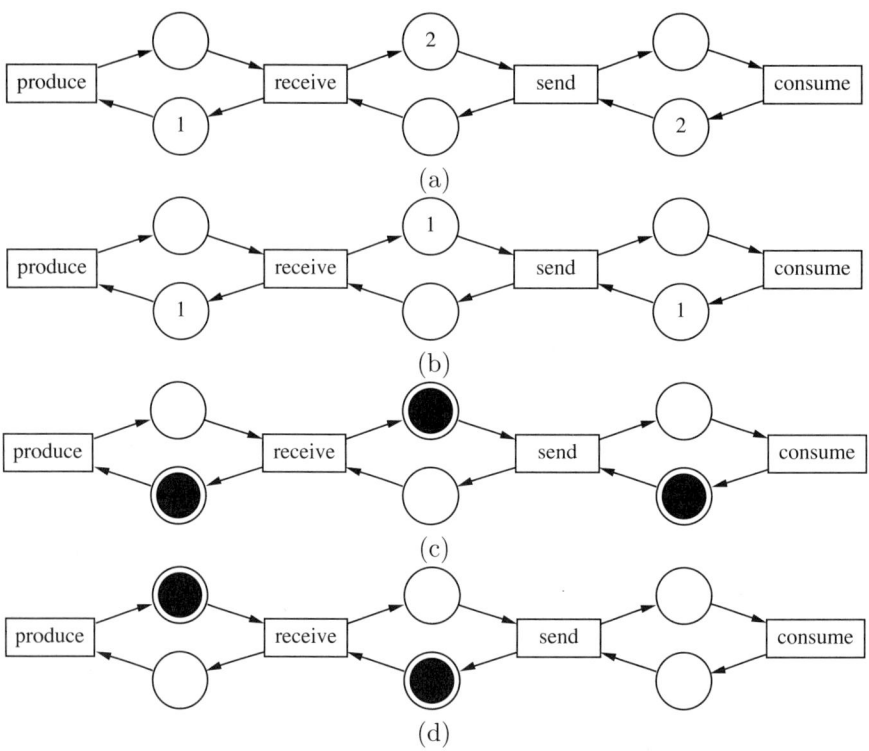

Fig. 4.15 (a) Two-bounded Petri net modeling a shop with shelf capacity of 2. (b) Safe Petri net modeling a shop with shelf capacity of 1. (c) Safe Petri net using tokens. (d) Petri net that is not live.

A Petri net is *Lk-live* if every transition in the net is Lk-live.

> **Example 4.3.5** The Petri net shown in Figure 4.16 describes a winery that produces wine until some point at which it chooses to deliver all its bottles to the shop. The patron must be signaled by both the winery and the shop as to when to get the wine. In this Petri net, the transition *consume* is dead, meaning that the poor patron will never get any wine. The transition *deliver* is L1-live, which means that wine will be delivered at most once. The transition *receive* is L2-live, which means that any finite number of bottles of wine can be received. Finally, the transition *produce* is L3-live, meaning that the winery could produce an infinite number of bottles of wine while delivering none of them.

When a Petri net is bounded, the number of reachable markings is finite, and they can all be found using the algorithm shown in Figure 4.17. This algorithm constructs a graph called a *reachability graph* (RG). In an RG, the vertices, Φ, are the markings and the edges, Γ, are the possible transition firings that lead the state of the Petri net between the two markings connected

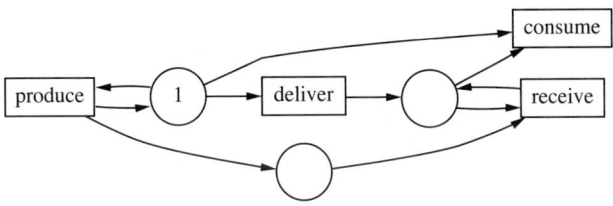

Fig. 4.16 Example in which *consume* is dead, *deliver* is L1-live, *receive* is L2-live, and *produce* is L3-live.

find_RG(Petri net $\langle P, T, F, M_0 \rangle$) {
    $M = M_0$;
    $T_e = \{t \in T | M \geq C_{\bullet t}\}$;
    $\Phi = \{M\}$;
    $\Gamma = \emptyset$;
    done = false;
    while ($\neg$ done) {
        t = select(T_e);
        if ($(T_e - \{t\} \neq \emptyset)$ then
            push(M, $T_e - \{t\}$);
        $M' = M - C_{\bullet t} + C_{t \bullet}$;
        if ($M' \notin \Phi$) then {
            $\Phi = \Phi \cup \{M'\}$;
            $\Gamma = \Gamma \cup \{(M, M')\}$;
            $M = M'$;
            $T_e = \{t \in T | M \geq C_{\bullet t}\}$;
        } else {
            $\Gamma = \Gamma \cup \{(M, M')\}$;
            if (stack is not empty) then
                (M, T_e) = pop();
            else
                done = true;
        }
    }
    return(Φ, Γ);
}

Fig. 4.17 Algorithm to find the reachability graph.

by the edge. For safe Petri nets, the vertices in this graph are labeled with the subset of the places included in the marking. The edges are labeled with the transition that fires to move the Petri net from one marking to the next.

Example 4.3.6 Consider the application of the *find_RG* algorithm to the Petri net shown in Figure 4.15(c). The set M is initially set to $\{p_2, p_3, p_6\}$, and T_e is initially $\{produce, send\}$. Let us assume that

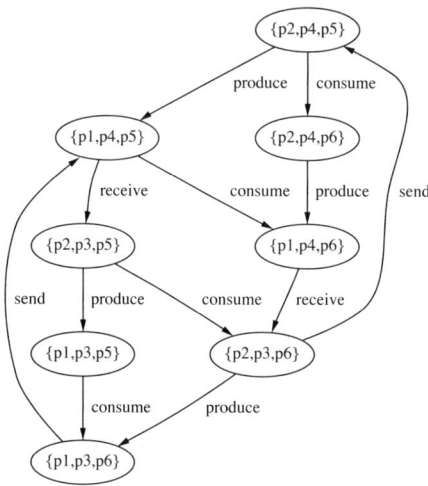

Fig. 4.18 Example reachability graph.

the *select* function chooses to fire *send*. The current marking and the remaining unfired transition, *produce*, are pushed onto the stack. The new marking is $M' = \{p_2, p_4, p_5\}$, which is not in Φ, so it is added along with a transition between these two markings to Γ. The new marking becomes the current marking, the new set of enabled transitions is $T_e = \{produce, consume\}$, and control loops back. At this point the *select* function chooses to fire *consume*, and it pushes the current marking and the remaining transition, *produce*, onto the stack. The new marking after firing *consume* is $\{p_2, p_4, p_6\}$. This marking is not found in Φ, so it is added, and a new set of enabled transitions is calculated to be $\{produce\}$. Next, the transition *produce* is fired. Since there are no remaining enabled transitions, there is nothing to push onto the stack. Firing *produce* leads to the marking $\{p_1, p_4, p_6\}$. This is again a new marking which is added, and the new set of enabled events is found to be $\{receive\}$. The *receive* transition is then fired, resulting in the marking $\{p_2, p_3, p_6\}$. This marking is found in Φ, so the marking $\{p_2, p_4, p_5\}$ and the enabled transition $\{produce\}$ are popped off the stack. This transition is fired, resulting in a new marking, $\{p_1, p_4, p_5\}$. In this new marking, the transition *receive* and *consume* are enabled. Firing *consume* results in the marking $\{p_1, p_4, p_6\}$, which is found in Φ. Therefore, we pop the marking $\{p_1, p_4, p_5\}$ and the set of enabled transition $\{receive\}$ off the stack, and we continue by firing *receive*. The rest of the markings are found using this algorithm, resulting in the RG shown in Figure 4.18.

There are several important classifications of Petri nets. Before getting into them, we first describe a few important terms. Two transitions t_1 and t_2 are said to be *concurrent* when there exists markings in which they are both

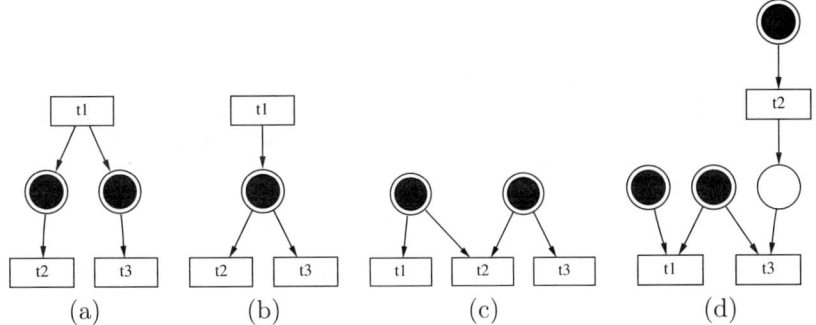

Fig. 4.19 (a) Petri net in which transitions t_2 and t_3 are concurrent. (b) Petri net in which transitions t_2 and t_3 are in conflict. (c) Petri net showing symmetric confusion. (d) Petri net showing asymmetric confusion.

enabled and they can fire in either order. Two transitions, t_1 and t_2, are said to be in *conflict* when the firing of one disables the other. In other words, t_1 can happen or t_2 can happen but not both. When concurrency and conflict are mixed, we get *confusion*.

Example 4.3.7 Transitions t_2 and t_3 are concurrent in Figure 4.19(a). Transitions t_2 and t_3 are in conflict in Figure 4.19(b). The Petri net in Figure 4.19(c) has *symmetric confusion* since t_1 and t_3 are concurrent, but they each conflict with t_2. The Petri net in Figure 4.19(d) has *asymmetric confusion* since t_1 and t_2 are concurrent, but t_1 conflicts with t_3 only if t_2 fires first.

The first Petri net classification is *state machines* (SM). A Petri net is a state machine if and only if every transition has exactly one place in its preset and one place in its postset (i.e., $\forall t \in T \,.\, |\bullet t| = |t \bullet| = 1$). In other words, state machines do not allow any concurrency, but they do allow conflict.

The second classification is *marked graphs* (MG). A Petri net is a marked graph if and only if every place has exactly one transition in its preset and one transition in its postset (i.e., $\forall p \in P \,.\, |\bullet p| = |p \bullet| = 1$). Marked graphs allow concurrency, but they do not allow conflict.

The third classification is *free-choice nets* (FC). A Petri net is free choice if and only if every pair of transitions that share a common place in their preset have only a single place in their preset. More formally,

$$\forall t, t' \in T, t \neq t' \,.\, \bullet t \cap \bullet t' \neq \emptyset \Rightarrow |\bullet t| = |\bullet t'| = 1$$

This can be written in an equivalent fashion on places as follows:

$$\forall p, p' \in P, p \neq p' \,.\, p \bullet \cap p' \bullet \neq \emptyset \Rightarrow |p \bullet| = |p' \bullet| = 1$$

Finally, a third equivalent form is:

$$\forall p \in P, \forall t \in T \,.\, (p, t) \in F \Rightarrow p\bullet = \{t\} \vee \bullet t = \{p\}$$

Free choice nets allow concurrency and conflict, but they do not allow confusion.

The fourth classification is *extended free-choice nets* (EFC). A Petri net is an extended free choice net if and only if every pair of places that share common transitions in their postset have exactly the same transitions in their postset. More formally,

$$\forall p, p' \in P \,.\, p \bullet \cap p' \bullet \neq \emptyset \Rightarrow p \bullet = p' \bullet$$

Extended free-choice nets also allow concurrency and conflict, but they do not allow confusion.

The fifth classification is *asymmetric choice nets* (AC). A Petri net is an asymmetric choice net if and only if for every pair of places that share common transitions in their postset, one has a subset of the transitions of the other. More formally,

$$\forall p, p' \in P \,.\, p \bullet \cap p' \bullet \neq \emptyset \Rightarrow p \bullet \subseteq p' \bullet \lor p' \bullet \subseteq p \bullet$$

Asymmetric choice nets allow *asymmetric confusion* but not *symmetric confusion*.

Example 4.3.8 The Petri nets shown in Figure 4.15(c) is not a state machine, but it is a marked graph. The Petri net shown in Figure 4.20(a) depicts a winery that sells to two separate shops which sell to two different patrons. Also, the winery is not allowed to produce another pair of wine bottles until both patrons have received the last production. This Petri net is a state machine, but it is not a marked graph. This Petri net is also a free-choice net. If the two shops sell to the same patron and the winery is allowed to produce more wine immediately after delivering the previous batch to the shop, it is represented using the Petri net shown in Figure 4.20(b). This net is not a state machine, a marked graph, or a free-choice net. The reason this net is not free choice is that the transitions *receive1* and *receive2* share a place in their preset, but they have two places in their preset. This pair of transitions does, however, satisfy the extended free-choice requirement since the places in their preset have the same transitions in their postset. The transitions *send1* and *send2* are neither free or extended free choice. It is, however, an asymmetric choice. Consider the places in the preset of *send1*. One has only *send1* in its postset while the other has both *send1* and *send2*, so one place has a subset of the transitions in its postset compared with the other. Similarly, this is true for the places in the preset of *send2*. The net in Figure 4.20(a) is therefore an asymmetric choice net. The Petri net shown in Figure 4.20(c) depicts a model where the winery can produce two bottles of wine at once which are sold to two separate shops that sell to two separate patrons. The winery, however, has the ability to recall the wine (perhaps some bad grapes). This Petri net falls into none of these classifications. Consider the two places in the preset of transitions *send1*, *send2*, and *recall*. Both places have the transition *recall* in their postset. However, they each have a distinct transition in their postset. This net has symmetric confusion.

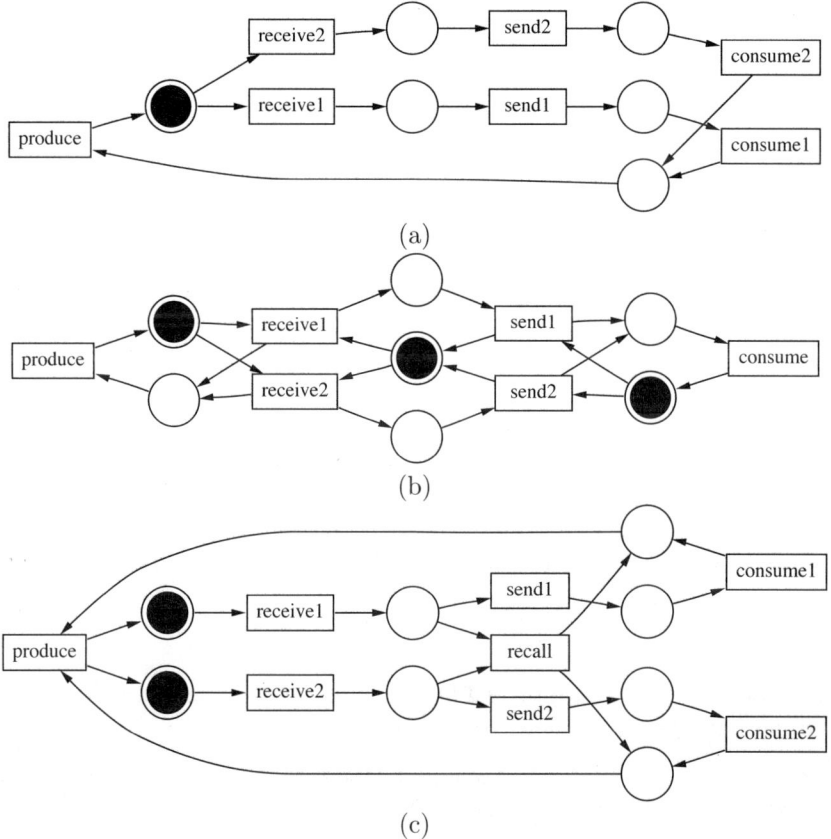

Fig. 4.20 (a) Free-choice Petri net. (b) Asymmetric choice net. (c) Petri net with symmetric confusion.

It is possible to check safety and liveness for certain restricted classes of Petri nets using the theorems given below.

Theorem 4.1 *A state machine is live and safe iff it is strongly connected and M_0 has exactly one token.*

Theorem 4.2 (Commoner, 1971) *A marked graph is live and safe iff it is strongly connected and M_0 places exactly one token on each simple cycle.*

A *siphon* is a nonempty subset of places, S, in which every transition having a postset place in S also has a preset place in S (i.e., $\bullet S \subseteq S \bullet$). If in some marking no place in S has a token, then in all future markings, no place in S will ever have a token. An example siphon is shown in Figure 4.21(a) in which the token count remains the same when firing $t1$, but reduced when $t2$ fires.

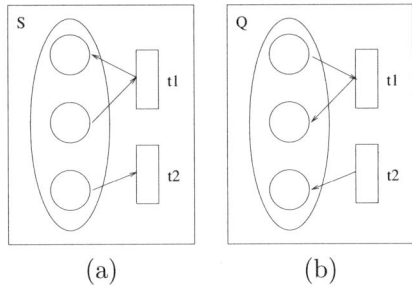

Fig. 4.21 (a) Example of a siphon. (b) Example of a trap.

A *trap* is a nonempty subset of places, Q, in which every transition having a preset place in Q also has a postset place in Q (i.e., $Q\bullet \subseteq \bullet Q$). If in some marking some place in Q has a token, then in all future markings some place in Q will have a token. An example trap is shown in Figure 4.21(b) in which the token count remains the same when firing $t1$, but increases when $t2$ fires.

Theorem 4.3 (Hack, 1972) *A free-choice net, N, is live iff every siphon in N contains a marked trap.*

Theorem 4.4 (Commoner, 1972) *An asymmetric choice net N is live if (but not only if) every siphon in N contains a marked trap.*

A *state machine* (SM) *component* of a net, N, is a subnet in which each transition has at most one place in its preset and one place in its postset and is generated by these places. The net generated by a set of places includes these places, all transitions in their preset and postset, and all connecting arcs. A net N is said to be covered by a set of SM-components when the set of components includes all places, transitions, and arcs from N.

Similarly, a *marked graph* (MG) *component* of a net, N, is a subnet in which each place has at most one transition in its preset and one transition in its postset and is generated by these transitions. The net generated by a set of transitions includes these transitions, all places in their preset and postset, and all connecting arcs. A net N is said to be covered by a set of MG-components when the set of components includes all places, transitions, and arcs from N.

The following theorems show that a live and safe free-choice net can be thought of as an interconnection of state machines or marked graphs.

Theorem 4.5 (Hack, 1972) *A live free-choice net, N, is safe iff N is covered by strongly connected SM-components each of which has exactly one token in M_0.*

Theorem 4.6 *If N is a live and safe free-choice net then N is covered by strongly connected MG-components.*

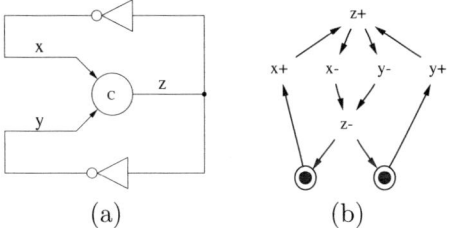

Fig. 4.22 (a) Simple C-element circuit. (b) STG model of the simple circuit.

4.3.2 Signal Transition Graphs

In order to use a Petri net to model asynchronous circuits, it is necessary to relate transitions to events on signal wires. There have been several variants of Petri nets that accomplish this, including *M-nets*, *I-nets*, and *change diagrams*. In this section we describe one of the most common variants, called a *signal transition graph* (STG). A STG is a labeled safe Petri net which is modeled with the 7-tuple $\langle P, T, F, M_0, N, s_0, \lambda_T \rangle$, where:

- $N = I \cup O$ is the set of signals where I is the set of input signals and O is the set of output signals.

- s_0 is the initial value for each signal in the initial state.

- $\lambda_T : T \to N \times \{+, -\}$ is the *transition labeling function*.

In a STG, each transition is labeled with either a rising transition, $s+$, or falling transition, $s-$. If it is labeled with $s+$, it indicates that this transition corresponds to a $0 \to 1$ transition on the signal wire s. If it is labeled with a $s-$, it indicates that the transition corresponds to a $1 \to 0$ transition on s. Note that as opposed to state machines, a STG imposes explicit restrictions on the environment's allowed behavior.

> **Example 4.3.9** An example STG model for the simple C-element circuit shown in Figure 4.22(a) is given in Figure 4.22(b). Note that implicit places (ones that have only a single transition in their preset and postset) are often omitted in STGs. In this STG, initially x and y are enabled to change from '0' to '1' (i.e., $x+$ and $y+$ are enabled). After both x and y change to '1' (in either order), z is enabled to change to '1'. This enables x and y to change from '1' to '0' (i.e., $x-$ and $y-$ are enabled). After they have both changed to '0', z can change back to '0'. This brings the system back to its initial state.

The allowable forms of STGs are often restricted to a synthesizable subset. For example, synthesis methods often restrict the STG to be *live* and *safe*. Some synthesis methods require STGs to be *persistent*. A STG is persistent if for all arcs $a* \to b*$, there exist other arcs that ensure that $b*$ fires before

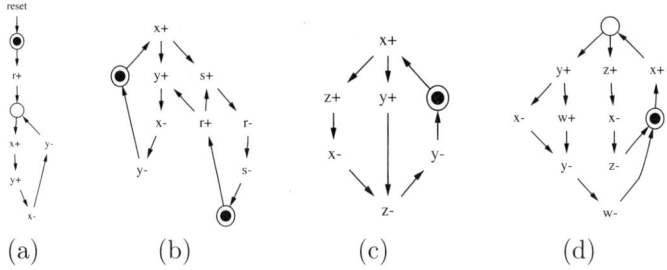

Fig. 4.23 (a) STG that is not live. (b) STG that is not safe. (c) STG that is not persistent. (d) STG that does not have single-cycle transitions.

the opposite transition of $a*$. Other methods require *single-cycle transitions*. A STG has single-cycle transitions if each signal name appears in exactly one rising and one falling transition.

> **Example 4.3.10** The STG in Figure 4.23(a) is not live since after $r+$ fires once, it can never fire again. The STG in Figure 4.23(b) is not safe. Consider the following sequences of transitions: $x+$, $r+$, $y+$, $x-$, $y-$. At this point, if $x+$ fires before $s+$, then the arc between $x+$ and $s+$ would receive a second token. The STG in Figure 4.23(c) is not persistent. Consider the following sequence of transitions: $x+$, $z+$, $x-$. At this point, there is still a token on the arc between $x+$ and $y+$ as $x-$ is firing. The STG in Figure 4.23(d) does not have single-cycle transitions since two transitions are labeled with $x-$.

None of these restrictions is actually a necessary requirement for a circuit implementation to exist. In fact, a circuit can be built with the behavior described by each of the STGs in Figure 4.23. However, the use of these restrictions can often simplify the synthesis algorithms, as we will see later.

In order to design a circuit from an STG, it is necessary to find its *state graph* (SG). A SG is modeled by the tuple $\langle S, \delta, \lambda_S \rangle$.

- S is the set of states.

- $\delta \subseteq S \times T \times S$ is the set of state transitions.

- $\lambda_S : S \rightarrow (N \rightarrow \{0,1\})$ is the *state labeling function*.

Each state s is labeled with a binary vector $\langle s(0), s(1), \ldots, s(n) \rangle$, where each $s(i)$ is either 0 or 1, indicating the value returned by λ_S. When the context is clear, we use $s(i)$ interchangeably with $\lambda_S(s)(i)$. For each state, we can determine the value that each signal is tending to. In particular, if in a state s_i, there exists a transition on signal u_i to another reachable state s_j [i.e., $\exists (s_i, t, s_j) \in \delta$. $\lambda_T(t) = u_i + \lor \lambda_T(t) = u_i-$], then the signal u_i is *excited*. If there does not exist such a transition, the signal u_i is in *equilibrium*. In either case, the value that each signal is tending to is called its *implied value*.

If the signal is excited, the implied value of u_i is $\overline{s(i)}$. If the signal is in equilibrium, the implied value of u_i is $s(i)$. The *implied state*, s' is labeled with a binary vector $\langle s'(0), s'(1), \ldots, s'(n) \rangle$ of the implied values. The function $X : S \rightarrow 2^N$ returns the set of excited signals in a given state [i.e., $X(s) = \{u_i \in S \mid s(i) \neq s'(i)\}$]. When $u_i \in X(s)$ and $s(i) = 0$, the state in the state diagram is annotated with an "R" in the corresponding location of the state vector to indicate that the signal u_i is excited to rise. When $u_i \in X(s)$ and $s(i) = 1$, the state is annotated with an "F" to indicate that the signal is excited to fall. The algorithm to find the SG from an STG is shown in Figure 4.24.

Example 4.3.11 Consider the application of the *find_SG* algorithm to the STG in Figure 4.22. The initial state for the initial marking is $\langle 000 \rangle$. In this state, $x+$ and $y+$ are enabled. Note that although $\lambda_S(s_0) = 000$, we annotate the state with the label $RR0$ to indicate that signals x and y are excited to rise. Let's assume that the algorithm chooses to fire $x+$. It then pushes the initial marking, $\langle 000 \rangle$, and $y+$ onto the stack. Next, it checks if firing $x+$ results in a safety violation, which it does not. It then updates the marking and changes the state to $\langle 100 \rangle$ (labeled $1R0$ in the figure). This state is new, so it adds it to the list of states, updates the state labeling function, and adds a state transition. It then determines that $y+$ is the only remaining enabled transition. Control loops back, and $y+$ is fired. There are no remaining transitions to push on the stack, so it skips to update the marking and state to $\langle 110 \rangle$. This state is again new, so the state is added to the set of states. In this state, $z+$ is enabled. Continuing in this way, the rest of the SG can be found, as shown in Figure 4.25.

For a SG to be well-formed, it must have a *consistent state assignment*. A SG has a consistent state assignment if for each state transition $(s_i, t, s_j) \in \delta$ exactly one signal changes value, and its value is consistent with the transition. In other words, a SG has a consistent state assignment if

$$\forall (s_i, t, s_j) \in \delta. \forall u \in N \quad . \quad (\lambda_T(t) \neq u * \land s_i(u) = s_j(u))$$
$$\lor \quad (\lambda_T(t) = u + \land s_i(u) = 0 \land s_j(u) = 1)$$
$$\lor \quad (\lambda_T(t) = u - \land s_i(u) = 1 \land s_j(u) = 0)$$

where "$*$" represents either "$+$" or "$-$". A STG produces a SG with a consistent state assignment if in any firing sequence the transitions of a signal strictly alternate between $+$'s and $-$'s.

Example 4.3.12 The STG in Figure 4.26(a) does not have a consistent state assignment. If the initial state is 000, then the sequence of transitions $x+$, $y+$, $z+$, $y-$, $z-$ takes it to the same marking, which now must be labeled 100. This is not a consistent labeling, so this STG produces a SG which does not have a consistent state assignment.

A SG is said to have a *unique state assignment* (USC) if no two different states (i.e., markings) have identical values for all signals [i.e., $\forall s_i, s_j \in S, s_i \neq$

```
find_SG(⟨P, T, F, M_0, N, s_0, λ_T⟩) {
  M = M_0;
  s = s_0;
  T_e = {t ∈ T | M ⊆ •t};
  S = {M};
  λ_S(M) = s;
  done = false;
  while (¬ done) {
    t = select(T_e);
    if (T_e − {t} ≠ ∅) then
      push(M, s, T_e − {t});
    if ((M − •t) ∩ t• ≠ ∅) then
      return(''STG is not safe.'');
    M' = (M − •t) ∪ t•;
    s' = s;
    if (λ_T(t) = u+) then
      s'(u) = 1;
    else if (λ_T(t) = u−) then
      s'(u) = 0;
    if (M' ∉ S) then {
      S = S ∪ {M'};
      λ_S(M') = s'
      δ = δ ∪ {(M, t, M')};
      M = M';
      s = s';
      T_e = {t ∈ T | M ⊆ •t};
    } else {
      if (λ_S(M') ≠ s') then
        return(''Inconsistent state assignment.'');
      if (stack is not empty) then
        (M, s, T_e) = pop();
      else
        done = true;
    }
  }
  return(⟨S, δ, λ_S⟩);
}
```

Fig. 4.24 Algorithm to find a state graph.

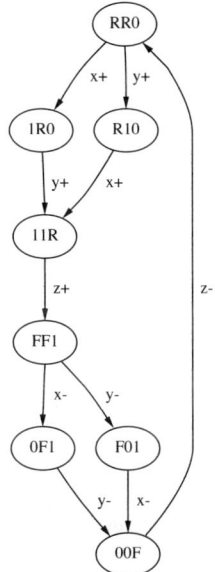

Fig. 4.25 State graph for C-element with state vector $\langle x, y, z \rangle$.

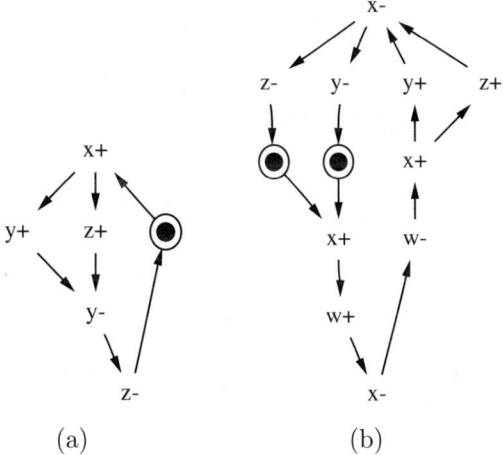

Fig. 4.26 (a) STG that does not have a consistent state assignment. (b) STG that does not have a unique state code.

$s_j \cdot \lambda(s_i) \neq \lambda(s_j)]$. Some synthesis methods are restricted to STGs that produce SGs with USC.

> **Example 4.3.13** The STG in Figure 4.26(b) produces a SG which does not have USC. In the initial marking all signals are '0' (i.e., the state vector is 0000). Consider the following sequence of transitions: $x+$, $w+$, $x-$, and $w-$. At the end of this sequence, the STG has a different marking, but all signals are again '0' (i.e., the state vector is 0000).

We conclude this section by returning to the passive/lazy-active wine shop example from the end of Section 4.2. Recall that this could not be modeled using an AFSM or even an XBM machine. This can be modeled as a STG as shown in Figure 4.27. Another example STG for the wine shop with two patrons from Section 4.2 is shown in Figure 4.28.

4.4 TIMED EVENT/LEVEL STRUCTURES

The major drawback of AFSMs is their inability to model arbitrary concurrency. The failing of Petri nets is their difficulty in expressing signal levels (see Figure 4.28). A *Timed event/level* (TEL) *structure* is a hybrid graphical representation method which is capable of modeling both arbitrary concurrency and signal levels. TEL structures support both event causality to specify sequencing and level causality to specify bit-value sampling. In this section we give an overview of TEL structures.

A TEL structure is modeled with a 6-tuple $T = \langle N, s_0, A, E, R, \# \rangle$, where:

1. N is the set of signals.
2. $s_0 = \{0,1\}^N$ is the initial state.
3. $A \subseteq N \times \{+,-\} \cup \$$ is the set of atomic *actions*.
4. $E \subseteq A \times (\mathcal{N}=\{0,1,2...\})$ is the set of *events*.
5. $R \subseteq E \times E \times \mathcal{N} \times (\mathcal{N} \cup \{\infty\}) \times (\text{b:}\{0,1\}^N \to \{0,1\})$ is the set of *rules*.
6. $R_O \subseteq R$ is the set of *initially marked rules*.
7. $\# \subseteq E \times E$ is the *conflict relation*.

The signal set, N, contains the signal wires in the specification. The state s_0 contains the initial value of each signal in N. The action set, A, contains for each signal, x, in N, a rising transition, $x+$, and a falling transition, $x-$. The set A also includes a *sequencing event*, \$, which is used to indicate an action that does not result in a signal transition. The event set, E, contains actions paired with occurrence indices (i.e., $\langle a, i \rangle$). Note that $\mathcal{N}$ represents the set of natural numbers. Rules represent causality between events. Each rule, r, is of the form $\langle e, f, l, u, b \rangle$, where:

TIMED EVENT/LEVEL STRUCTURES   117

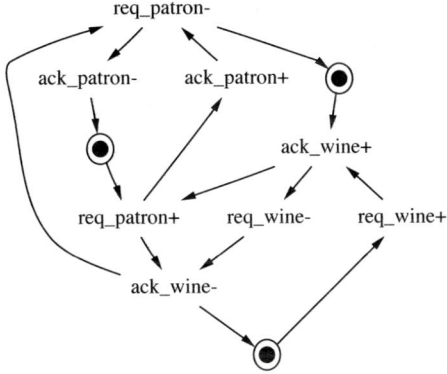

Fig. 4.27 STG for the reshuffled passive/lazy-active wine shop.

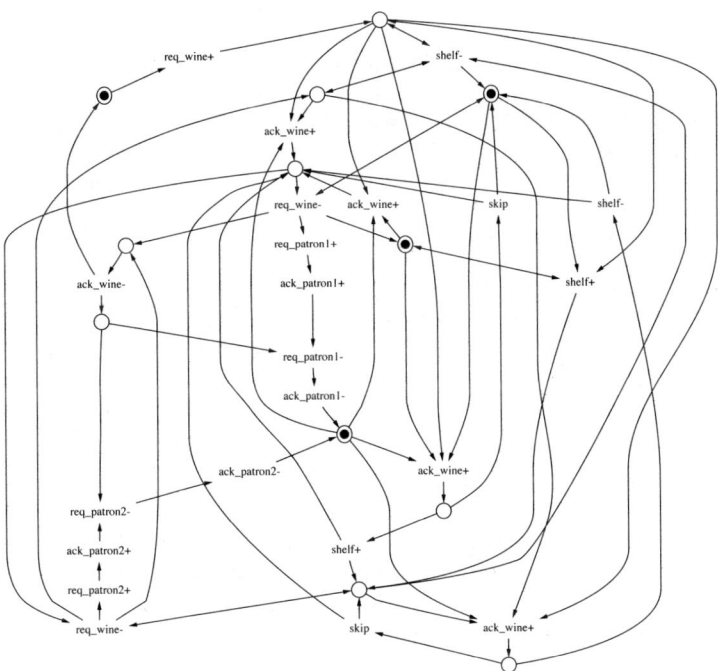

Fig. 4.28 STG for the wine shop with two patrons.

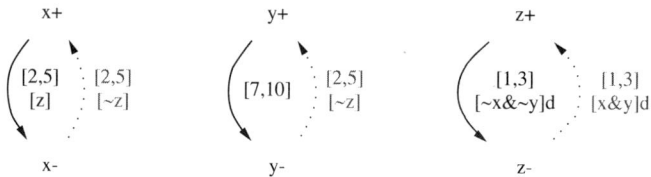

Fig. 4.29 TEL structure for a C-element and its environment.

1. e = enabling event.
2. f = enabled event.
3. $[l, u]$ = bounded timing constraint.
4. b = a boolean function over the signals in N.

> **Example 4.4.1** Figure 4.29 shows the TEL structure for a C-element and its environment. The dotted arcs represent initially marked rules, and the initial state is 000. The nodes in these graphs are events, and the arcs are rules. Each rule is annotated with a bounded timing constraint and a level expression. If the level expression is equal to **true**, it is omitted from the graph. If the rule is disabling, the level expression is terminated with a "d." Otherwise, the rule is nondisabling. There are three processes. The first represents the behavior of the signal x. The signal x rises 2 to 5 time units after z goes low, and it falls 2 to 5 time units after z rises. The behavior of y is a little different. The signal y again rises 2 to 5 time units after z goes low, but it falls 7 to 10 time units later regardless of the value of z.

A rule is *enabled* if its enabling event has occurred and its boolean function is true in the current state. A rule is *satisfied* if it has been enabled at least l time units. A rule becomes *expired* when it has been enabled u time units. Excluding conflicts, an event cannot occur until every rule enabling it is satisfied, and it must occur before every rule enabling it has expired. Each rule is defined to be *disabling* or *nondisabling*. If a rule is disabling and its boolean condition becomes false after it has become enabled, the rule ceases to be enabled and must wait until the condition is true again before it can fire. If a rule is nondisabling, the fact that the boolean condition has become false is ignored, and the rule can fire as soon as its lower bound is met. For the purposes of verification, if a rule becomes disabled, this may indicate that the enabled event has a hazard which is considered a failure.

> **Example 4.4.2** Let us first consider the behavior of the TEL structure in Figure 4.29, ignoring timing. In the initial state, $x+$ and $y+$ are enabled. Note that $z+$ is not enabled because the expression $[x\&y]d$ is false. Let us assume that $x+$ fires first, followed by $y+$. At this point, $y-$ and $z+$ are enabled. If $y-$ fires first, the expression $[x\&y]d$ becomes

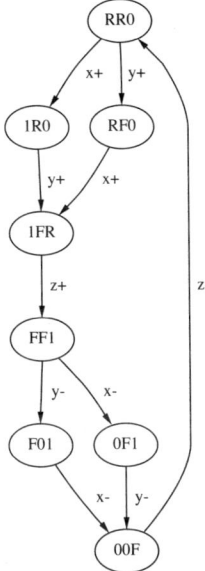

Fig. 4.30 SG for a timed C-element circuit.

disabled. This indicates that the C-element has a hazard and may glitch. Let us now consider the behavior using the timing constraints. Again in the initial state $x+$ and $y+$ are enabled. After 2 time units, the rules $\langle x-, x+, [2,5], [\sim z] \rangle$ and $\langle y-, y+, [2,5], [\sim z] \rangle$ are satisfied and can fire. By 5 time units, these rules become expired and must fire. Let's assume that $y+$ fires at time 2 (this is the worst-case scenario), which enables $y-$. After 3 more time units, $x+$ is forced to fire, enabling $z+$. After 3 more time units, we know for sure that $z+$ has fired. Therefore, it takes at most 6 time units after $y+$ fires before $z+$ fires. Since it takes at least 7 time units after y rises before y falls, we know that z does not glitch. The SG for this timed C-element circuit is shown in Figure 4.30. Note the subtle difference between this SG and the one in Figure 4.25. In this SG, $y-$ is enabled to fall sooner: namely, in states 010 (i.e., $RF0$) and 110 (i.e., $1FR$). Due to timing, however, y cannot actually fall until state 111 (i.e., $FF1$).

The conflict relation, #, is used to model disjunctive behavior and choice. When two events e and e' are in conflict (denoted $e \# e'$), this specifies that either e or e' can occur but not both. Taking the conflict relation into account, if two rules have the same enabled event and conflicting enabling events, only one of the two mutually exclusive enabling events needs to occur to cause the enabled event. This models a form of disjunctive causality. Choice is modeled when two rules have the same enabling event and conflicting enabled events. In this case, only one of the enabled events can occur.

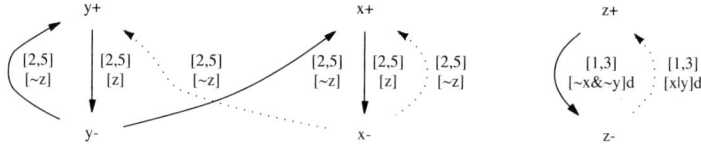

Fig. 4.31 TEL structure with conflict: namely, $x + \#y+$, $x + \#y-$, $x - \#y+$, and $x - \#y-$.

Example 4.4.3 The TEL structure shown in Figure 4.31 has the following conflicts: $x + \#y+$, $x + \#y-$, $x - \#y+$, and $x - \#y-$. In the initial state, $x+$ and $y+$ are both enabled. Even though in each case only one of its two rules is initially marked, they are still enabled since the enabling events of these rules, $x-$ and $y-$, are in conflict. In other words, only one of $x-$ or $y-$ is necessary for $x+$ or $y+$ to be enabled. Let us assume that $y+$ fires first. In the new state, $x+$ is no longer enabled since $x+$ and $y+$ are in conflict, and $z+$ is now enabled since the expression $[x|y]$ has become true. After $z+$ fires, $y-$ can fire followed by $z-$. In this state, $x+$ and $y+$ are again both enabled.

We conclude this chapter with the TEL structure for the shop process from the wine shop with two patrons from Section 4.2.3, which is shown in Figure 4.32. This TEL structure has numerous conflicts. Namely, all events in the set { $ack\_wine-/1$, $req\_patron1+$, $req\_patron1-$, $\$/2$ } conflict with all events in the set { $ack\_wine-/2$, $req\_patron2+$, $req\_patron2-$, $\$/3$ }. One important advantage of TEL structures is a more direct correlation with the handshaking-level specification. When you compare the STG for this example to the TEL structure, it is quite clear that the TEL structure has a more direct correspondence to the handshaking-level specification.

4.5 SOURCES

The Huffman flow table was introduced in [170, 172]. Burst mode was originally developed by Stevens [365], and it was applied successfully to a number of industrial designs by Davis's group at Hewlett-Packard [89, 99, 100, 101]. Nowick constrained and formalized burst mode into its current form, and he proposed the unique entry point and maximal set property. He also proved that any burst-mode machine satisfying these properties have a hazard-free gate-level implementation, and he developed the first hazard-free synthesis procedure [301]. Yun et al. introduced extended burst-mode machines [422].

Petri nets were introduced by Petri [312]. A couple of good surveys of Petri nets are found in [280, 311]. Seitz introduced the Petri net variant *machine nets* (M-nets) for the specification of asynchronous circuits [344]. An M-net is essentially a labeled safe Petri net in which each transition is

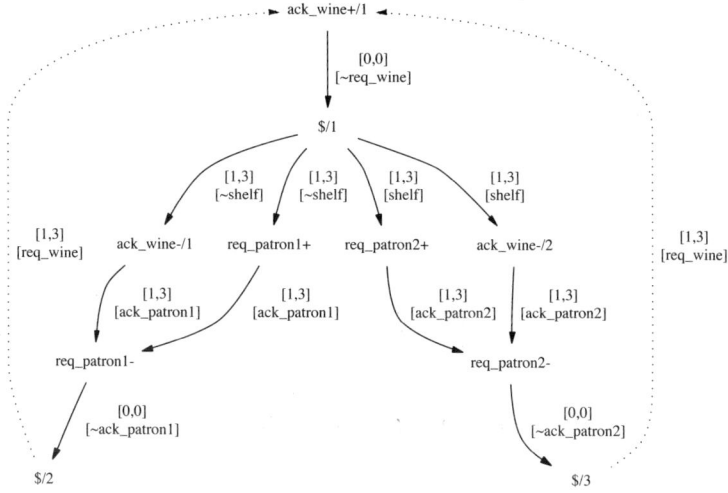

Fig. 4.32 TEL structure for the wine shop with two patrons.

labeled with either a signal name, s, or its complement, s'. If it is labeled with a signal name, it indicates that this transition corresponds to a $0 \rightarrow 1$ transition on the corresponding signal wire. If it is labeled with a signal complement, it indicates that the transition corresponds to a $1 \rightarrow 0$ transition. Molnar et al. [274] introduced another Petri net variant, *interface nets* (I-nets), for asynchronous interface specifications. An I-net is a safe Petri net with transitions labeled with interface signal names. The I-net describes the allowed interface behavior. *Change diagrams* were introduced by Varshavsky [393]. Change diagrams are quite similar to STGs, but they are capable of specifying initial behavior and a limited form of disjunctive causality. In particular, a change diagram is composed of three different types of arcs. There are *strong precedence arcs*, which model conjunctive (AND) causality. There are also *weak precedence arcs*, which model disjunctive (OR) causality. In this case, tokens only need to be present on a single incoming arc for the transition to be enabled. The last type of arcs are the *disengageable strong precedence arcs*. These arcs behave just like strong precedence arcs the first time they are encountered. They are then disengaged from the graph and are not considered in subsequent cycles. They are useful for specifying initial startup behavior. The signal transition graph (STG) was introduced by Chu [82, 83] and concurrently by Rosenblum and Yakovlev [329]. Chu also developed a method to translate AFSMs to STGs [84]. Muller and Bartky introduced the idea of a state graph [279].

TEL structures were introduced by Belluomini and Myers [34]. Timing has also been added to Petri nets by Ramchandani [320] and others.

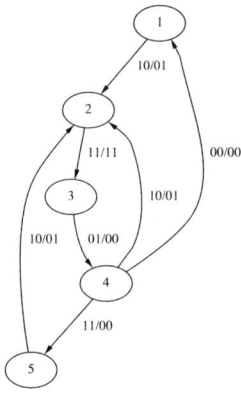

Fig. 4.33 AFSM for Problem 4.1.

Problems

4.1 Asynchronous Finite State Machines
The AFSM for a *quick-return linkage circuit* (QRL) is shown in Figure 4.33.

4.1.1. Find the flow table for the AFSM.

4.1.2. Draw a burst-mode state machine for the QRL circuit.

4.2 Burst-Mode State Machines
For each BM state diagram in Figure 4.34, determine whether:

4.2.1. It has the maximal set property.

4.2.2. It represents a BM machine.

4.3 Burst-Mode to Flow Table
Translate the BM machine shown in Figure 4.35(a) into a flow table.

4.4 Extended Burst-Mode State Machines
For the extended burst-mode state machines in Figure 4.36, determine whether they satisfy the maximal set property, and if not, explain why not.

4.5 Extended Burst-Mode to Flow Table
Translate the XBM machine shown in Figure 4.35(b) into a flow table.

4.6 Petri net Properties
For the Petri nets in Figure 4.37, determine whether they have the following properties:

4.6.1. k-bounded, and if so, for what value of k? Is it safe?

4.6.2. Live, and if not, classify each transition's degree of liveness.

4.7 Petri net Classifications
For the Petri nets in Figure 4.37, determine whether they fall into the following classifications:

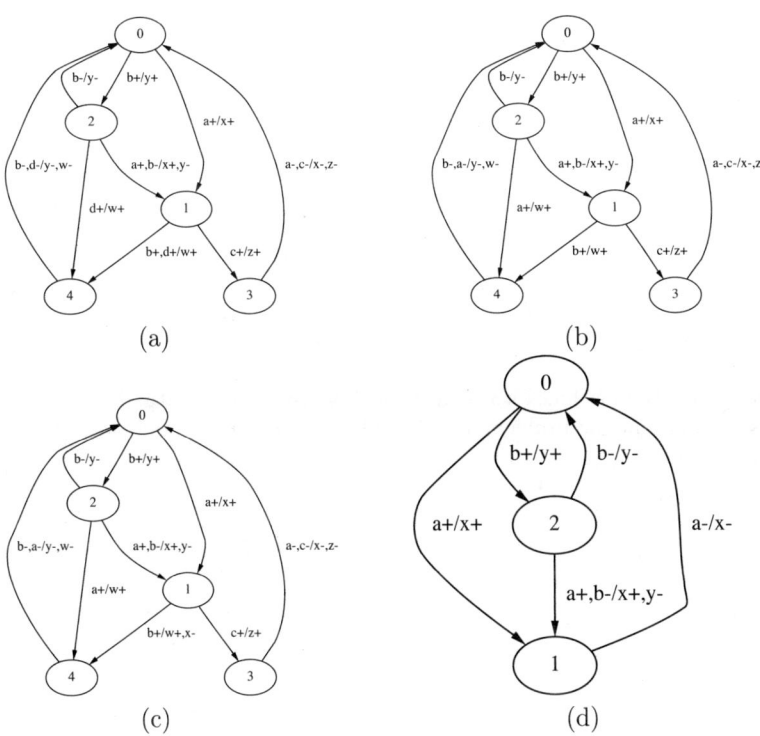

Fig. 4.34 BM machines for Problem 4.2.

124   GRAPHICAL REPRESENTATIONS

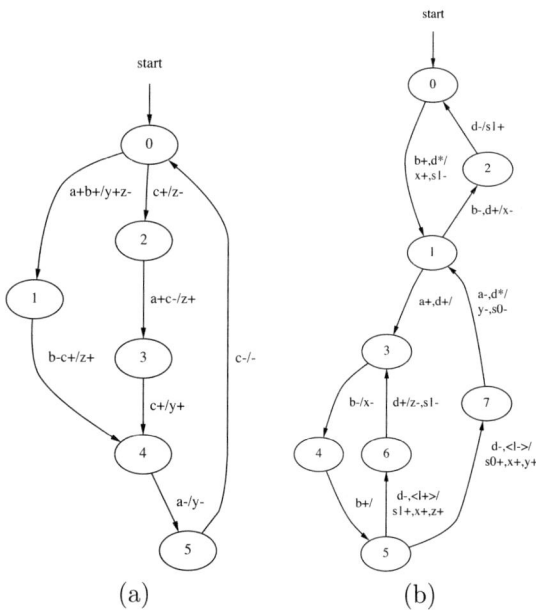

Fig. 4.35 (a) BM machine for Problem 4.3. Note that $abc = 000$ and $yz = 01$ initially. (b) XBM machine for Problem 4.5. Note that $abdl = 000-$ and $s0s1xyz = 01000$ initially.

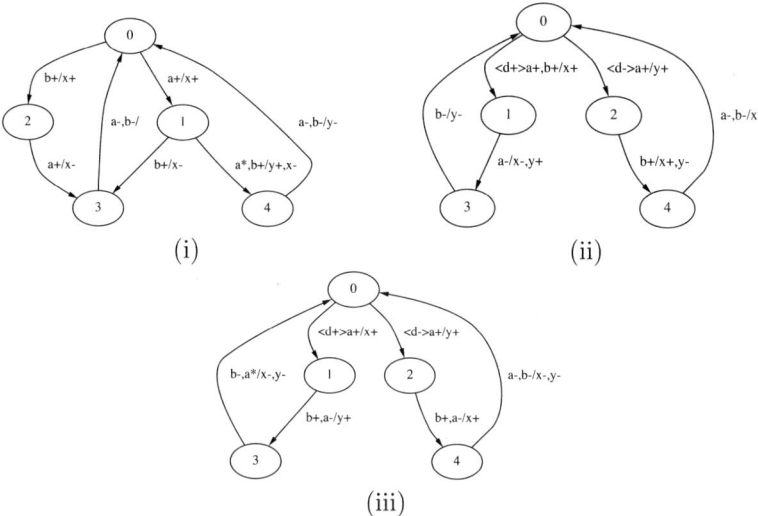

Fig. 4.36 Extended burst-mode machines for Problem 4.4.

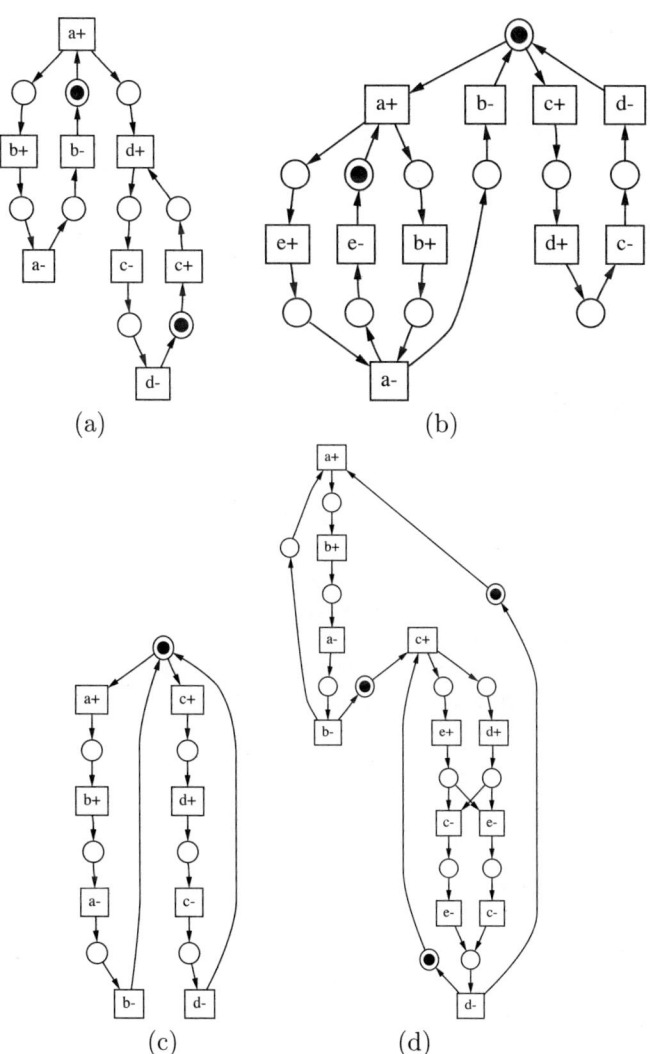

Fig. 4.37 Petri nets for Problems 4.6, 4.7, and 4.8.

126 GRAPHICAL REPRESENTATIONS

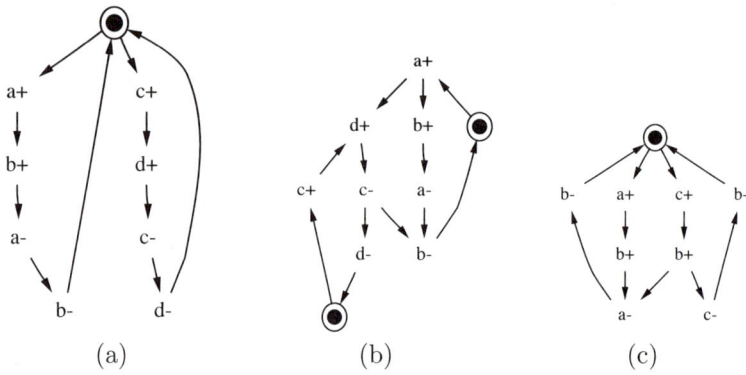

Fig. 4.38 Signal transition graphs for problems 4.9 and 4.10.

- **4.7.1.** State machine
- **4.7.2.** Marked graph
- **4.7.3.** Free-choice net
- **4.7.4.** Extended free-choice net
- **4.7.5.** Asymmetric choice net

4.8 Reachability Graphs
For the Petri net in Figure 4.37(b), determine its reachability graph.

4.9 Signal Transition Graphs
For the signal transition graphs in Figure 4.38, state which of the following properties are satisfied or not, and if not, say why:
- **4.9.1.** Live
- **4.9.2.** Safe
- **4.9.3.** Persistent
- **4.9.4.** Single-cycle transitions

4.10 State Graphs
For the signal transition graphs in Figure 4.38, find their state graphs and determine if they have the following properties, and if not, say why:
- **4.10.1.** Consistent state assignment
- **4.10.2.** Unique state assignment

4.11 Signal Transition Graphs
For the signal transition graphs in Figure 4.39, state which of the following properties are satisfied or not, and if not, say why:
- **4.11.1.** Live
- **4.11.2.** Safe
- **4.11.3.** Persistent
- **4.11.4.** Single-cycle transitions

4.12 State Graphs

For the signal transition graphs in Figure 4.39, find their state graphs and determine if they have the following properties, and if not, say why:

4.12.1. Consistent state assignment

4.12.2. Unique state assignment

4.13 Signal Transition Graphs

For the signal transition graphs in Figure 4.40, state which of the following properties are satisfied or not, and if not, say why:

4.13.1. Live

4.13.2. Safe

4.13.3. Persistent

4.13.4. Single-cycle transitions

4.14 State Graphs

For the signal transition graphs in Figure 4.40, find their state graphs and determine if they have the following properties, and if not, say why:

4.14.1. Consistent state assignment

4.14.2. Unique state assignment

4.15 State Graphs

Find the state graph from the STG representation of the quick return linkage (QRL) circuit shown in Figure 4.41(a).

4.16 State Graphs

Assuming that all signals are initially low, find the SG from the STG in Figure 4.41(b).

4.17 State Graphs

Assuming that all signals are initially low, find the SG from the TEL structure in Figure 4.42. The events $a+$, $a-$, $b+$, and $b-$ all conflict with each other.

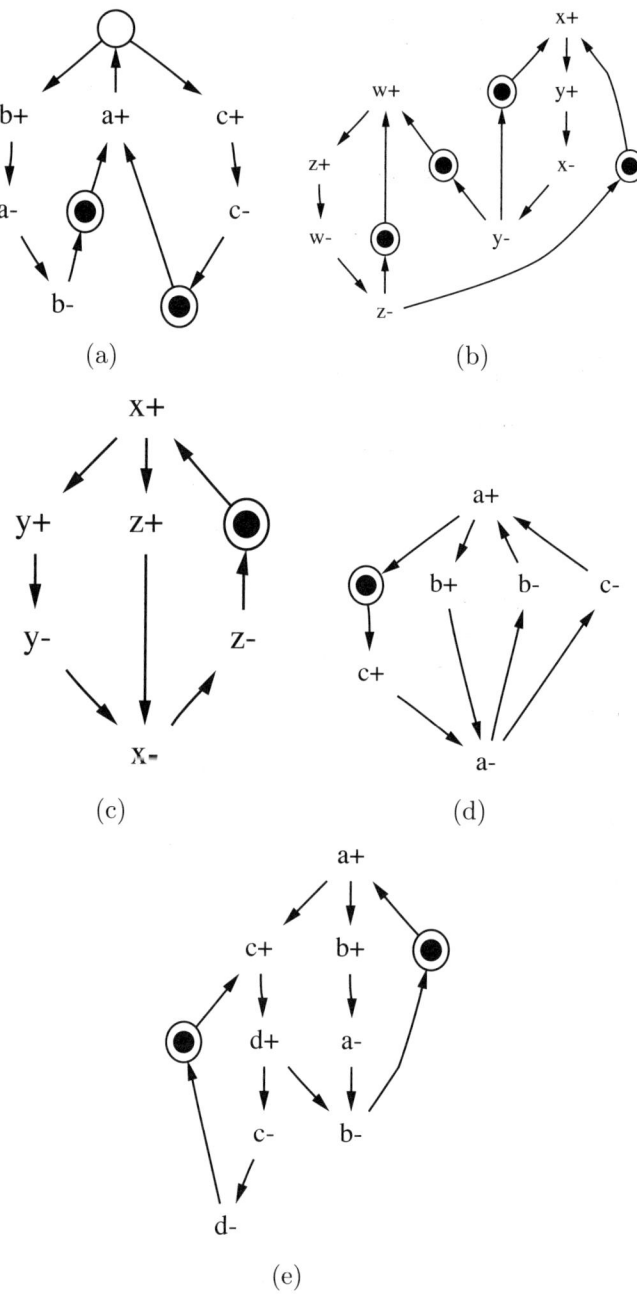

Fig. 4.39 Signal transition graphs for problems 4.11 and 4.12.

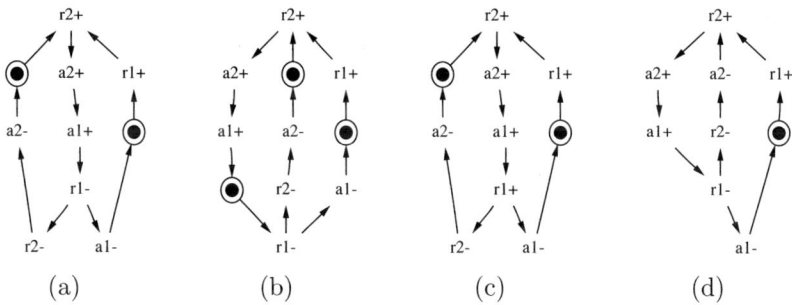

Fig. 4.40 Signal transition graphs for Problems 4.13 and 4.14.

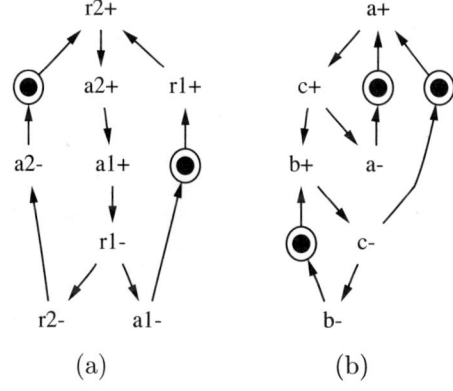

Fig. 4.41 (a) QRL circuit for Problem 4.15. (b) STG for Problem 4.16.

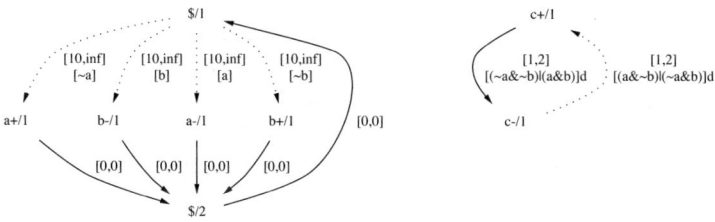

Fig. 4.42 TEL for Problem 4.17.

4.18 Translation

For the following handshaking expansion, draw graphical representations using the following methods:

- BM machine
- STG
- TEL structure

```
shopPA_dual_rail:process
begin
  guard_or(bottle0,'1',bottle1,'1');
  if bottle0 = '1' then assign(shelf0,'1',1,2);
  elsif bottle1 = '1' then assign(shelf1,'1',1,2);
  end if;
  guard(ack_patron,'1');
  assign(ack_wine,'1',1,2,shelf0,'0',1,2,shelf1,'0',1,2);
  guard_and(bottle0,'0',bottle1,'0',ack_patron,'0');
  assign(ack_wine,'0',1,2);
end process;
```

5
Huffman Circuits

Never do today what you can put off till tomorrow. Delay may give clearer light as to what is best to be done.
—Aaron Burr

Delay is preferable to error.
—Thomas Jefferson

Feedback is your friend.
—Professor Doyle at my Caltech Frosh Camp

In this chapter we introduce the Huffman school of thought to the synthesis of asynchronous circuits. *Huffman circuits* are designed using a traditional asynchronous state machine approach. As depicted in Figure 5.1, an asynchronous state machine has primary inputs, primary outputs, and fed-back state variables. The state is stored in the feedback loops and thus may need delay elements along the feedback path to prevent state changes from occurring too rapidly. The design of Huffman circuits begins with a specification given in a flow table which may have been derived from an AFSM, BM, or XBM machine. The goal of the synthesis procedure is to derive a correct circuit netlist which has been optimized according to design criteria such as area, speed, or power. The approach taken for the synthesis of synchronous state machines is to divide the synthesis problem into three steps. The first step is *state minimization*, in which compatible states are merged to produce a simpler flow table. The second step is *state assignment* in which a binary encoding is assigned to each state. The third step is *logic minimization* in

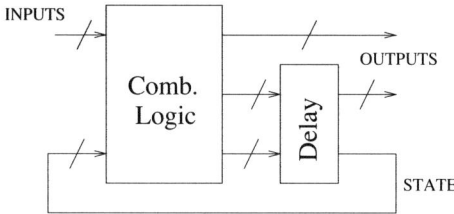

Fig. 5.1 Huffman circuit model.

which an optimized netlist is derived from an encoded flow table. The design of Huffman circuits can follow the same three-step process, but each step must be modified to produce correct circuits under an asynchronous timing model.

Huffman circuits are typically designed under the *bounded gate and wire delay* model. Under this model, circuits are guaranteed to work regardless of gate and wire delays as long as a bound on the delays is known. In order to design correct Huffman circuits, it is also necessary to put some constraints on the behavior of the environment, namely when inputs are allowed to change. There are a number of different restrictions on inputs that have been proposed, each resulting in variations in the synthesis procedure. The first is *single-input change* (SIC), which states that only one input is allowed to change at a time. In other words, each input change must be separated by a minimum time interval. If this minimum time interval is set to be the maximum delay for the circuit to stabilize, the restriction is called single-input change *fundamental mode*. This is quite restrictive, though, so another approach is to allow *multiple-input changes* (MIC). Again, if input changes are allowed only after the circuit stabilizes, this mode of operation is called multiple-input change fundamental mode. We first introduce each of the synthesis steps under the single-input change fundamental mode restriction, and later the synthesis methods are extended to support a limited form of multiple-input changes: extended burst mode.

5.1 SOLVING COVERING PROBLEMS

The last step of state minimization, state assignment, and logic synthesis is to solve a *covering problem*. For this reason, we begin this chapter by describing an algorithm for solving covering problems. Informally, a covering problem exists whenever you must select a set of choices with minimum cost which satisfy a set of constraints. The classic example is when deriving a minimum two-level sum-of-products cover of a logic function, one must choose the minimum number of prime implicants that cover all the minterms of a given function.

More formally, the set of choices is represented using a Boolean vector, $\mathbf{x} = (x_1, \ldots, x_n)$. If a Boolean variable, x_i, is set to 1, this indicates that this choice has been included in the solution. When $x_i = 0$, this indicates that this choice has not been included in the solution. A covering problem can now be expressed as a product-of-sums f where each product (or *clause*) represents a constraint that must be satisfied. Each clause is the sum of those choices that would satisfy the constraint. We are also given a cost function:

$$\text{cost}(\mathbf{x}) = \sum_{i=1}^{n} w_i x_i \tag{5.1}$$

where n is the number of choices and w_i is the cost of each choice. The goal now is to solve the covering problem by finding the assignment of the x_i's which results in the minimum cost.

Example 5.1.1 An example product-of-sums formulation of a covering problem is shown below. Note that any solution must include x_1, must not include x_2, must either not include x_3 or include x_4, etc.

$$f = x_1 \overline{x_2} (\overline{x_3} + x_4)(\overline{x_3} + x_4 + x_5 + x_6)(\overline{x_1} + x_4 + x_5 + x_6)(\overline{x_4} + x_1 + x_6)(\overline{x_5} + x_6)$$

When choices appear only in their positive form (i.e., uncomplemented), it is a *unate covering problem*. When any choice appears in both its positive and negative form (i.e., complemented), it is a *binate covering problem*. In this section we consider the more general case of the binate covering problem, but the solution clearly applies to both.

The function f is represented using a *constraint matrix*, $\mathbf{A}$, which includes a column for each x_i variable and a row for every clause. Each entry of the matrix a_{ij} is "–" if the variable x_i does not appear in the clause, "0" if it appears complemented, and "1" if it appears uncomplemented. The ith row of $\mathbf{A}$ is denoted a_i while the jth column is denoted by A_j.

Example 5.1.2 The constraint matrix for the covering problem from Example 5.1.1 is shown below.

$$\mathbf{A} = \begin{array}{c} \\ \\ \\ \\ \\ \\ \\ \end{array} \begin{array}{cccccc} x_1 & x_2 & x_3 & x_4 & x_5 & x_6 \\ \left[\begin{array}{cccccc} 1 & - & - & - & - & - \\ - & 0 & - & - & - & - \\ - & - & 0 & 1 & - & - \\ - & - & 0 & 1 & 1 & 1 \\ 0 & - & - & 1 & 1 & 1 \\ 1 & - & - & 0 & - & 1 \\ - & - & - & - & 0 & 1 \end{array}\right] \end{array} \begin{array}{c} 1 \\ 2 \\ 3 \\ 4 \\ 5 \\ 6 \\ 7 \end{array}$$

The binate covering problem is to find an assignment to the Boolean variables, $\mathbf{x}$, of minimum cost such that for every row a_i either

1. $\exists j \,.\, (a_{ij} = 1) \wedge (x_j = 1)$; or
2. $\exists j \,.\, (a_{ij} = 0) \wedge (x_j = 0)$.

```
bcp(A, x, b) {
  (A, x) = reduce(A, x);        /* Find essentials / apply dominance.*/
  L = lower_bound(A, x);        /* Compute lower bound for this matrix.*/
  if (L ≥ cost(b)) then return(b);          /* Check against best.*/
  if (terminalCase(A)) then {   /* Check if solved or infeasible.*/
    if (A has no rows) return(x);    /* Matrix is fully reduced.*/
    else return(b);             /* Matrix is infeasible return best.*/
  }
  c = choose_column(A);         /* Select column to branch on.*/
  $x_c$ = 1;                      /* Include selected column in solution.*/
  $A^1$ = select_column(A, c);    /* Use c in solution.*/
  $x^1$ = bcp($A^1$, x, b)        /* Recurse with c in solution.*/
  if (cost($x^1$)<cost(b)) then {             /* If better record it.*/
    b = $x^1$;
    if (cost(b)= L) return(b);  /* If lower bound, return.*/
  }
  $x_c$ = 0;                     /* Exclude selected column from the solution.*/
  $A^0$ = remove_column(A, c);    /* Do not use c in solution.*/
  $x^0$ = bcp($A^0$, x, b)        /* Recurse with c not in solution.*/
  if (cost($x^0$)<cost(b)) then b = $x^0$;    /* If better, record it.*/
  return(b);
}
```

Fig. 5.2 Branch-and-bound algorithm for binate covering.

A branch-and-bound algorithm for solving the binate covering problem is shown in Figure 5.2. It takes as input a constraint matrix, **A**, the current partial solution vector **x**, and the best solution found so far, **b**. The *bcp* algorithm is invoked with $\mathbf{x} = \mathbf{0}$ and $\mathbf{b} = no\_solution$. The cost of $no\_solution$ is defined to be infinite. The *bcp* algorithm returns a vector representing a minimal cover. In the rest of this section we describe this algorithm in detail.

5.1.1 Matrix Reduction Techniques

The first step of the *bcp* algorithm is to reduce the matrix using a number of reduction techniques, shown in Figure 5.3. The first deals with *essential rows*. A row a_i of **A** is essential when there exists exactly one j such that a_{ij} is not equal to "−". This corresponds to a clause which consists of only a single literal. If the literal is x_j (i.e., $a_{ij} = 1$), the variable is *essential*, and the solution **x** is updated such that $x_j = 1$. The matrix **A** is reduced by removing the column corresponding to the essential literal, A_j, and all rows a_k which are solved by selecting this column (i.e., $a_{kj} = 1$). If the literal is $\overline{x_j}$ (i.e., $a_{ij} = 0$), the variable is *unacceptable*. The matrix **A** is reduced by removing the column corresponding to the unacceptable literal, A_j, and all rows a_k which are solved by excluding this column (i.e., $a_{kj} = 0$). In other

```
reduce(A,x) {
  do {
    A' = A;
    (A,x) = find_essential_rows(A,x);
    A = delete_dominating_rows(A);
    (A,x) = delete_dominated_columns(A,x);
  } while (A ≠ ∅ and A ≠ A');  /* Repeat until no improvement.*/
  return(A,x);
}
```

Fig. 5.3 Reduce algorithm.

words, the variable is set to the value of the literal, the column corresponding to the variable is removed, and any row where the variable has the same value is removed.

Example 5.1.3 The constraint matrix shown in Example 5.1.2 has two essential rows, rows 1 and 2. Row 1 implies that x_1 is an essential variable and must be set to 1. Setting x_1 to 1 allows us to remove rows 1 and 6. The new constraint matrix after x_1 is selected is shown below.

$$\mathbf{A} = \begin{array}{c} \\ \\ \\ \\ \\ \end{array} \begin{array}{c} x_2 \quad x_3 \quad x_4 \quad x_5 \quad x_6 \\ \left[\begin{array}{ccccc} 0 & - & - & - & - \\ - & 0 & 1 & - & - \\ - & 0 & 1 & 1 & 1 \\ - & - & 1 & 1 & 1 \\ - & - & - & 0 & 1 \end{array}\right] \begin{array}{c} 2 \\ 3 \\ 4 \\ 5 \\ 7 \end{array} \end{array}$$

Row 2 implies that x_2 is an unacceptable variable and must be set to 0. The new constraint matrix after x_2 is excluded is shown below.

$$\mathbf{A} = \begin{array}{c} \\ \\ \\ \\ \end{array} \begin{array}{c} x_3 \quad x_4 \quad x_5 \quad x_6 \\ \left[\begin{array}{cccc} 0 & 1 & - & - \\ 0 & 1 & 1 & 1 \\ - & 1 & 1 & 1 \\ - & - & 0 & 1 \end{array}\right] \begin{array}{c} 3 \\ 4 \\ 5 \\ 7 \end{array} \end{array}$$

The second reduction is *row dominance*. A row a_k dominates another row a_i if it has all the 1's and 0's of a_i. In other words, a row a_k dominates another row a_i if for each column A_j of $\mathbf{A}$, one of the following is true:

- $a_{ij} = -$

- $a_{ij} = a_{kj}$

Dominating rows can be removed without affecting the set of solutions.

Example 5.1.4 In the matrix from Example 5.1.3, row 4 dominates row 3, so row 4 can be removed. The resulting constraint matrix is:

$$\mathbf{A} = \begin{matrix} & x_3 & x_4 & x_5 & x_6 & \\ & \begin{bmatrix} 0 & 1 & - & - \\ - & 1 & 1 & 1 \\ - & - & 0 & 1 \end{bmatrix} & \begin{matrix} 3 \\ 5 \\ 7 \end{matrix} \end{matrix}$$

The third reduction is *column dominance*. A column A_j dominates another column A_k if for each clause a_i of A, one of the following is true:

- $a_{ij} = 1$.
- $a_{ij} = -$ and $a_{ik} \neq 1$.
- $a_{ij} = 0$ and $a_{ik} = 0$.

Dominated columns can be removed without affecting the existence of a solution. Note that when you remove a column, the associated variable is set to 0. This means that if any rows include that column with a 0 entry, they can be removed as well.

Example 5.1.5 In the matrix from Example 5.1.4, column x_6 dominates columns x_3 and x_5. Note that by eliminating the column associated with x_3 and x_5, rows 3 and 7 are also removed. The new matrix is shown below.

$$\mathbf{A} = \begin{matrix} & x_4 & x_6 & \\ & \begin{bmatrix} 1 & 1 \end{bmatrix} & 5 \end{matrix}$$

At this point, either x_4 or x_6 can be selected to solve the covering problem. Therefore, there are two possible minimal solutions: either $x_1 = x_4 = 1$ with $x_2 = x_3 = x_5 = x_6 = 0$ or $x_1 = x_6 = 1$ with $x_2 = x_3 = x_4 = x_5 = 0$. If either assignment is plugged into the initial product-of-sums given above, it is clear that they yield a 1 for f.

Assuming that an assignment of 1 to any variable has weight 1, then either solution found in our example is a minimum-cost solution. If the weights are not equal, it is necessary to also check the weights of the columns before removing dominated columns. Namely, if the weight of the dominating column, w_j, is greater than the weight of the dominated column, w_k, then x_k should not be removed.

Example 5.1.6 Consider the constraint matrix below with $w_1 = 3$, $w_2 = 1$, and $w_3 = 1$.

$$\mathbf{A} = \begin{matrix} & x_1 & x_2 & x_3 & \\ & \begin{bmatrix} 1 & 1 & - \\ - & 0 & 1 \end{bmatrix} & \begin{matrix} 1 \\ 2 \end{matrix} \end{matrix}$$

In this example, column x_1 dominates x_2, and selecting x_1 solves this matrix. However, x_1 has a higher weight than x_2, so we do not remove x_2. In fact, the lowest-cost solution is $x_2 = x_3 = 1$ and $x_1 = 0$, which has cost 2, while $x_1 = 1$ and $x_2 = x_3 = 0$ would have cost 3.

5.1.2 Bounding

After the reduction steps described above, the matrix may or may not be solved. If it is solved, the cost of the solution can be determined by Equation 5.1. The reduced matrix may have a *cyclic core* and thus may not be completely solved. At this point, it is worthwhile to test whether or not a good solution can be derived from the partial solution found up to this point. This is accomplished by determining a lower bound on the final cost, starting with the current partial solution. If the cost of the current solution or lower bound on the partial solution is greater than or equal to the cost of the best solution found so far, the previous best solution is returned.

Finding an exact lower bound is as difficult as solving the covering problem, so we need to use a heuristic algorithm. A satisfactory heuristic method to determine the lower bound is to find a *maximal independent set* (MIS) of rows. Two rows are independent when it is not possible to satisfy both by setting a single variable to 1. By this definition, any row which contains a complemented variable is dependent on any other clause, so we must ignore these rows. A heuristic algorithm to find a maximal independent set to compute a lower bound is shown in Figure 5.4.

> **Example 5.1.7** The length of the shortest row in the cyclic constraint matrix below is 2, so the algorithm chooses row 1 to be part of our maximal independent set. Row 1 intersects rows 2, 7, and 10. Row 3 is also length 2, so it is added to the independent set. Row 3 intersects rows 4, 5, 8, and 9. This leaves only row 6. Therefore, the maximal independent set is $\{1, 3, 6\}$ and the lower bound is 3. This constraint matrix can be solved by selecting three columns: x_2, x_3, and x_4.

$$\mathbf{A} = \begin{bmatrix}
1 & 1 & - & - & - & - & - & - & - \\
1 & - & 1 & - & - & - & - & - & - \\
- & - & - & 1 & 1 & - & - & - & - \\
- & - & - & 1 & - & 1 & - & - & - \\
- & - & 1 & - & 1 & 1 & - & - & - \\
- & - & 1 & - & - & - & 1 & - & - \\
- & 1 & - & - & - & - & 1 & - & - \\
- & - & - & 1 & - & - & - & 1 & - \\
- & - & - & 1 & - & - & - & - & 1 \\
- & 1 & - & - & - & - & - & 1 & 1
\end{bmatrix}
\begin{matrix} x_1 & x_2 & x_3 & x_4 & x_5 & x_6 & x_7 & x_8 & x_9 \end{matrix}
\quad \begin{matrix} 1 \\ 2 \\ 3 \\ 4 \\ 5 \\ 6 \\ 7 \\ 8 \\ 9 \\ 10 \end{matrix}$$

5.1.3 Termination

The next step determines if $\mathbf{A}$ has been reduced to a terminal case. If $\mathbf{A}$ has no more rows, all the constraints have been satisfied by the current solution, $\mathbf{x}$, and the algorithm has reached a terminal case. Another possible terminating case is when there does not exist any solution for a given constraint matrix.

```
lower_bound(A,x) {
    MIS = ∅
    A = delete_rows_with_complemented_variables(A);
    do {
        i =choose_shortest_row(A);
        MIS = MIS ∪ {i};        /* Add row to maximal independent set.*/
        A =delete_intersecting_rows(A,i);
    } while (A ≠ ∅);            /* Repeat until matrix empty.*/
    return(|MIS| + cost(x));    /* Size of set plus partial solution.*/
}
```

Fig. 5.4 Lower-bound algorithm.

Example 5.1.8 Consider the following formula:

$$f = (x_1 + x_2)(\overline{x_1} + x_2)(x_1 + \overline{x_2})(\overline{x_1} + \overline{x_2})$$

The constraint matrix is shown below.

$$\mathbf{A} = \begin{array}{c} x_1 \ x_2 \\ \left[\begin{array}{cc} 1 & 1 \\ 0 & 1 \\ 1 & 0 \\ 0 & 0 \end{array}\right] \begin{array}{c} 1 \\ 2 \\ 3 \\ 4 \end{array}\end{array}$$

There is clearly no assignment to x_1 or x_2 that makes this formula true. In this case, the function *terminalCase* detects this, and the previous best solution is returned.

5.1.4 Branching

If **A** has not been reduced to a terminal case, the matrix is said to be *cyclic*. To find an exact minimal solution, it is necessary to consider different possible alternatives at this point. The first step is to determine a column to branch on. A column which intersects many short rows should be preferred. This is based on the assumption that shorter rows have a lower chance to be covered. This can be accomplished by assigning a weight to each row that is inversely proportional to the *row length* (i.e., the number of 1's in the row) and by summing the weights of all the rows covered by a column in order to determine the value of that column. The column with the highest value is chosen for case splitting.

Example 5.1.9 Consider the constraint matrix from Example 5.1.7. The row weights are shown in Table 5.1. The weights associated with the columns for this example are shown in Table 5.2. Clearly, the best choice is column x_4.

Next, the variable x_c associated with the column selected for branching is set to 1, the constraint matrix is reduced, and *bcp* is called recursively.

Table 5.1 Row weights for Example 5.1.7.

Row	Weight
1	1/2
2	1/2
3	1/2
4	1/2
5	1/3
6	1/2
7	1/2
8	1/2
9	1/2
10	1/3

Table 5.2 Column weights for Example 5.1.7.

Column	Weight
x_1	1.00
x_2	1.33
x_3	1.33
x_4	2.00
x_5	0.83
x_6	0.83
x_7	1.00
x_8	0.83
x_9	0.83

Example 5.1.10 The result after selecting x_4 is shown below.

$$\mathbf{A} = \begin{bmatrix} x_1 & x_2 & x_3 & x_5 & x_6 & x_7 & x_8 & x_9 \\ 1 & 1 & - & - & - & - & - & - \\ 1 & - & 1 & - & - & - & - & - \\ - & - & 1 & 1 & 1 & - & - & - \\ - & - & 1 & - & - & 1 & - & - \\ - & 1 & - & - & - & 1 & - & - \\ - & 1 & - & - & - & - & 1 & 1 \end{bmatrix} \begin{matrix} 1 \\ 2 \\ 5 \\ 6 \\ 7 \\ 10 \end{matrix}$$

The result returned, $\mathbf{x}^1$, is then checked to see if it is better than the best solution found so far. If so, it becomes the new best solution. If it also meets the lower bound, L, we have found a minimal solution, and it can be returned. If not, we remove x_c from consideration by setting it to 0, reduce the constraint matrix, and call *bcp* again.

Example 5.1.11 The result after removing x_4 is shown below.

$$\mathbf{A} = \begin{bmatrix} x_1 & x_2 & x_3 & x_5 & x_6 & x_7 & x_8 & x_9 \\ 1 & 1 & - & - & - & - & - & - \\ 1 & - & 1 & - & - & - & - & - \\ - & - & - & 1 & - & - & - & - \\ - & - & - & - & 1 & - & - & - \\ - & - & 1 & 1 & 1 & - & - & - \\ - & - & 1 & - & - & 1 & - & - \\ - & 1 & - & - & - & 1 & - & - \\ - & - & - & - & - & - & 1 & - \\ - & - & - & - & - & - & - & 1 \\ - & 1 & - & - & - & - & 1 & 1 \end{bmatrix} \begin{matrix} 1 \\ 2 \\ 3 \\ 4 \\ 5 \\ 6 \\ 7 \\ 8 \\ 9 \\ 10 \end{matrix}$$

The result $\mathbf{x}^0$ is then compared with the previous best, and if it is lower cost, it becomes the new best solution. Finally, the best solution is returned.

5.2 STATE MINIMIZATION

A Huffman flow table is used to describe the sequence of outputs that should be provided for any possible sequence of inputs. There may be many such flow tables to represent the same behavior. The original flow table that is given as a specification may often contain more rows, or states, than is necessary to represent this behavior. To build a Huffman circuit for a given flow table, it is necessary to encode each state using state variables. If the number of states to be encoded is s and the number of state variables used is n, then n must be greater than or equal to $\lceil \log_2 s \rceil$. If we can reduce the number of states, we can also reduce the number of state variables needed to encode the states. Therefore, the first step in the synthesis process for Huffman circuits is to minimize the number of states in the flow table by finding the flow table with the smallest number of rows which produces the same desired behavior.

In this section we describe a procedure to find a flow table with the minimum number of rows to represent a desired behavior. The procedure first identifies all *compatible pairs* of states. These are states which can potentially be merged. Next, it finds all *maximal compatibles*. These are the largest sets of states which are all pairwise compatible. As illustrated later, it may not be possible to find a minimal solution using only maximal compatibles. Therefore, the procedure extends this list to a set of *prime compatibles*. These sets of states are potentially smaller than the maximal compatibles. A minimal solution can always be found using the prime compatibles. Finally, a covering problem is setup where the prime compatibles are the potential solutions and the states are what needs to be covered. In this section we describe state minimization for single-input change fundamental mode, which is the same as for synchronous state machines. Modifications needed for XBM machines are described later.

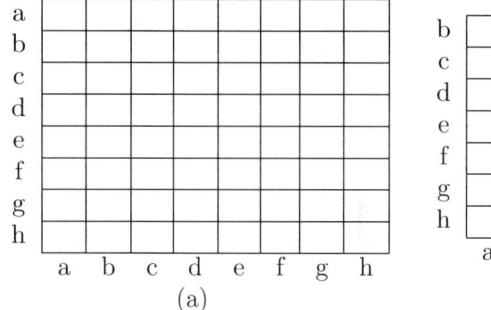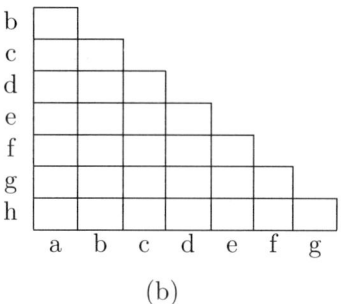

Fig. 5.5 Example pair chart.

5.2.1 Finding the Compatible Pairs

The first step in state minimization is to determine all compatible pairs. To do this, we make use of a *pair chart* (also known as a *compatibility table*), shown in Figure 5.5. Since compatibility is a reflexive and symmetric relation, the top of the chart is redundant and does not need to be considered. The reduced pair chart is depicted in Figure 5.5(b).

The first step to fill in the pair chart is to go through each pair of states, (u,v), in turn, and check if they are *unconditionally compatible*. Two states u and v are unconditionally compatible when they are *output compatible* and for each input produce the same next state when they are both specified. Two states are output compatible when for each input in which they are both specified to produce an output, they produce the same output. When two states u and v are unconditionally compatible, the corresponding (u,v) entry is marked with the symbol $\sim$.

> **Example 5.2.1** Consider the flow table shown in Figure 5.6. States a and b are output compatible since with input x_1, they both output a 0, and for all other inputs the output is unspecified in either a or b. They are not, however, unconditionally compatible because on input x_3 they produce different next states, namely d and a, respectively. Rows b and c, however, are both output compatible and unconditionally compatible. Rows a and g are also unconditionally compatible even though they have different next states on input x_6 because the next states are a and g (exactly those being considered). The pair chart after the first step is shown in Figure 5.7.

The second step is to mark each pair of states which are *incompatible*. When two states u and v are not output compatible, the states are incompatible. When two states u and v are incompatible, the (u,v) entry is marked with the symbol $\times$.

	x_1	x_2	x_3	x_4	x_5	x_6	x_7
a	a,0	–	d,0	e,1	b,0	a,–	–
b	b,0	d,1	a,–	–	a,–	a,1	–
c	b,0	d,1	a,1	–	–	–	g,0
d	–	e,–	–	b,–	b,0	–	a,–
e	b,–	e,–	a,–	–	b,–	e,–	a,1
f	b,0	c,–	–,1	h,1	f,1	g,0	–
g	–	c,1	–	e,1	–	g,0	f,0
h	a,1	e,0	d,1	b,0	b,–	e,–	a,1

Fig. 5.6 Huffman flow table used to illustrate state minimization.

	a	b	c	d	e	f	g
b							
c		∼					
d							
e				∼			
f							
g	∼						
h				∼			

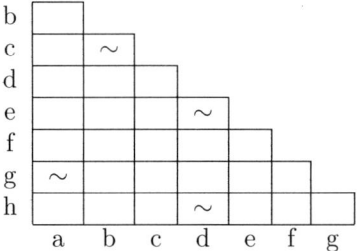

Fig. 5.7 Pair chart after marking unconditional compatibles.

	a	b	c	d	e	f	g
b							
c	×	∼					
d							
e			×	∼			
f	×	×		×			
g	∼	×			×		
h	×	×	×	∼		×	×

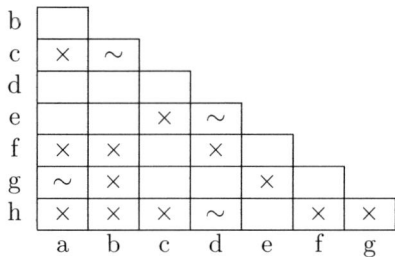

Fig. 5.8 Pair chart after marking output incompatible states.

Example 5.2.2 States a and c are incompatible since for input x_3, a outputs a 0 and c outputs a 1. The pair chart after the second step is shown in Figure 5.8.

The third step is to fill in the remaining entries with the pairs of next states which differ. These state pairs are only *conditionally compatible.* If their differing next states are merged into a new state during state minimization, these states become compatible. When two states u and v are compatible only when states s and t are merged, the (u,v) entry is marked with s,t.

	a	b	c	d	e	f	g
b	a,d						
c	×	~					
d	b,e	a,b d,e	d,e a,g				
e	a,b a,d	d,e a,b a,e	×	~			
f	×	×	c,d	×	c,e b,f e,g		
g	~	×	c,d f,g	c,e b,e a,f	×	e,h	
h	×	×	×	~	a,b a,d	×	×

Fig. 5.9 Pair chart after marking conditional compatibles.

	a	b	c	d	e	f	g
b	a,d						
c	×	~					
d	b,e	a,b d,e	d,e a,g				
e	a,b a,d	d,e a,b a,e	×	~			
f	×	×	c,d	×	×		
g	~	×	c,d f,g	×	×	e,h	
h	×	×	×	~	a,b a,d	×	×

Fig. 5.10 Final pair chart.

Example 5.2.3 States a and b go to states d and a on input x_3, so they are only compatible if d and a are compatible. The pair chart after the third step is shown in Figure 5.9.

The final step is to check each pair of conditional compatibles, and if any pair of next states are known to be incompatible, the pair of states are are also incompatible. Again, in this case, the (u,v) entry is marked with the symbol ×.

Example 5.2.4 For states d and g to be compatible, it must be the case that states c and e are merged. However, we know that states c and e are incompatible. Therefore, we also know that states d and g are incompatible. The pair chart after the fourth and final step is shown in Figure 5.10.

5.2.2 Finding the Maximal Compatibles

After finding the compatible pairs, the next step is to find larger sets of states that can be covered by a single state of some table. Note that if S is a compatible, any subset of S is also a compatible. A *maximal compatible* is a compatible that is not a subset of any larger compatible. From the set of maximal compatibles, it is possible to determine all other compatibles. In this subsection we present two approaches to finding the maximal compatibles.

The first approach uses the compatible pairs found in the preceding subsection. Begin by initializing a *compatible list* (*c*-list) with the compatible pairs in the rightmost column having at least one non-× entry in the compatibility table. Examine the columns from right to left, and perform the following steps:

1. Set S_i to the set of states in column i which do not contain an ×.
2. Intersect S_i with each member of the current *c*-list.
3. If the intersection has more than one member, add to the *c*-list an entry composed of the intersection unioned with i.
4. Before moving to the next column, remove duplicated entries and those that are a subset of other entries.
5. Finally, add pairs which consist of i and any members of S_i that did not appear in any of the intersections.
6. Repeat until the left of the pair chart has been reached.

The final *c*-list plus any individual states not contained in any other member make up the maximal compatibles.

Example 5.2.5 The procedure applied to the compatible pairs from the pair chart in Figure 5.10 is shown below.

First step: $\quad c = \{fg\}$
$S_e = h$: $\quad c = \{fg, eh\}$
$S_d = eh$: $\quad c = \{fg, deh\}$
$S_c = dfg$: $\quad c = \{cfg, deh, cd\}$
$S_b = cde$: $\quad c = \{cfg, deh, bde, bcd\}$
$S_a = bdeg$: $\quad c = \{cfg, deh, bcd, abde, ag\}$

As an alternative approach, we can use the incompatible pairs, which can readily be found from the compatibility table. If s_i and s_j have been found to be incompatible, we know that no maximal compatible can include both s_i and s_j. In general, every group of states that does not include a pair of incompatible states is in fact a compatible set of states. We can now write a Boolean formula that gives the conditions for a set of states to be compatible. For each state s_i, $x_i = 1$ means that s_i is in the set. Therefore, given that the states s_i and s_j are incompatible, the clause $(\overline{x_i} + \overline{x_j})$ can be used to express that a valid compatible set cannot include both states. If we form a conjunction of clauses for each incompatible pair, we can express the conditions necessary for a set of states to be compatible. Next, we convert this product-of-sums into a sum-of-products by multiplying it out and eliminating absorbed terms (i.e., terms which contain a subset of the literals of other terms). Each term of the resulting sum-of-products defines a maximal compatible set. The states which correspond to variables that do not occur in the term make up the maximal compatible.

Example 5.2.6 The function derived for the incompatible pairs from the pair chart in Figure 5.10 is shown below.

$$(\bar{a}+\bar{c})(\bar{a}+\bar{f})(\bar{a}+\bar{h})(\bar{b}+\bar{f})(\bar{b}+\bar{g})(\bar{b}+\bar{h})(\bar{c}+\bar{e})$$
$$(\bar{c}+\bar{h})(\bar{d}+\bar{f})(\bar{d}+\bar{g})(\bar{e}+\bar{f})(\bar{e}+\bar{g})(\bar{f}+\bar{h})(\bar{g}+\bar{h})$$

After multiplying out the function above, we get the following sum-of-products:

$$\bar{a}\bar{b}\bar{d}\bar{e}\bar{h} + \bar{a}\bar{b}\bar{c}\bar{f}\bar{g} + \bar{a}\bar{e}\bar{f}\bar{g}\bar{h} + \bar{c}\bar{f}\bar{g}\bar{h} + \bar{b}\bar{c}\bar{d}\bar{e}\bar{f}\bar{h}$$

The set of maximal compatibles implied by this function are

$$cfg, deh, bcd, abde, ag$$

which match those found by the first method.

5.2.3 Finding the Prime Compatibles

Recall that some pairs of states are compatible only if other pairs are merged into a single state. In other words, the selection of one compatible may imply that another must be selected. The set of compatibles implied by each compatible is called its *class set*. The implied compatibles must be selected to guarantee *closure* of the solution. Assume that C_1 and C_2 are two compatibles and Γ_1 and Γ_2 are their respective class sets. If $C_1 \subset C_2$, it may appear that C_2 is clearly better, but if $\Gamma_1 \subset \Gamma_2$, C_1 may actually be the better choice. The selection of C_2 may actually imply extra compatibles that make it difficult to use. Therefore, the best set of compatibles may include ones that are not maximal. To address this problem, we introduce the notion of *prime compatibles*. A compatible C_1 is prime if and only if there does not exist $C_2 \supset C_1$ such that $\Gamma_2 \subseteq \Gamma_1$. The following theorem states that an optimum solution can always be found using only prime compatibles.

Theorem 5.1 (Grasselli, 1965) *The members of at least one minimal covering are prime compatibility classes.*

An algorithm to find the prime compatibles is shown in Figure 5.11. This algorithm takes as input the compatibility table and the list of maximal compatibles, and it returns the set of prime compatibles. The algorithm begins by initializing *done*, a variable to keep track of compatibles that have been processed, to the empty set. It then loops through the list of compatibles beginning with the largest. The loop finds all compatibles of size k and adds them to the list of prime compatibles, P. It then considers each prime of size k. If it has a nonempty class set, the algorithm attempts to find a smaller compatible which has a smaller class set. Any such compatible found is added to the set of prime compatibles. This process continues until all prime compatibles have been considered in turn.

```
prime_compatibles(C, M){
    done = ∅;                              /* Initialize already computed set.*/
    for (k = |largest(M)|; k ≥ 1; k − −) { /* Loop largest to smallest.*/
        foreach (q ∈ M; |q| = k) enqueue(P, q); /* Queue all of size k.*/
        foreach (p ∈ P; |p| = k) {         /* Consider all primes of size k.*/
            if (class_set(C, p) = ∅) then continue;  /* If empty, skip.*/
            foreach (s ∈ max_subsets(p)) {  /* Check all maximal subsets.*/
                if (s ∈ done) then continue;  /* If computed, skip.*/
                Γ_s = class_set(C, s);       /* Find subset's class set.*/
                prime = true;                /* Initialize prime as true.*/
                foreach (q ∈ P; |q| ≥ k) {   /* Check all larger primes.*/
                    if (s ⊂ q) then {        /* If contained in prime, check it.*/
                        Γ_q = class_set(C, q);  /* Compute class set.*/
                        if (Γ_q ⊆ Γ_s) then {   /* If smaller, not prime.*/
                            prime = false;
                            break;
                        }
                    }
                }
                if (prime = 1) then enqueue(P, s);  /* If prime, queue it.*/
                done = done ∪ {s};           /* Mark as computed.*/
            }
        }
    }
    return(P);                              /* Return prime compatibles.*/
}
```

Fig. 5.11 Algorithm to find prime compatibles.

Example 5.2.7 For our running example, the size of the largest maximal compatible is 4, and it is *abde*. It is the only maximal compatible of size 4, so $P = \{abde\}$. Next, each prime p of size 4 is considered. The class set of the prime p, Γ_p, is found by examining the compatibility table. If the class set is empty, this means that the compatible is unconditionally compatible, so no further computation on this prime is necessary. In this case, p is set equal to $\{abde\}$, and its class set is determined by considering each pair. First, for (a,b) to be compatible, (a,d) must be merged, but this pair is included in *abde*. Next, for (a,d) to be compatible, (b,e) must be merged, which is again included in *abde*. Similarly, (a,e) requires (a,b) and (a,d), which are included. The pair (b,d) requires (a,b) and (d,e) which are included. Finally, (d,e) is unconditionally compatible. This means that the class set of *abde* is the empty set. Since this is the only prime of size 4, we move to primes of size 3.

There are three maximal compatibles of size 3, *bcd*, *cfg*, and *deh*. Let us consider *bcd* first. In this case, (b,c) are unconditionally compatible, but (b,d) requires (a,b) and (d,e) to be merged. Also, (c,d) requires (d,e)

STATE MINIMIZATION

Table 5.3 Prime compatibles and their class sets.

	Prime compatibles	Class set
1	$abde$	$\emptyset$
2	bcd	$\{(a,b),(d,e),(a,g)\}$
3	cfg	$\{(c,d),(e,h)\}$
4	deh	$\{(a,b),(a,d)\}$
5	bc	$\emptyset$
6	cd	$\{(d,e),(a,g)\}$
7	cf	$\{(c,d)\}$
8	cg	$\{(c,d),(f,g)\}$
9	fg	$\{(e,h)\}$
10	dh	$\emptyset$
11	ag	$\emptyset$
12	f	$\emptyset$

and (a,g) to be merged. Therefore, the class set for bcd is $\{(a,b), (d,e), (a,g)\}$. Since the class set is nonempty, the subsets of size 2, $\{bc, bd, cd\}$, must each be checked as a potential prime.

First, *done* is checked to see if it has already been processed. In this case, none of them have. Next, the class set is found for each potential prime compatible. The class set for bc is empty. Next, each prime q which is bigger than this compatible is checked. The only prime found is bcd. In this case, bc has a smaller class set than bcd, so bc is added to the list of primes and to *done*. Next, bd is checked, which has a class set of $\{(a,b),(d,e)\}$. The prime $abde$ found earlier is a superset of bd, and it has an empty class set. Therefore, bd is not a prime compatible, so it is only added to *done*. The last potential prime to be considered is cd, which has a class set of $\{(d,e), (a,g)\}$. This one is only a subset of bcd, so cd is a prime compatible.

Next, cfg is considered, which has a class set of $\{(c,d), (e,h)\}$. In this case, all three subsets of size 2 are prime compatibles. Finally, deh is examined, which has a class set of $\{(a,b),(a,d)\}$. In this case, de is discarded even though it has an empty class set because it is a subset of $abde$, which also has an empty class set. The compatible eh is also discarded because its class set is the same as the one for deh. Finally, the compatible dh is prime, since its class set is empty.

Next, each of the primes of size 2 is considered. The first two, ag and bc, are skipped because they have empty class sets. For the prime cd, c is not prime because it is part of bc, which has an empty class set. Similarly, d is not prime because it is part of $abde$, which also has an empty class set. For the prime cf, f is prime, since it is not part of any other prime with an empty class set. At this point, all primes have been found, and they are tabulated in Table 5.3.

5.2.4 Setting Up the Covering Problem

A collection of prime compatibles forms a valid solution when it is a *closed cover*. A collection of compatibles is a *cover* when all states are contained in some compatible in the set. A collection is *closed* when all compatibles implied by some compatible in the collection are contained in some other compatible in the collection. The variable $c_i = 1$ when the ith prime compatible is a member of the solution. Using the c_i variables, it is possible to write a Boolean formula that represents the conditions for a solution to be a closed cover. The formula is a product-of-sums where each product is a *covering* or *closure constraint*.

There is one covering constraint for each state. The product is simply a disjunction of the prime compatibles that include the state. In other words, for the covering constraint to yield 1, one of the primes that includes the state must be in the solution. There is a closure constraint for each implied compatible for each prime compatible. In other words, for the closure constraint to yield 1, either the prime compatible is not included in the solution or some other prime compatible is included which satisfies the implied compatible. The closure constraints imply that the covering problem is binate.

Example 5.2.8 The covering constraint for state a is

$$(c_1 + c_{11})$$

Similarly, the entire set of covering constraints can be constructed to produce the following product-of-sums:

$$(c_1 + c_{11})(c_1 + c_2 + c_5)(c_2 + c_3 + c_5 + c_6 + c_7 + c_8)$$
$$(c_1 + c_2 + c_4 + c_6 + c_{10})(c_1 + c_4)(c_3 + c_7 + c_9 + c_{12})$$
$$(c_3 + c_8 + c_9 + c_{11})(c_4 + c_{10})$$

The prime bcd requires the following states to be merged: (a,b), (a,g), (d,e). Therefore, if we include bcd in the cover (i.e., c_2), we must also select compatibles which will merge these other state pairs. For instance, $abde$ is the only prime compatible that merges a and b. Therefore, we have a closure constraint of the form

$$c_2 \Rightarrow c_1$$

The prime ag is the only one that merges states a and g, so we also need a closure constraint of the form

$$c_2 \Rightarrow c_{11}$$

Finally, primes $abde$ and deh both merge states d and e, so the resulting closure constraint is

$$c_2 \Rightarrow (c_1 + c_4)$$

Converting the implication into disjunctions, we can express the complete set of closure constraints for bcd as follows:

$$(\overline{c_2} + c_1)(\overline{c_2} + c_{11})(\overline{c_2} + c_1 + c_4)$$

After finding all covering and closure constraints, we obtain the following formula:

$$(c_1 + c_{11})(c_1 + c_2 + c_5)(c_2 + c_3 + c_5 + c_6 + c_7 + c_8)$$
$$\cdot (c_1 + c_2 + c_4 + c_6 + c_{10})(c_1 + c_4)(c_3 + c_7 + c_9 + c_{12})$$
$$\cdot (c_3 + c_8 + c_9 + c_{11})(c_4 + c_{10})(\overline{c_2} + c_1)(\overline{c_2} + c_{11})(\overline{c_2} + c_1 + c_4)$$
$$\cdot (\overline{c_3} + c_2 + c_6)(\overline{c_3} + c_4)(\overline{c_4} + c_1)(\overline{c_4} + c_1)(\overline{c_6} + c_{11})(\overline{c_6} + c_1 + c_4)$$
$$\cdot (\overline{c_7} + c_2 + c_6)(\overline{c_8} + c_2 + c_6)(\overline{c_8} + c_3 + c_9)(\overline{c_9} + c_4) \quad = \quad 1$$

To find a minimal number of prime compatibles that include all the states has now been formulated into a binate covering problem. We can rewrite the equation above as a constraint matrix:

$$\mathbf{A} = \begin{array}{c} \\ \end{array} \begin{array}{c} c_1 \; c_2 \; c_3 \; c_4 \; c_5 \; c_6 \; c_7 \; c_8 \; c_9 \; c_{10} \; c_{11} \; c_{12} \\ \left[\begin{array}{cccccccccccc} 1 & - & - & - & - & - & - & - & - & - & 1 & - \\ 1 & 1 & - & - & 1 & - & - & - & - & - & - & - \\ - & 1 & 1 & - & 1 & 1 & 1 & 1 & - & - & - & - \\ 1 & 1 & - & 1 & - & 1 & - & - & - & 1 & - & - \\ 1 & - & - & 1 & - & - & - & - & - & - & - & - \\ - & - & 1 & - & - & - & 1 & - & 1 & - & - & 1 \\ - & - & 1 & - & - & - & - & 1 & 1 & - & 1 & - \\ - & - & - & 1 & - & - & - & - & - & 1 & - & - \\ 1 & 0 & - & - & - & - & - & - & - & - & - & - \\ - & 0 & - & - & - & - & - & - & - & - & 1 & - \\ 1 & 0 & - & 1 & - & - & - & - & - & - & - & - \\ - & 1 & 0 & - & - & 1 & - & - & - & - & - & - \\ - & - & 0 & 1 & - & - & - & - & - & - & - & - \\ 1 & - & - & 0 & - & - & - & - & - & - & - & - \\ 1 & - & - & 0 & - & - & - & - & - & - & - & - \\ - & - & - & - & - & 0 & - & - & - & - & 1 & - \\ 1 & - & - & 1 & - & 0 & - & - & - & - & - & - \\ - & 1 & - & - & - & 1 & 0 & - & - & - & - & - \\ - & 1 & - & - & - & 1 & - & 0 & - & - & - & - \\ - & - & 1 & - & - & - & - & 0 & 1 & - & - & - \\ - & - & - & 1 & - & - & - & - & 0 & - & - & - \end{array} \right] \begin{array}{c} 1 \\ 2 \\ 3 \\ 4 \\ 5 \\ 6 \\ 7 \\ 8 \\ 9 \\ 10 \\ 11 \\ 12 \\ 13 \\ 14 \\ 15 \\ 16 \\ 17 \\ 18 \\ 19 \\ 20 \\ 21 \end{array} \end{array}$$

Rows 4, 11, and 17 dominate row 5, so rows 4, 11, and 17 can be removed. Also, row 14 dominates row 15 (they are identical), so row 14 can be removed. The resulting constraint matrix is shown below.

150   HUFFMAN CIRCUITS

$$\mathbf{A} = \begin{bmatrix}
c_1 & c_2 & c_3 & c_4 & c_5 & c_6 & c_7 & c_8 & c_9 & c_{10} & c_{11} & c_{12} \\
1 & - & - & - & - & - & - & - & - & - & 1 & - \\
1 & 1 & - & - & 1 & - & - & - & - & - & - & - \\
- & 1 & 1 & - & 1 & 1 & 1 & 1 & - & - & - & - \\
1 & - & - & 1 & - & - & - & - & - & - & - & - \\
- & - & 1 & - & - & - & 1 & - & 1 & - & - & 1 \\
- & - & 1 & - & - & - & - & 1 & 1 & - & 1 & - \\
- & - & - & 1 & - & - & - & - & - & 1 & - & - \\
1 & 0 & - & - & - & - & - & - & - & - & - & - \\
- & 0 & - & - & - & - & - & - & - & - & 1 & - \\
- & 1 & 0 & - & - & 1 & - & - & - & - & - & - \\
- & - & 0 & 1 & - & - & - & - & - & - & - & - \\
1 & - & - & 0 & - & - & - & - & - & - & - & - \\
- & - & - & - & 0 & - & - & - & - & - & 1 & - \\
- & 1 & - & - & - & 1 & 0 & - & - & - & - & - \\
- & 1 & - & - & - & 1 & - & 0 & - & - & - & - \\
- & - & 1 & - & - & - & - & 0 & 1 & - & - & - \\
- & - & - & 1 & - & - & - & - & 0 & - & - & -
\end{bmatrix} \begin{matrix} 1 \\ 2 \\ 3 \\ 5 \\ 6 \\ 7 \\ 8 \\ 9 \\ 10 \\ 12 \\ 13 \\ 15 \\ 16 \\ 18 \\ 19 \\ 20 \\ 21 \end{matrix}$$

Next, we compute a lower bound. In doing so, we ignore all rows which include a 0 entry (i.e., rows 9 – 21). The length of the shortest remaining row is 2, so we choose row 1 to be part of our maximal independent set. Row 1 intersects with rows 2, 5, and 7. The shortest remaining row is row 8, which we add to our independent set. This row does not intersect any remaining row. The next row chosen is row 6. This row intersects row 3, leaving no rows left. Therefore, our maximal independent set is $\{1, 6, 8\}$, which gives a lower bound of 3. This is clearly not a terminal case, so we must select a variable to branch on. The column for variable c_1 has a weight of 1.33, which is best, so we set c_1 to 1 and remove all rows that it intersects to get the table below, and we recursively call the *bcp* algorithm.

$$\mathbf{A} = \begin{bmatrix}
c_2 & c_3 & c_4 & c_5 & c_6 & c_7 & c_8 & c_9 & c_{10} & c_{11} & c_{12} \\
1 & 1 & - & 1 & 1 & 1 & 1 & - & - & - & - \\
- & 1 & - & - & - & 1 & - & 1 & - & - & 1 \\
- & 1 & - & - & - & - & 1 & 1 & - & 1 & - \\
- & - & 1 & - & - & - & - & - & 1 & - & - \\
0 & - & - & - & - & - & - & - & - & 1 & - \\
1 & 0 & - & - & 1 & - & - & - & - & - & - \\
- & 0 & 1 & - & - & - & - & - & - & - & - \\
- & - & - & 0 & - & - & - & - & - & 1 & - \\
1 & - & - & - & 1 & 0 & - & - & - & - & - \\
1 & - & - & - & 1 & - & 0 & - & - & - & - \\
- & 1 & - & - & - & - & 0 & 1 & - & - & - \\
- & - & 1 & - & - & - & - & 0 & - & - & -
\end{bmatrix} \begin{matrix} 3 \\ 6 \\ 7 \\ 8 \\ 10 \\ 12 \\ 13 \\ 16 \\ 18 \\ 19 \\ 20 \\ 21 \end{matrix}$$

This table can be reduced a bit. Column c_4 dominates column c_{10}, so c_{10} can be removed. This makes c_4 an essential variable, since it is now the only possible solution for row 8. Therefore, we set c_4 to 1 and remove intersecting rows. Now, c_9 dominates c_{12}, so c_{12} can be removed. The resulting cyclic matrix is shown below.

$$\mathbf{A} = \begin{bmatrix} c_2 & c_3 & c_5 & c_6 & c_7 & c_8 & c_9 & c_{11} \\ 1 & 1 & 1 & 1 & 1 & 1 & - & - \\ - & 1 & - & - & 1 & - & 1 & - \\ - & 1 & - & - & - & 1 & 1 & 1 \\ 0 & - & - & - & - & - & - & 1 \\ 1 & 0 & - & 1 & - & - & - & - \\ - & - & - & 0 & - & - & - & 1 \\ 1 & - & - & 1 & 0 & - & - & - \\ 1 & - & - & 1 & - & 0 & - & - \\ - & 1 & - & - & - & 0 & 1 & - \end{bmatrix} \begin{matrix} 3 \\ 6 \\ 7 \\ 10 \\ 12 \\ 16 \\ 18 \\ 19 \\ 20 \end{matrix}$$

Next, we calculate the lower bound for this matrix. We ignore rows 10, 12, 16, 18, 19, and 20 because they all contain a 0 entry. This leaves rows 3, 6, and 7, which are not independent, since they all intersect in column c_3. Therefore, the lower bound of this matrix is 1. When we add that to the partial solution (i.e., $\{c_1, c_4\}$), we get a lower bound of 3. Column c_3 has a value of 0.75, which is the highest, so it is selected to branch on. After setting c_3 to 1 and removing intersecting rows, we get the following matrix:

$$\mathbf{A} = \begin{bmatrix} c_2 & c_5 & c_6 & c_7 & c_8 & c_9 & c_{11} \\ 0 & - & - & - & - & - & 1 \\ 1 & - & 1 & - & - & - & - \\ - & - & 0 & - & - & - & 1 \\ 1 & - & 1 & 0 & - & - & - \\ 1 & - & 1 & - & 0 & - & - \end{bmatrix} \begin{matrix} 10 \\ 12 \\ 16 \\ 18 \\ 19 \end{matrix}$$

In this matrix, rows 18 and 19 dominate row 12, so rows 18 and 19 can be removed. Column c_{11} dominates columns c_5, c_7, c_8, and c_9, so they can be removed. The resulting cyclic matrix is shown below.

$$\mathbf{A} = \begin{bmatrix} c_2 & c_6 & c_{11} \\ 0 & - & 1 \\ 1 & 1 & - \\ - & 0 & 1 \end{bmatrix} \begin{matrix} 10 \\ 12 \\ 16 \end{matrix}$$

The lower bound of this matrix is 1, which when added to the length of the current partial solution, $\{c_1, c_3, c_4\}$, we get 4. We select c_2, which results in the following matrix:

$$\mathbf{A} = \begin{bmatrix} c_6 & c_{11} \\ - & 1 \\ 0 & 1 \end{bmatrix} \begin{matrix} 10 \\ 16 \end{matrix}$$

In this matrix, c_{11} is essential and selecting it solves the matrix. The final solution is $\{c_1, c_2, c_3, c_4, c_{11}\}$, which becomes our new best solution.

It is, however, larger than our lower bound, 4, so we try again removing c_2 to produce the following matrix:

$$\mathbf{A} = \begin{bmatrix} c_6 & c_{11} \\ 1 & - \\ 0 & 1 \end{bmatrix} \begin{matrix} 12 \\ 16 \end{matrix}$$

To solve this matrix, we must select both c_6 and c_{11}, so we find another solution $\{c_1, c_3, c_4, c_6, c_{11}\}$ of size 5. We then back up further, to the point where we selected c_3 and consider not selecting it. The resulting constraint matrix is shown below:

$$\mathbf{A} = \begin{bmatrix} c_2 & c_5 & c_6 & c_7 & c_8 & c_9 & c_{11} \\ 1 & 1 & 1 & 1 & 1 & - & - \\ - & - & - & 1 & - & 1 & - \\ - & - & - & - & 1 & 1 & 1 \\ 0 & - & - & - & - & - & 1 \\ - & - & 0 & - & - & - & 1 \\ 1 & - & 1 & 0 & - & - & - \\ 1 & - & 1 & - & 0 & - & - \\ - & - & - & - & 0 & 1 & - \end{bmatrix} \begin{matrix} 3 \\ 6 \\ 7 \\ 10 \\ 16 \\ 18 \\ 19 \\ 20 \end{matrix}$$

This matrix is cyclic, and we select to branch on c_9. The matrix after selecting c_9 is shown below.

$$\mathbf{A} = \begin{bmatrix} c_2 & c_5 & c_6 & c_7 & c_8 & c_{11} \\ 1 & 1 & 1 & 1 & 1 & - \\ 0 & - & - & - & - & 1 \\ - & - & 0 & - & - & 1 \\ 1 & - & 1 & 0 & - & - \\ 1 & - & 1 & - & 0 & - \end{bmatrix} \begin{matrix} 3 \\ 10 \\ 16 \\ 18 \\ 19 \end{matrix}$$

In this matrix c_5 dominates c_7 and c_8, so the columns corresponding to c_7 and c_8 as well as rows 18 and 19 can be removed. Now, c_5 dominates columns c_2 and c_6, which results in those columns being removed along with rows 10 and 16. Finally, c_5 has become essential to cover row 3, so it must be selected solving the matrix. The solution found is $\{c_1, c_4, c_5, c_9\}$ which with size 4 is better than the previous best, which is size 5. At this point, the algorithm goes back and removes c_9 from the solutions, and we obtain another solution of size 5: $\{c_1, c_2, c_4, c_7, c_{11}\}$. At this point, we would recurse back and consider removing c_1 from the solution, resulting in the following matrix:

In this matrix, c_{11} is essential and c_2 and c_4 are unacceptable. After setting c_{11} to 1 and c_2 and c_4 to 0, we get

$$\mathbf{A} = \begin{bmatrix} - & 1 & - & - & - & - & - & - \\ 1 & 1 & 1 & 1 & 1 & - & - & - \\ - & - & - & - & - & - & - & - \\ 1 & - & - & 1 & - & 1 & - & 1 \\ - & - & - & - & - & - & 1 & - \\ 0 & - & 1 & - & - & - & - & - \\ 0 & - & - & - & - & - & - & - \\ - & - & 1 & 0 & - & - & - & - \\ - & - & 1 & - & 0 & - & - & - \\ 1 & - & - & - & 0 & 1 & - & - \\ - & - & - & - & - & 0 & - & - \end{bmatrix} \begin{matrix} 2 \\ 3 \\ 5 \\ 6 \\ 8 \\ 12 \\ 13 \\ 18 \\ 19 \\ 20 \\ 21 \end{matrix}$$

with columns $c_3\ c_5\ c_6\ c_7\ c_8\ c_9\ c_{10}\ c_{12}$.

In this matrix, row 5 is dominated by all others, so all rows but row 5 are removed. All remaining columns mutually dominate, so the resulting terminal case is shown below.

$$\mathbf{A} = \begin{bmatrix} c_3 \\ - \end{bmatrix} \quad 5$$

There is no solution to this matrix, so this implies that c_1 must be part of any valid solution. At this point, the *bcp* algorithm terminates, returning the best solution found: $\{c_1, c_4, c_5, c_9\}$.

	x_1	x_2	x_3	x_4	x_5	x_6	x_7
1	1,0	{1,4},1	1,0	1,1	1,0	1,1	1,1
4	1,1	{1,4},0	1,1	{1,5},0	{1,5},0	{1,4},–	1,1
5	{1,5},0	{1,4},1	1,1	–	1,–	1,1	9,0
9	{1,5},0	5,1	–,1	4,1	9,1	9,0	9,0

Fig. 5.12 Reduced Huffman flow table.

	x_1	x_2	x_3	x_4	x_5	x_6	x_7
1	1,0	1,1	1,0	1,1	1,0	1,1	1,1
4	1,1	1,0	1,1	1,0	1,0	1,–	1,1
5	1,0	1,1	1,1	–	1,–	1,1	9,0
9	1,0	5,1	–,1	4,1	9,1	9,0	9,0

Fig. 5.13 Final reduced Huffman flow table.

5.2.5 Forming the Reduced Flow Table

After finding a minimal solution, it is necessary to derive the reduced flow table. There is a row in the new flow table for each compatible selected. The entries in this row are found by combining the entries from all states contained in the compatible. If in any state the value of the output is specified, the value of the output in the merged state takes that value (note that by construction of the compatibles there can be no conflict in specified output values). For the next states, the reduced machine must go to a merged state, which includes all the next states for all the states contained in this compatible.

> **Example 5.2.9** For compatible 1, *abde*, under input x_1, we find that *a* and *b* are the only possible next states. The only compatible which contains both of these states is compatible 1. Next, consider input x_2 for compatible 1. The only possible next states are *d* and *e*, which appear in both compatible 1, *abde*, and compatible 4, *deh*. Therefore, we have a choice of which next state to use. The reduced flow table reflecting these choices is shown in Figure 5.12. One possible final reduced flow table is shown in Figure 5.13.

5.3 STATE ASSIGNMENT

After finding a minimum row flow table representation of an asynchronous finite state machine, it is now necessary to encode each row using a unique binary code. In synchronous design, a correct encoding can be assigned arbitrarily using n bits for a flow table with 2^n rows or less. In asynchronous design, more care must be taken to ensure that a circuit can be built that is independent of signal delays.

	x_1	x_2	x_3	x_4
a	a	b	d	c
b	c	b	b	b
c	c	d	b	c
d	a	d	d	b

(a)

	$y_1 y_2$	$y_1 y_2 y_3$
a	00	000
b	01	011
c	10	110
d	11	101

(b)

Fig. 5.14 (a) Simple Huffman flow table. (b) Two potential state assignments.

We first define a few terms. When the codes representing the present state and next state are the same, the circuit is *stable*. When the codes differ in a single bit location, the circuit is *in transition* from the present state to the next state. When the codes differ in multiple bit positions, the circuit is *racing* from the present state to the next state. If the circuit is racing and there exists a possibility where differences in delays can cause it to reach a different stable state than the one intended, the race is *critical*. All other races are *noncritical*. A transition from state s_i to state s_j is *direct* (denoted $[s_i,s_j]$) when all state variables are excited to change at the same time. When all state transitions are direct, the state assignment is called a *minimum-transition-time state assignment*. A flow table in which each unstable state leads directly to a stable state is called a *normal flow table*. A direct transition $[s_i,s_j]$ races critically with another direct transition $[s_k,s_l]$ when unequal delays can cause these two transitions to pass through a common state. A state assignment for a Huffman circuit is correct when it is free of critical races.

Example 5.3.1 A simple flow table is shown in Figure 5.14(a) and two potential state assignments for this machine are shown in Figure 5.14(b). Consider the machine in state b under input x_2. When the input changes to x_1, the machine is excited to enter state c. Using the first code, the machine is going from state 01 to state 10, which means that y_1 and y_2 are racing. Unless the two bits switch at about the same time, it is likely that the machine will momentarily pass through state 00 or state 11. In either case, the machine would then become excited to go to state a and the machine would end up in the wrong state. On the other hand, using the second code, the machine would be excited to go from state 011 to 110. Even though y_1 and y_3 are racing, y_2 is stable 1, keeping the machine from intersecting state 000 or 101.

5.3.1 Partition Theory and State Assignment

In this subsection we introduce *partition theory* and a theorem that relates it to critical race free state assignment. First, a partition π on a set S is a set of subsets of S such that their pairwise intersection is empty. The disjoint subsets of π are called *blocks*. A partition is *completely specified* if the union of the subsets is S. Otherwise, the partition is said to be *incompletely specified*. Elements of S which do not appear in π are said to be *unspecified*.

A state assignment composed of n state variables $y_1, \ldots, y_n$ is composed of the τ-partitions $\tau_1, \ldots, \tau_n$ induced by their respective state variable. In other words, each state coded with a 0 in bit position y_1 is in one block of the partition τ_1, while those coded with a 1 are in the other block. Since each partition is created by a single variable, each partition can be composed of only one or two blocks. The order in which the blocks appear, or which one is assigned a 0 or 1, does not matter. This means that once we find one critical race free state assignment, any formed by complementing a state variable in each state or reordering the state variables is also a valid assignment.

Example 5.3.2 The partitions induced by the two state codes from Figure 5.14(b) are shown below.

First state code:

$$\tau_1 = \{ab; cd\}$$
$$\tau_2 = \{ac; bd\}$$

Second state code:

$$\tau_1 = \{ab; cd\}$$
$$\tau_2 = \{ad; bc\}$$
$$\tau_3 = \{ac; bd\}$$

A partition π_2 is less than or equal to another partition π_1 (denoted $\pi_2 \leq \pi_1$) if and only if all elements specified in π_2 are also specified in π_1 and each block of π_2 appears in a unique block of π_1. A *partition list* is a collection of partitions of the form $\{s_p, s_q; s_r, s_s\}$ or $\{s_p, s_q; s_t\}$, where $[s_p, s_q]$ and $[s_r, s_s]$ are transitions in the same column and s_t is a stable state also in the same column. Since these partitions are composed of exactly two blocks, they are also called *dichotomies*. A state assignment for a normal flow table is a minimum transition time assignment free of critical races if and only if each partition in the partition list is $\leq$ some τ_i. These conditions are expressed more formally in the following theorem.

Theorem 5.2 (Tracey, 1966) *A row assignment allotting one y-state per row can be used for direct transition realization of normal flow tables without critical races if, and only if, for every transition* $[s_i, s_j]$:

1. *If* $[s_m, s_n]$ *is another transition in the same column, then at least one y-variable partitions the pair* $\{s_i, s_j\}$ *and the pair* $\{s_m, s_n\}$ *into separate blocks.*

2. *If* s_k *is a stable state in the same column then at least one y-variable partitions the pair* $\{s_i, s_j\}$ *and the state* s_k *into separate blocks.*

3. *For* $i \neq j$, s_i *and* s_j *are in separate blocks of at least one y-variable partition.*

Example 5.3.3 The partition list for our example in Figure 5.14(a) is shown below.

$$\pi_1 = \{ad; bc\}$$
$$\pi_2 = \{ab; cd\}$$
$$\pi_3 = \{ad; bc\}$$
$$\pi_4 = \{ac; bd\}$$

It is clear that the partition induced by the first code is not sufficient since it does not address partitions π_1 and π_3. The second state code includes a partition that is less than or equal to each of these, so it is a valid code.

5.3.2 Matrix Reduction Method

The partition list can be converted into a *Boolean matrix*. Each partition in the partition list forms a row in this matrix and each column represents a state. The entries are annotated with a 0 if the corresponding state is a member of the first block of the partition, with a 1 if the state is a member of the second block, and a $-$ if the state does not appear in the partition.

Example 5.3.4 Consider the flow table shown in Figure 5.15. The partition list is given below.

$$\pi_1 = \{ab; cf\}$$
$$\pi_2 = \{ae; cf\}$$
$$\pi_3 = \{ac; de\}$$
$$\pi_4 = \{ac; bf\}$$
$$\pi_5 = \{bf; de\}$$
$$\pi_6 = \{ad; bc\}$$
$$\pi_7 = \{ad; ce\}$$
$$\pi_8 = \{ac; bd\}$$
$$\pi_9 = \{ac; ef\}$$
$$\pi_{10} = \{bd; ef\}$$

This partition list is converted into the Boolean matrix shown in Figure 5.16.

The state assignment problem has now been reduced to finding a Boolean matrix C with a minimum number of rows such that each row in the original Boolean matrix constructed from the partition list is covered by some row of C. The rows of this reduced matrix represent the two-block τ-partitions. The columns of this matrix represent one possible state assignment. The number of rows therefore is the same as the number of state variables needed in the assignment.

	x_1	x_2	x_3	x_4
a	a,0	c,1	d,0	c,1
b	a,0	f,1	c,1	b,0
c	f,1	c,1	c,1	c,1
d	-,-	d,0	d,0	b,0
e	a,0	d,0	c,1	e,1
f	f,1	f,1	-,-	e,1

Fig. 5.15 More complex Huffman flow table.

	a	b	c	d	e	f
π_1	0	0	1	–	–	1
π_2	0	–	1	–	0	1
π_3	0	–	0	1	1	–
π_4	0	1	0	–	–	1
π_5	–	0	–	1	1	0
π_6	0	1	1	0	–	–
π_7	0	–	1	0	1	–
π_8	0	1	0	1	–	–
π_9	0	–	0	–	1	1
π_{10}	–	0	–	0	1	1

Fig. 5.16 Boolean matrix.

5.3.3 Finding the Maximal Intersectibles

One approach to minimizing the size of the Boolean matrix would be to find two rows that intersect, and replace those two rows with their *intersection*. Two rows of a Boolean matrix, R_i and R_j, have an intersection if R_i and R_j agree wherever both R_i and R_j are specified. The intersection is formed by creating a row which has specified values taken from either R_i or R_j. Entries where neither R_i or R_j are specified are left unspecified. A row, R_i, *includes* another row, R_j, when R_j agrees with R_i wherever R_i is specified. A row, R_i, *covers* another row, R_j, if R_i includes R_j or R_i includes the complement of R_j (denoted $\overline{R_j}$).

Example 5.3.5 Consider the Boolean matrix shown in Figure 5.16. Rows 1 and 2 intersect, and their intersection is 001 – 01. Also, rows 3, 4, 8, and 9 intersect to form the row 010111. The complement of row 5 and and row 6 intersect. Recall that the assignment of 0 to the left partition and 1 to the right is arbitrary. Finding the complement of the row effectively reverses this decision. The intersection of these two rows is 011001. Finally, rows 7 and 10 intersect to form the new row 001011. The results are summarized in Figure 5.17(a). There are no remaining rows that can be intersected, and the state assignment

	a	b	c	d	e	f
(π_1,π_2)	0	0	1	–	0	1
$(\pi_3,\pi_4,\pi_8,\pi_9)$	0	1	0	1	1	1
$(\overline{\pi_5},\pi_6)$	0	1	1	0	0	1
(π_7,π_{10})	0	0	1	0	1	1

(a)

	$y_1y_2y_3y_4$
a	0000
b	0110
c	1011
d	-100
e	0101
f	1111

(b)

Fig. 5.17 (a) Reduced Boolean matrix. (b) Corresponding state assignment.

	a	b	c	d	e	f
(π_1,π_7,π_{10})	0	0	1	0	1	1
$(\pi_2,\overline{\pi_5},\pi_6)$	0	1	1	0	0	1
$(\pi_3,\pi_4,\pi_8,\pi_9)$	0	1	0	1	1	1

(a)

	$y_1y_2y_3$
a	000
b	011
c	110
d	001
e	101
f	111

(b)

Fig. 5.18 (a) Minimal Boolean matrix. (b) Corresponding state assignment.

that it implies requires four state variables (one for each row), as shown in Figure 5.17(b). However, the matrix shown in Figure 5.18(a) also covers the partition while requiring only three state variables, as shown in Figure 5.18(b).

The problem is that the order in which we combine the intersections can affect the final result. In order to find a minimum row reduced matrix, it is necessary to find all possible row intersections. If a set of partitions, π_i, π_j, ..., π_k, has an intersection, it is called *intersectible*. An intersectible may be enlarged by adding a partition π_l if and only if π_l has an intersection with every element in the set. An intersectible which cannot be enlarged further is called a *maximal intersectible*.

The first step in finding the reduced matrix is to find all pairwise intersectibles. For each pair of rows R_i and R_j, one should check whether R_i and R_j have an intersection. It is also necessary to check whether R_i and $\overline{R_j}$ have an intersection since the resulting row would also cover R_i and R_j. If there are n partitions to cover, this implies the need to consider 2n *ordered partitions*. The following theorem can be used to reduce the number of ordered partitions considered.

	π_1	π_2	π_3	π_4	π_5	π_6	π_7	π_8	π_9	π_{10}	$\overline{\pi_5}$
π_2	~										
π_3	X	X									
π_4	X	X	~								
π_5	X	X	~	X							
π_6	X	~	X	X	X						
π_7	~	X	X	X	X	~					
π_8	X	X	~	~	X	X	X				
π_9	X	X	~	~	X	X	X	~			
π_{10}	~	X	X	X	X	X	~	X	~		
$\overline{\pi_5}$	X	~	X	~	X	~	X	X	X	X	
$\overline{\pi_{10}}$	X	X	X	X	X	X	X	~	X	X	X

Fig. 5.19 Pair chart for state assignment example.

Theorem 5.3 (Unger, 1969) *Let D be a set of ordered partitions derived from some set of unordered partitions. For some state s, label as p_1, p_2, etc. the members of D having s in their left sets, and label as q_1, q_2, etc. the members of D that do not contain s in either set. Then a minimal set of maximal intersectibles covering each member of D or its complement can be found by considering only the ordered partitions labeled as p's or q's. (The complements of the p's can be ignored.)*

Example 5.3.6 Consider the example shown in Figure 5.16. If we select $s = a$, then we only need to consider $\pi_1, \ldots, \pi_{10}$ and $\overline{\pi_5}$ and $\overline{\pi_{10}}$.

To find the pairwise intersectibles, we can now construct a pair chart for the remaining ordered partitions.

Example 5.3.7 The pairwise intersectibles from Figure 5.16 are found using the pair chart shown in Figure 5.19, and they are listed below.

$(\pi_1, \pi_2)(\pi_1, \pi_7)(\pi_1, \pi_{10})(\pi_2, \pi_6)(\pi_2, \overline{\pi_5})(\pi_3, \pi_4)(\pi_3, \pi_5)(\pi_3, \pi_8)(\pi_3, \pi_9)$
$(\pi_4, \pi_8)(\pi_4, \pi_9)(\pi_4, \overline{\pi_5})(\pi_6, \pi_7)(\pi_6, \overline{\pi_5})(\pi_7, \pi_{10})(\pi_8, \pi_9)(\pi_8, \overline{\pi_{10}})(\pi_9, \pi_{10})$

The second step is to use the set of pairwise intersectibles to derive all maximal intersectibles. The approach taken is the same as the one described earlier to find maximal compatibles.

STATE ASSIGNMENT   161

Table 5.4 Maximal intersectibles for Example 5.3.4.

x_1	(π_1, π_2)
x_2	(π_1, π_7, π_{10})
x_3	$(\pi_2, \pi_6, \overline{\pi_5})$
x_4	$(\pi_3, \pi_4, \pi_8, \pi_9)$
x_5	(π_3, π_5)
x_6	$(\pi_4, \overline{\pi_5})$
x_7	(π_6, π_7)
x_8	$(\pi_8, \overline{\pi_{10}})$
x_9	(π_9, π_{10})

Example 5.3.8 Here is the derivation of the maximal intersectibles from the intersectibles from Example 5.3.7:

First step: $\quad c = \{(\pi_9, \pi_{10})\}$

$S_{\pi_8} = \pi_9, \overline{\pi_{10}}: \quad c = \{(\pi_9, \pi_{10}), (\pi_8, \pi_9), (\pi_8, \overline{\pi_{10}})\}$

$S_{\pi_7} = \pi_{10}: \quad c = \{(\pi_9, \pi_{10}), (\pi_8, \pi_9), (\pi_8, \overline{\pi_{10}}), (\pi_7, \pi_{10})\}$

$S_{\pi_6} = \pi_7, \overline{\pi_5}: \quad c = \{(\pi_9, \pi_{10}), (\pi_8, \pi_9), (\pi_8, \overline{\pi_{10}}), (\pi_7, \pi_{10}),$
$\qquad (\pi_6, \pi_7), (\pi_6, \overline{\pi_5})\}$

$S_{\pi_4} = \pi_8, \pi_9, \overline{\pi_5}: \quad c = \{(\pi_9, \pi_{10}), (\pi_8, \overline{\pi_{10}}), (\pi_7, \pi_{10}), (\pi_6, \pi_7),$
$\qquad (\pi_6, \overline{\pi_5}), (\pi_4, \pi_8, \pi_9), (\pi_4, \overline{\pi_5})\}$

$S_{\pi_3} = \pi_4, \pi_5, \pi_8, \pi_9: \quad c = \{(\pi_9, \pi_{10}), (\pi_8, \overline{\pi_{10}}), (\pi_7, \pi_{10}), (\pi_6, \pi_7),$
$\qquad (\pi_6, \overline{\pi_5}), (\pi_4, \overline{\pi_5}), (\pi_3, \pi_4, \pi_8, \pi_9),$
$\qquad (\pi_3, \pi_5)\}$

$S_{\pi_2} = \pi_6, \overline{\pi_5}: \quad c = \{(\pi_9, \pi_{10}), (\pi_8, \overline{\pi_{10}}), (\pi_7, \pi_{10}), (\pi_6, \pi_7),$
$\qquad (\pi_4, \overline{\pi_5}), (\pi_3, \pi_4, \pi_8, \pi_9), (\pi_3, \pi_5),$
$\qquad (\pi_2, \pi_6, \overline{\pi_5})\}$

$S_{\pi_1} = \pi_2, \pi_7, \pi_{10}: \quad c = \{(\pi_9, \pi_{10}), (\pi_8, \overline{\pi_{10}}), (\pi_6, \pi_7), (\pi_4, \overline{\pi_5}),$
$\qquad (\pi_3, \pi_4, \pi_8, \pi_9), (\pi_3, \pi_5), (\pi_2, \pi_6, \overline{\pi_5}),$
$\qquad (\pi_1, \pi_7, \pi_{10}), (\pi_1, \pi_2)\}$

The final set of maximal intersectibles is shown in Table 5.4.

5.3.4 Setting Up the Covering Problem

The last step is to select a minimum number of maximal intersectibles such that each row of the partition matrix is covered by one intersectible. This again results in a covering problem, which can be solved as above. In this covering problem, the clauses are now the partitions, while the literals are the maximal intersectibles. Note there are no implications in this problem, so the covering problem is unate.

	x_1	x_2	x_3	x_4
000	000,0	011,1	100,0	011,1
110	000,0	111,1	011,1	110,0
011	111,1	011,1	011,1	011,1
100	-,-	100,0	100,0	110,0
101	000,0	100,0	011,1	101,1
111	111,1	111,1	-,-	111,1

Fig. 5.20 Huffman flow table from Figure 5.15 after state assignment.

Example 5.3.9 The constraint matrix for the maximal intersectibles from Table 5.4 is shown below.

$$\mathbf{A} = \begin{bmatrix} 1 & 1 & - & - & - & - & - & - & - \\ 1 & - & 1 & - & - & - & - & - & - \\ - & - & - & 1 & 1 & - & - & - & - \\ - & - & - & 1 & - & 1 & - & - & - \\ - & - & 1 & - & 1 & 1 & - & - & - \\ - & - & 1 & - & - & - & 1 & - & - \\ - & 1 & - & - & - & - & 1 & - & - \\ - & - & - & 1 & - & - & - & 1 & - \\ - & - & - & 1 & - & - & - & - & 1 \\ - & 1 & - & - & - & - & - & 1 & 1 \end{bmatrix} \begin{matrix} \pi_1 \\ \pi_2 \\ \pi_3 \\ \pi_4 \\ \pi_5 \\ \pi_6 \\ \pi_7 \\ \pi_8 \\ \pi_9 \\ \pi_{10} \end{matrix}$$

with columns x_1 x_2 x_3 x_4 x_5 x_6 x_7 x_8 x_9.

This matrix cannot be reduced, so we find a maximal independent set of rows, { 1, 3, 6 }, which shows that the lower bound is 3. We then select column x_4 to branch on, reduce the matrix, and initiate recursion on the following matrix:

$$\mathbf{A} = \begin{bmatrix} 1 & 1 & - & - & - & - & - & - \\ 1 & - & 1 & - & - & - & - & - \\ - & - & 1 & 1 & 1 & - & - & - \\ - & - & 1 & - & - & 1 & - & - \\ - & 1 & - & - & - & 1 & - & - \\ - & 1 & - & - & - & - & 1 & 1 \end{bmatrix} \begin{matrix} \pi_1 \\ \pi_2 \\ \pi_5 \\ \pi_6 \\ \pi_7 \\ \pi_{10} \end{matrix}$$

with columns x_1 x_2 x_3 x_5 x_6 x_7 x_8 x_9.

In this matrix, column x_3 dominates columns x_5 and x_6. Column x_2 dominates columns x_8 and x_9. Now, π_5 and π_{10} are essential rows, implying the need to include x_3 and x_2 in the solution. This solves the matrix. Therefore, the solution $(\pi_3, \pi_4, \pi_8, \pi_9)$, $(\pi_2, \pi_6, \overline{\pi_5})$, and (π_1, π_7, π_{10}) has been found. Since its size, 3, matches the lower bound we found earlier, we are done. The resulting state code is the transpose of the Boolean matrix shown in Figure 5.18. The Huffman flow table after state assignment is shown in Figure 5.20.

	a	b	c	d	e	f
π_2	0	–	1	–	0	1
π_3	0	–	0	1	1	–
π_4	0	1	0	–	–	1
π_5	–	0	–	1	1	0
π_6	0	1	1	0	–	–
π_8	0	1	0	1	–	–
π_9	0	–	0	–	1	1

Fig. 5.21 Boolean matrix.

5.3.5 Fed-Back Outputs as State Variables

Previously, we ignored the values of the outputs during state assignment. It may be possible, however, to feed back the outputs and use them as state variables. To do this, we determine in each state under each input the value of each output upon entry. This information can be used to satisfy some of the necessary partitions. By reducing the number of partitions that must be satisfied, we can reduce the number of explicit state variables that are needed. We will illustrate this through an example.

Example 5.3.10 Consider again the flow table from Figure 5.15. The value of the output is always 0 upon entering states a, b, and d, and it is always 1 upon entering states c, e, and f. Previously, we found the following partition: $\{ab; cf\}$. If we consider the fed-back output as a state variable, it satisfies this partition since it is 0 in states a and b and 1 in states c and f. Using the output as a state variable reduces the partition list to the one given below which is converted into the Boolean matrix shown in Figure 5.21.

$$\pi_2 = \{ae, cf\}$$
$$\pi_3 = \{ac, de\}$$
$$\pi_4 = \{ac, bf\}$$
$$\pi_5 = \{bf, de\}$$
$$\pi_6 = \{ad, bc\}$$
$$\pi_8 = \{ac, bd\}$$
$$\pi_9 = \{ac, ef\}$$

Using Theorem 5.3 and state a, we determine that the only complement that needs to be considered is $\overline{\pi_5}$. The pairwise intersectibles from Figure 5.21 are found using the pair chart shown in Figure 5.22 and are listed below.

$$(\pi_2, \pi_6)(\pi_2, \overline{\pi_5})(\pi_3, \pi_4)(\pi_3, \pi_5)(\pi_3, \pi_8)(\pi_3, \pi_9)(\pi_4, \pi_8)$$
$$(\pi_4, \pi_9)(\pi_4, \overline{\pi_5})(\pi_6, \overline{\pi_5})(\pi_8, \pi_9)$$

	π_2	π_3	π_4	π_5	π_6	π_8	π_9
π_3	×						
π_4	×	∼					
π_5	×	∼	×				
π_6	∼	×	×	×			
π_8	×	∼	∼	×	×		
π_9	×	∼	∼	×	×	∼	
$\overline{\pi_5}$	∼	×	∼	×	∼	×	×

Fig. 5.22 Pair chart for output state assignment example.

Table 5.5 Maximal intersectibles for Example 5.3.4 using outputs as state variables.

x_1	$(\pi_2, \pi_6, \overline{\pi_5})$
x_2	$(\pi_3, \pi_4, \pi_8, \pi_9)$
x_3	(π_3, π_5)
x_4	$(\pi_4, \overline{\pi_5})$

Here is the derivation of the maximal intersectibles:

First step: $c = \{(\pi_8, \pi_9)\}$
$S_{\pi_6} = \overline{\pi_5}$: $c = \{(\pi_8, \pi_9), (\pi_6, \overline{\pi_5}),\}$
$S_{\pi_4} = \pi_8, \pi_9, \overline{\pi_5}$: $c = \{(\pi_4, \pi_8, \pi_9), (\pi_4, \overline{\pi_5}), (\pi_6, \overline{\pi_5})\}$
$S_{\pi_3} = \pi_4, \pi_5, \pi_8, \pi_9$: $c = \{(\pi_3, \pi_4, \pi_8, \pi_9), (\pi_3, \pi_5), (\pi_4, \overline{\pi_5}), (\pi_6, \overline{\pi_5})\}$
$S_{\pi_2} = \pi_6, \overline{\pi_5}$: $c = \{(\pi_3, \pi_4, \pi_8, \pi_9), (\pi_3, \pi_5), (\pi_4, \overline{\pi_5}),$
$(\pi_2, \pi_6, \overline{\pi_5})\}$

The final set of maximal intersectibles is shown in Table 5.5.

The constraint matrix for the maximal intersectibles from Table 5.5 is shown below.

$$\mathbf{A} = \begin{bmatrix} 1 & - & - & - \\ - & 1 & 1 & - \\ - & 1 & - & 1 \\ 1 & - & 1 & 1 \\ 1 & - & - & - \\ - & 1 & - & - \\ - & 1 & - & - \end{bmatrix} \begin{matrix} \pi_2 \\ \pi_3 \\ \pi_4 \\ \pi_5 \\ \pi_6 \\ \pi_8 \\ \pi_9 \end{matrix}$$

with column headers $x_1\ x_2\ x_3\ x_4$.

Row π_2 is essential, so x_1 must be part of the solution. Row π_8 is also essential, so x_2 must be part of the solution. This solves the entire matrix and the final solution is $(\pi_2, \pi_6, \overline{\pi_5})$ and $(\pi_3, \pi_4, \pi_8, \pi_9)$.

The new flow table after state assignment is shown in Figure 5.23. Note that only two state variables are needed. When only these two variables are considered, it appears that states b and f have the same code. Also, states d and e appear to have the same state code. The fed-back output, however, serves to disambiguate these states.

	x_1	x_2	x_3	x_4
00	00,0	01,1	10,0	01,1
11	00,0	11,1	01,1	11,0
01	11,1	01,1	01,1	01,1
10	–,–	10,0	10,0	11,0
10	00,0	10,0	01,1	10,1
11	11,1	11,1	–,–	11,1

Fig. 5.23 Huffman flow table from Figure 5.15 after state assignment using outputs as fed-back state variables.

5.4 HAZARD-FREE TWO-LEVEL LOGIC SYNTHESIS

After finding a critical race free state assignment, the next step of the design process is to synthesize the logic for each next state and output signal. The traditional synchronous approach to logic synthesis for FSMs would be to derive a *sum-of-products* (SOP) implementation for each next state and output signal. For asynchronous FSMs, care must be taken to avoid *hazards* in the SOP implementation. A circuit has a hazard when there exists an assignment of delays such that a *glitch*, an unwanted signal transition, can occur. Hazards must be avoided in asynchronous design since they may be misinterpreted by other parts of the circuit as a valid signal transition causing erroneous behavior. In this section we describe a method for hazard-free two-level logic synthesis under the SIC fundamental-mode assumption.

5.4.1 Two-Level Logic Minimization

An *incompletely specified Boolean function* f of n variables $x_1, x_2, \ldots, x_n$ is a mapping: $f : \{0,1\}^n \rightarrow \{0, 1, -\}$. Each element m of $\{0,1\}^n$ is called a *minterm*. The value of a variable x_i in a minterm m is given by $m(i)$. The *ON-set* of f is the set of minterms which return 1. The *OFF-set* of f is the set of minterms which return 0. The *don't care (DC)-set* of f is the set of minterms which return $-$.

A *literal* is either the variable, x_i, or its complement, $\overline{x_i}$. The literal x_i evaluates to 1 in the minterm m when $m(i) = 1$. The literal $\overline{x_i}$ evaluates to 1 when $m(i) = 0$. A *product* is a conjunction (AND) of literals. A product evaluates to 1 for a given minterm if each literal evaluates to 1 in the minterm, and the product is said to *contain* the minterm. A set of minterms which can be represented with a product is called a *cube*. A product Y contains another product X (i.e., $X \subseteq Y$) if the minterms contained in X are a subset of those in Y. The *intersection* of two products is the set of minterms contained in both products. A *sum-of-products* (SOP) is a set of products that are disjunctively combined. In other words, a SOP contains a minterm when one of the products in the SOP contains the minterm.

166 HUFFMAN CIRCUITS

	wx						wx			
	00	01	11	10			00	01	11	10
00	1	1	1	1		00	1	1	1	1
01	0	1	1	–		01	0	1	1	–
11	0	1	1	0		11	0	1	1	0
10	0	–	0	0		10	0	–	0	0
		(a)						(b)		

(with yz labels on the left)

Fig. 5.24 (a) Karnaugh map for small two-level logic minimization example. (b) Minimal two-level SOP cover.

An *implicant* of a function is a product that contains no minterms in the OFF-set of the function. A *prime implicant* is an implicant which is contained by no other implicant. A *cover* of a function is a SOP which contains the entire ON-set and none of the OFF-set. A cover may optionally include part of the DC-set. The two-level logic minimization problem is to find a minimum-cost cover of the function. Ignoring hazards, a minimal cover is always composed only of prime implicants, as stated in the following theorem.

Theorem 5.4 (Quine, 1952) *A minimal SOP must always consist of a sum of prime implicants if any definition of cost is used in which the addition of a single literal to any formula increases the cost of the formula.*

Example 5.4.1 Consider a function of four variables, w, x, y, and z, depicted in the Karnaugh map in Figure 5.24(a). The minterms are divided as follows:

$$\text{ON-set} = \{\bar{w}\bar{x}\bar{y}\bar{z}, \bar{w}x\bar{y}\bar{z}, wx\bar{y}\bar{z}, w\bar{x}\bar{y}\bar{z}, \bar{w}x\bar{y}z, wx\bar{y}z, \bar{w}xyz, wxyz\}$$
$$\text{OFF-set} = \{\bar{w}\bar{x}\bar{y}z, \bar{w}\bar{x}yz, w\bar{x}yz, \bar{w}\bar{x}y\bar{z}, wxy\bar{z}, w\bar{x}y\bar{z}\}$$
$$\text{DC-set} = \{w\bar{x}\bar{y}z, \bar{w}xy\bar{z}\}$$

The product $\bar{y}z$ is not an implicant since it contains the OFF-set minterm $\bar{w}\bar{x}\bar{y}z$. The product $w\bar{y}z$ is an implicant since it does not intersect the OFF-set. It is not, however, a prime implicant, since it is contained in the implicant $w\bar{y}$, which is prime. The minimal SOP cover for this function is

$$f = \bar{y}\bar{z} + xz$$

which is depicted in Figure 5.24(b).

5.4.2 Prime Implicant Generation

For functions of fewer than four variables, prime implicants can easily be found using a Karnaugh map. For functions of more variables, the Karnaugh map method quickly becomes too tedious. A better approach is Quine's tabular method, but this method requires that all minterms be listed explicitly at the beginning. In this subsection we briefly explain a recursive procedure based on *consensus* and *complete sums*.

The consensus theorem states that $xy + \bar{x}z = xy + \bar{x}z + yz$. The product yz is called the consensus for xy and $\bar{x}z$. A complete sum is defined to be a SOP formula composed of all the prime implicants. The following results are very useful for finding complete sums.

Theorem 5.5 (Blake, 1937) *A SOP is a complete sum if and only if:*

1. *No term includes any other term.*

2. *The consensus of any two terms of the formula either does not exist or is contained in some term of the formula.*

Theorem 5.6 (Blake, 1937) *If we have two complete sums f_1 and f_2, we can obtain the complete sum for $f_1 \cdot f_2$ using the following two steps:*

1. *Multiply out f_1 and f_2 using the following properties:*

   - $x \cdot x = x$ (idempotent)
   - $x \cdot (y + z) = xy + xz$ (distributive)
   - $x \cdot \bar{x} = 0$ (complement)

2. *Eliminate all terms absorbed by some other term (i.e., $a + ab = a$).*

Based upon this result and Boole's expansion theorem, we can define a recursive procedure for finding the complete sum for a function f:

$$\mathrm{cs}(f) = \mathrm{abs}([x_1 + \mathrm{cs}(f(0, x_2, \ldots, x_n))] \cdot [\bar{x_1} + \mathrm{cs}(f(1, x_2, \ldots, x_n))])$$

where $abs(f)$ removes absorbed terms from f.

Example 5.4.2 The ON-set and DC-set of the function depicted in Figure 5.24(a) can be expressed using the following formula:

$$f(w, x, y, z) = \bar{y}\bar{z} + xz + w\bar{x}\bar{y}z + \bar{w}xy\bar{z}$$

which is clearly not a complete sum. A complete sum can be found using the recursive procedure as follows:

$$
\begin{aligned}
f(w, x, y, z) &= \bar{y}\bar{z} + xz + w\bar{x}\bar{y}z + \bar{w}xy\bar{z} \\
f(w, x, y, 0) &= \bar{y} + \bar{w}xy \\
f(w, x, 0, 0) &= 1 \\
f(w, x, 1, 0) &= \bar{w}x \\
\mathrm{cs}(f(w, x, y, 0)) &= \mathrm{abs}((y + 1)(\bar{y} + \bar{w}x)) = \bar{y} + \bar{w}x \\
f(w, x, y, 1) &= x + w\bar{x}\bar{y} \\
f(w, 0, y, 1) &= w\bar{y}
\end{aligned}
$$

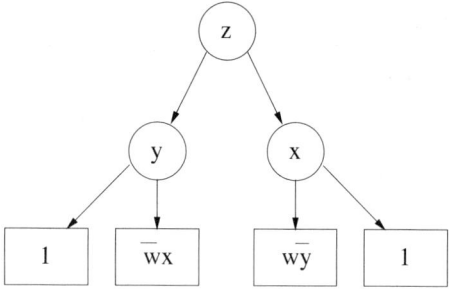

Fig. 5.25 Recursion tree for prime generation example.

$$
\begin{aligned}
f(w,1,y,1) &= 1 \\
\mathrm{cs}(f(w,x,y,1)) &= \mathrm{abs}((x+w\bar{y})(\bar{x}+1)) = x + w\bar{y} \\
\mathrm{cs}(f(w,x,y,z)) &= \mathrm{abs}((z+\bar{y}+\bar{w}x)(\bar{z}+x+w\bar{y})) \\
&= \mathrm{abs}(xz + w\bar{y}z + \bar{y}\bar{z} + x\bar{y} + w\bar{y} + \bar{w}x\bar{z} + \bar{w}x) \\
&= xz + \bar{y}\bar{z} + x\bar{y} + w\bar{y} + \bar{w}x
\end{aligned}
$$

The recursion tree for this example is shown in Figure 5.25.

5.4.3 Prime Implicant Selection

In order to select the minimal number of prime implicants necessary to cover the function, we need to solve a covering problem. We can create a constraint matrix where the rows are the minterms in the function and the columns are the prime implicants.

Example 5.4.3 The constraint matrix for the example shown in Figure 5.24(a) is given below.

	xz	$\bar{y}\bar{z}$	$x\bar{y}$	$w\bar{y}$	$\bar{w}x$
$\bar{w}\bar{x}\bar{y}\bar{z}$	–	1	–	–	–
$\bar{w}x\bar{y}\bar{z}$	–	1	1	–	1
$wx\bar{y}\bar{z}$	–	1	1	1	–
$w\bar{x}\bar{y}\bar{z}$	–	1	–	1	–
$\bar{w}x\bar{y}z$	1	–	1	–	1
$wx\bar{y}z$	1	–	1	1	–
$\bar{w}xyz$	1	–	–	–	1
$wxyz$	1	–	–	–	–

The rows corresponding to the minterms $\bar{w}\bar{x}\bar{y}\bar{z}$ and $wxyz$ are essential. This implies that $\bar{y}\bar{z}$ and xz are *essential primes*. Selecting these two primes solves the entire covering problem, so they make up our minimal SOP cover.

5.4.4 Combinational Hazards

For asynchronous design, the two-level logic minimization problem is complicated by the fact that there can be no hazards in the SOP implementation. Let us consider the design of a function f to implement either an output or next-state variable. Under the SIC model, when an input changes, the circuit moves from one minterm m_1 to another minterm m_2, where the two minterms differ in value in exactly one variable, x_i. During this transition, there are four possible transitions that f can make.

1. If $f(m_1) = 0$ and $f(m_2) = 0$, then f is making a *static $0 \to 0$ transition*.
2. If $f(m_1) = 1$ and $f(m_2) = 1$, then f is making a *static $1 \to 1$ transition*.
3. If $f(m_1) = 0$ and $f(m_2) = 1$, then f is making a *dynamic $0 \to 1$ transition*.
4. If $f(m_1) = 1$ and $f(m_2) = 0$, then f is making a *dynamic $1 \to 0$ transition*.

In order to design a hazard-free SOP cover, we must consider each of these cases in turn. First, if during a static $0 \to 0$ transition, the cover of f due to differences in delays momentarily evaluate to 1, we say that there exists a *static 0-hazard*. In a SOP cover of a function, no product term is allowed to include either m_1 or m_2 since they are members of the OFF-set. Therefore, the only way this can occur is if some product includes both x_i and $\overline{x_i}$. Clearly, such a product is not useful since it contains no minterms. If we exclude such product terms from the cover, the SOP cover can never produce a static 0-hazard. This result is summarized in the following theorem.

Theorem 5.7 (McCluskey, 1965) *A circuit has a static 0-hazard between the adjacent minterms m_1 and m_2 that differ only in x_j iff $f(m_1) = f(m_2) = 0$, there is a product term, p_i, in the circuit that includes x_j and $\overline{x_j}$, and all other literals in p_i have value 1 in m_1 and m_2.*

If during a static $1 \to 1$ transition, the cover of f can momentarily evaluate to 0, we say that there exists a *static 1-hazard*. In a SOP cover, consider the case where there is one product p_1 which contains m_1 but not m_2 and another product p_2 which contains m_2 but not m_1. The cover includes both m_1 and m_2, but there can be a static 1-hazard. If p_1 is implemented with a faster gate than p_2, the gate for p_1 can turn off faster than the gate for p_2 turns on, which can lead to the cover momentarily evaluating to a 0. In order to eliminate all static 1-hazards, for each possible transition $m_1 \to m_2$, there must exist a product in the cover that includes both m_1 and m_2. This result is summarized in the following theorem.

Theorem 5.8 (McCluskey, 1965) *A circuit has a static 1-hazard between adjacent minterms m_1 and m_2 where $f(m_1) = f(m_2) = 1$ iff there is no product term that has the value 1 in both m_1 and m_2.*

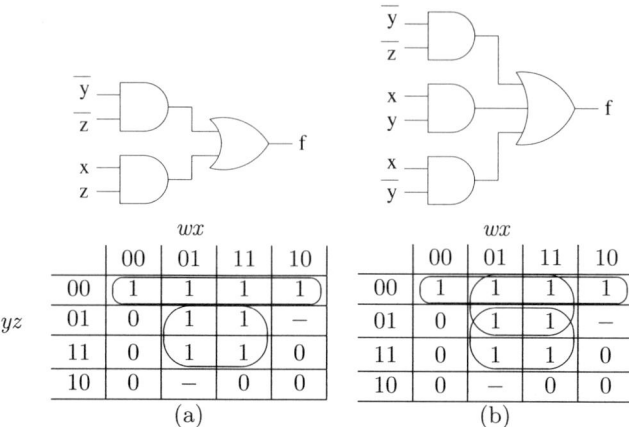

Fig. 5.26 (a) Circuit with static 1-hazard. (b) Circuit without static 1-hazard.

Example 5.4.4 Consider the transition from $\overline{w}x\overline{y}\overline{z}$ to $\overline{w}x\overline{y}z$. The function should maintain a constant value of 1 during this transition, so it is a static $1 \to 1$ transition. Consider the implementation shown in Figure 5.26(a). If the gate implementing $\overline{y}\overline{z}$ changes to 0 faster than xz changes to 1, then it is possible that the output of the function can momentarily yield a 0. The result is a static 1-hazard. If we include the prime $x\overline{y}$ in the cover as shown in Figure 5.26(b), this hazard is eliminated since this product yields a 1 during the transition, holding the output at a constant 1.

The next type of transition is the dynamic $0 \to 1$ transition. If during a $0 \to 1$ transition, the cover can change from 0 to 1 back to 0 and finally stabilize at 1, we say that the cover has a *dynamic $0 \to 1$ hazard*. Again, assuming no useless product terms (ones that include both x_i and $\overline{x_i}$), this is impossible under the SIC assumption. No product in the cover is allowed to include m_1 since it is in the OFF-set. Any product in the cover that includes m_2 will turn on monotonically. Similarly, there are no *dynamic $1 \to 0$ hazards*. This result is summarized in the following theorem.

Theorem 5.9 (McCluskey, 1965) *A SOP circuit has a dynamic hazard between adjacent minterms m_1 and m_2 that differ only in x_j iff $f(m_1) \neq f(m_2)$, the circuit has a product term p_i that contains x_j and $\overline{x_j}$, and all other literals of p_i have value 1 in m_1 and m_2.*

A simple, inefficient approach to produce a hazard-free SOP cover under SIC operation is to include all prime implicants in the cover. This eliminates static 1-hazards for input transitions, since the two minterms m_1 and m_2 are distance 1 apart; they must be included together in some prime. This means that an implicant exists which is made up of all literals that are equal in both m_1 and m_2. This implicant must be part of some prime implicant.

Example 5.4.5 For the example shown in Figure 5.24(a), the following cover is guaranteed to be hazard-free under the SIC assumption:

$$f = xz + \overline{y}\,\overline{z} + x\overline{y} + w\overline{y} + \overline{w}x$$

This approach is clearly inefficient, so a better approach is to reformulate the covering problem. The traditional SOP covering problem is to find the minimum number of prime implicants that cover all minterms in the ON-set. Instead, let us form an implicant out of each pair of states m_1 and m_2 involved in a static $1 \rightarrow 1$ transition which includes each literal that is the same value in both m_1 and m_2. The covering problem is now to find the minimum number of prime implicants that cover each of these *transition cubes*.

Example 5.4.6 For the example shown in Figure 5.24(a), let's assume that there exists a static $1 \rightarrow 1$ transition between each pair of minterms that are *distance 1 apart* (i.e., differ in a single literal). The resulting constraint matrix would be

	xz	$\overline{y}\,\overline{z}$	$x\overline{y}$	$w\overline{y}$	$\overline{w}x$
$\overline{w}\,\overline{y}\,z$	–	1	–	–	–
$\overline{x}\,\overline{y}\,z$	–	1	–	–	–
$\overline{w}x\overline{y}$	–	–	1	–	1
$wx\overline{y}$	–	–	1	1	–
$w\overline{y}z$	–	1	–	1	–
$x\overline{y}z$	1	–	1	–	–
$\overline{w}xz$	1	–	–	–	1
wxz	1	–	–	–	–

Again, xz and $\overline{y}\,\overline{z}$ are essential, and they must be included in the cover, leading to the following reduced constraint matrix:

	$x\overline{y}$	$w\overline{y}$	$\overline{w}x$
$\overline{w}x\overline{y}$	1	–	1
$wx\overline{y}$	1	1	–

The prime $x\overline{y}$ dominates the others, so the final hazard-free cover is

$$f = xz + \overline{y}\,\overline{z} + x\overline{y}$$

5.5 EXTENSIONS FOR MIC OPERATION

In Section 5.4, we restricted the class of circuits to those which only allowed a single input to change at a time. In other words, if these machines are specified as XBM machines, each input burst is allowed to include only a single transition. In this section we extend the synthesis method to allow multiple input changes. This allows us to synthesize any XBM machine which satisfies the maximal set property.

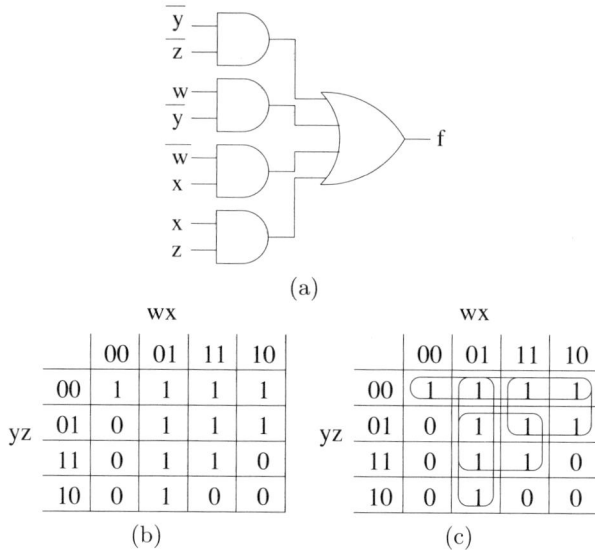

Fig. 5.27 (a) Hazardous circuit implementation. (b) Karnaugh map for small two-level logic minimization example. (c) Minimal two-level SOP cover.

5.5.1 Transition Cubes

Transitions in the MIC case begin in one minterm m_1 and end in another m_2, where the values of multiple variables may change during the transition. The minterm m_1 is called the *start point* and m_2 is called the *end point* of the transition. The smallest cube that contains both m_1 and m_2 is called the *transition cube*, and it is denoted $[m_1, m_2]$. This cube includes all possible minterms that a machine may pass through starting in m_1 and ending in m_2. The transition cube can also be represented with a product which contains a literal for each variable x_i in which $m_1(i) = m_2(i)$. An *open transition cube* $[m_1, m_2)$ includes all minterms in $[m_1, m_2]$ except those in m_2. An open transition cube usually must be represented using a set of products.

Example 5.5.1 Consider the transition $[\overline{w}\,x\,\overline{y}\,\overline{z},\ w\,x\,\overline{y}\,z]$ in the Karnaugh map in Figure 5.27(b). The transition cube for this transition is $x\,\overline{y}$, and it includes the four minterms that may be passed through during this transition. As another example, consider the open transition cube $[\overline{w}\,x\,y\,z,\ w\,x\,y\,\overline{z})$. This transition is represented by two products $\overline{w}\,x\,y$ and $x\,y\,z$.

5.5.2 Function Hazards

If a function f does not change monotonically during a multiple-input change, f has a *function hazard* for that transition. A function f contains a function hazard during a transition from m_1 to m_2 if there exists an m_3 and m_4 such that:

1. $m_3 \neq m_1$ and $m_4 \neq m_2$.

2. $m_3 \in [m_1, m_2]$ and $m_4 \in [m_3, m_2]$.

3. $f(m_1) \neq f(m_3)$ and $f(m_4) \neq f(m_2)$.

If $f(m_1) = f(m_2)$, it is a *static function hazard*, and if $f(m_1) \neq f(m_2)$, it is a *dynamic function hazard*.

> **Example 5.5.2** Consider the transition $[\overline{w}\,\overline{x}\,\overline{y}\,\overline{z},\ \overline{w}\,x\,\overline{y}\,z]$ in the Karnaugh map in Figure 5.27(b), which has a transition cube $\overline{w}\,\overline{y}$ with $m_3 = m_4 = \overline{w}\,\overline{x}\,\overline{y}\,z$. This transition has a static function hazard since $f(\overline{w}\,\overline{x}\,\overline{y}\,\overline{z}) = f(\overline{w}\,x\,\overline{y}\,z) = 1$ and $f(\overline{w}\,\overline{x}\,\overline{y}\,z) = 0$. The transition $[\overline{w}\,x\,y\,\overline{z},\ w\,\overline{x}\,y\,z]$ has a dynamic function hazard where $m_3 = w\,x\,y\,\overline{z}$ and $m_4 = w\,x\,y\,z$, since $f(m_1) = 1$, $f(m_3) = 0$, $f(m_4) = 1$, and $f(m_2) = 0$.

The following theorem states that if a transition has a function hazard, there does not exist an implementation of the function which avoids the hazard during this transition.

Theorem 5.10 (Eichelberger, 1965) *If a Boolean function, f, contains a function hazard for the input change m_1 to m_2, it is impossible to construct a logic gate network realizing f such that the possibility of a hazard pulse occurring for this transition is eliminated.*

Fortunately, as explained later, the synthesis method for XBM machines never produces a design with a transition that has a function hazard.

5.5.3 Combinational Hazards

Allowing multiple inputs to change besides introducing the potential of function hazards also complicates the elimination of combinational hazards. Even in the SIC case, if we use a minimum transition time state assignment, we must deal with static combinational hazards. Since after the input(s) change and the output and next state logic are allowed to stabilize (under the fundamental mode assumption), multiple changing next-state variables may be fed back to the input of the FSM. Again, the circuit moves from one minterm m_1 to another minterm m_2, but this time, multiple state variables may be changing concurrently. If we assume that we have only *normal flow tables* (all unstable states lead directly to stable states), and outputs are changed only in unstable states, then the only possible transitions are static ones. In other words, we restrict the set of allowable flow tables such that fed-back state variable changes cannot produce further output or next-state variable changes. Therefore, for state variable changes, we only need to consider static hazards.

HUFFMAN CIRCUITS

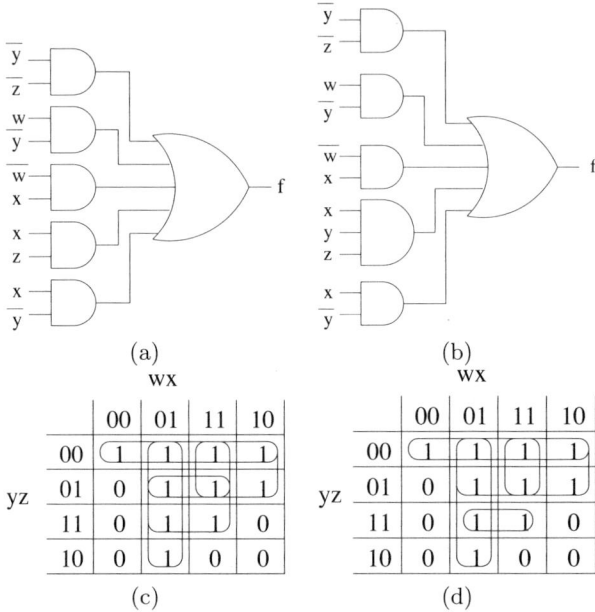

Fig. 5.28 (a) Static hazard-free circuit. (b) Dynamic hazard-free circuit. (c) Static hazard-free cover. (d) Dynamic hazard-free cover.

The results for static hazards in the MIC case are similar to those for the SIC case. Again, there can be no static 0-hazards. Static 1-hazards are a little more interesting. Since multiple variables may be changing concurrently, the cover may pass through other minterms along the way between m_1 and m_2. To be free of static 1-hazards, it is necessary that a single product in the cover include all these minterms. In other words, each transition cube, $[m_1, m_2]$, where $f(m_1) = f(m_2) = 1$, must be contained in some product in the cover to eliminate static 1-hazards. These transition cubes, therefore, are called *required cubes*.

Example 5.5.3 The circuit shown in Figure 5.27(a) is free of static 1-hazards under the SIC model. However, for the transition $[\overline{w}\,x\,\overline{y}\,z, w\,x\,\overline{y}\,z]$ there does not exist a product in the cover that includes this entire transition. In order to eliminate the static 1-hazard for this transition, it is necessary to add the product $x\,\overline{y}$. Now, all prime implicants are included in the cover. The resulting circuit is shown in Figure 5.28(a).

The following theorem states that a cover composed of all the prime implicants is guaranteed to be static hazard-free for every function hazard-free transition.

Theorem 5.11 (Eichelberger, 1965) *A SOP realization of f (assuming no product terms with complementary literals) will be free of all static logic hazards iff the realization contains all the prime implicants of f.*

The results for dynamic hazards are a bit more complicated. For each dynamic $1 \to 0$ transition $[m_1, m_2]$, if a product in the SOP cover intersects $[m_1, m_2]$ (i.e., it includes a minterm from the transition), it must also include the start point, m_1. If it does not include the start point, then if this product term is slow, it may turn on after the other product terms have turned off, causing a glitch during the $1 \to 0$ transition. For each dynamic $0 \to 1$ transition $[m_1, m_2]$, if a product in the SOP cover intersects $[m_1, m_2]$, it must also include the end point, m_2. In this case, the illegally intersecting product term may turn on and off quickly before the other product terms hold the function on. The result would be a glitch on the output. Since the transition cubes for dynamic $1 \to 0$ and $0 \to 1$ transitions must be carefully intersected, they are called *privileged cubes*. These results are summarized in the following theorem.

Theorem 5.12 (Bredeson, 1972) *Let f be a function which contains no dynamic function hazard for the transition $[m_1, m_2]$ where $f(m_1) = 0$ and $f(m_2) = 1$, and let $m_3 \in [m_1, m_2]$ where $f(m_3) = 1$. A SOP realization of f (assuming no product terms with complementary literals) will contain no dynamic logic hazard for transition $[m_1, m_2]$ (or $[m_2, m_1]$) iff for any m_3 which is covered by an implicant α in the SOP cover, it is also true that α covers m_2.*

The end point of the transition cube for a dynamic $0 \to 1$ transition is also a required cube. The *transition subcubes* for each dynamic $1 \to 0$ transition are required cubes. The transition subcubes for $1 \to 0$ transition $[m_1, m_2]$ are all cubes of the form $[m_1, m_3]$ such that $f(m_3) = 1$. Note that as an optimization you can eliminate any subcube contained in another.

Example 5.5.4 The transition $[\overline{w}\,x\,\overline{y}\,\overline{z},\ \overline{w}\,\overline{x}\,\overline{y}\,z]$ is a dynamic $1 \to 0$ transition from the Karnaugh map in Figure 5.28(c). The cubes required for this transition are $\overline{w}\,x\,\overline{y}$ and $\overline{w}\,\overline{y}\,\overline{z}$. For this transition, the product xz illegally intersects this transition, since it does not contain the start point. On the other hand, the cube that we added to eliminate the static 1-hazard, $x\,\overline{y}$, includes the start point, so it does not illegally intersect this transition. The result is that we must reduce the prime xz to xyz to eliminate this problem. Therefore, to eliminate dynamic hazards, it may be necessary to include nonprime implicants in the cover. In fact, if we also have a static transition $[\overline{w}\,x\,\overline{y}\,z, wxyz]$, then there would be no solution since xz is the only product that could include this transition, and it would cause a dynamic hazard. Fortunately, this situation can always be avoided for XBM machines.

(a)

	x=0	x=1
y=0	0	0
y=1	0	0

(b)

	x=0	x=1
y=0	1	1
y=1	1	1

(c)

	x=0	x=1
y=0	0	0
y=1	0	1

(d)

	x=0	x=1
y=0	1	1
y=1	1	0

(e)

	x=0	x=1
y=0	0	1
y=1	0	0

(f)

	x=0	x=1
y=0	1	0
y=1	1	1

(g)

	x=0	x=1
y=0	0	1
y=1	1	1

(h)

	x=0	x=1
y=0	1	0
y=1	0	0

Fig. 5.29 In each example, consider the transition cube $[\overline{x}\,\overline{y}, x\,y]$: (a) legal BM $0 \to 0$ transition, (b) legal BM $1 \to 1$ transition, (c) legal BM $0 \to 1$ transition, (d) legal BM $1 \to 0$ transition, (e) illegal BM $0 \to 0$ transition, (f) illegal BM $1 \to 1$ transition, (g) illegal BM $0 \to 1$ transition, (h) illegal BM $1 \to 0$ transition.

5.5.4 Burst-Mode Transitions

If we begin with a legal BM machine specification, the types of transitions possible are restricted. Namely, a function may change value only after every transition in the input burst has occurred. A transition $[m_1, m_2]$ for a function f is a *burst-mode transition* if for every minterm $m_i \in [m_1, m_2]$, $f(m_i) = f(m_1)$.

> **Example 5.5.5** Consider the Karnaugh maps shown in Figure 5.29 with the transition $[\overline{x}\,\overline{y},\ x\,y]$. The transitions in Figure 5.29(a), (b), (c), and (d) are legal burst-mode transitions. Those in Figure 5.29(e), (f), (g), and (h) are illegal since they do not maintain a constant value throughout $[\overline{x}\,\overline{y},\ x\,y)$.

If a function f has only burst-mode transitions, it is free of function hazards. Also, BM machines are free of dynamic $0 \to 1$ hazards. Finally, for any legal BM machine, there exists a hazard-free cover for each output and next-state variable before state minimization. These results are summarized in the following three theorems.

Theorem 5.13 (Nowick, 1993) *If f has a BM transition $[m_1, m_2]$, then f is free of function hazards for that transition.*

Theorem 5.14 (Nowick, 1993) *If f has a $0 \to 1$ BM transition in $[m_1, m_2]$, then a SOP implementation is free of logic hazards for this transition.*

Theorem 5.15 (Nowick, 1993) *Let G be any BM specification, let z be any output variable of G, let F be an unminimized flow table synthesized from G using an arbitrary state assignment, and let f_z be the output function for z in table F. Then the set of required cubes for f_z is a hazard-free cover.*

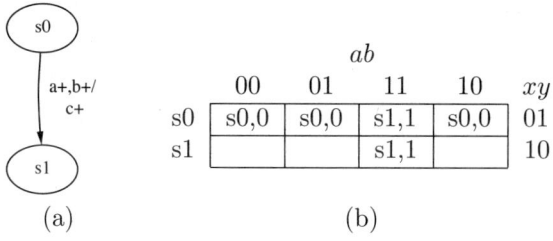

Fig. 5.30 (a) Simple burst-mode transition. (b) Corresponding flow table.

Example 5.5.6 Consider the simple burst-mode transition and corresponding flow table shown in Figure 5.30. For this burst-mode transition, the transition cube $[\bar{a}\,\bar{b}\,\bar{x}\,y,\,a\,b\,\bar{x}\,y]$ is a dynamic $0 \to 1$ transition for output c and next-state variable X and a dynamic $1 \to 0$ transition for next-state variable Y. This implies that $a\,b\,\bar{x}\,y$ is a required cube for c and X, while $a\,\bar{b}\,\bar{x}\,y$ and $\bar{a}\,b\,\bar{x}\,y$ are required cubes for Y. The cube $\bar{x}\,y$ is a priveledged cube for c, X, and Y. The transition cube $[a\,b\,\bar{x}\,y,\,a\,b\,x\,\bar{y}]$ is a static $1 \to 1$ transition for output c and next-state variable X and a static $0 \to 0$ transition for next-state variable Y. This implies that $a\,b$ is a required cube for c and X.

5.5.5 Extended Burst-Mode Transitions

When inputs are allowed to change nonmonotonically during multiple-input changes as in an XBM machine, we need to generalize the notion of transition cubes to allow the start and end points to be cubes rather than simply minterms. In the *generalized transition cube* $[c_1, c_2]$, the cube c_1 is called the *start cube* and c_2 is called the *end cube*. The *open generalized transition cube*, $[c_1, c_2)$, is defined to be all minterms in $[c_1, c_2]$ excluding those in c_2 (i.e., $[c_1, c_2) = [c_1, c_2] - c_2$).

In an XBM machine, some signals are rising, some are falling, and others are levels which can change nonmonotonically. Rising and falling signals change monotonically (i.e., at most once in a legal transition cube). Level signals must hold the same value in c_1 and c_2, where the value is either a constant (0 or 1) or a don't care (−). Level signals, if they are don't cares, may change nonmonotonically. In an XBM machine, the types of transitions are again restricted such that each function may change value only after the completion of an input burst. A generalized transition $[c_1, c_2]$ for a function f is an *extended burst-mode transition* if for every minterm $m_i \in [c_1, c_2)$, $f(m_i) = f(c_1)$ and for every minterm $m_i \in c_2$, $f(m_i) = f(c_2)$. The following theorem states that if a function has only extended burst-mode transitions, then it is function hazard-free.

		xl						xl			
		00	01	11	10			00	01	11	10
	00	1	1	1	1		00	0	0	0	0
yz	01	1	1	1	1		01	0	0	1	1
	11	1	1	0	0		11	0	0	1	0
	10	1	1	0	0		10	0	0	0	0
		(a)						(b)			

Fig. 5.31 (a) Example extended burst-mode transition. (b) Example illustrating dynamic $0 \rightarrow 1$ hazard issues.

Theorem 5.16 (Yun, 1999) *Every extended burst-mode transition is function hazard-free.*

> **Example 5.5.7** Consider the Karnaugh map shown in Figure 5.31(a), where x, y, and z are transition signals and l is a level signal. During the generalized transition $[\bar{x} - \bar{y}-, x - y-]$, x and y will rise, z is a directed don't-care (assume that it is rising), and l is a level signal which may change value arbitrarily. This transition is an extended burst-mode transition, since f is always 1 in $[\bar{x} - \bar{y}-, x - y-)$ and f is always 0 while in the cube $x - y-$.
>
> **Example 5.5.8** Consider the generalized transition $[\bar{x} - \bar{y}-, x - yz]$ shown in Figure 5.31(a). In this transition, x, y, and z will rise. This transition, however, is not an extended burst-mode transition, since in $[\bar{x} - \bar{y}-, x - yz)$, f can be either 1 or 0.

The *start subcube*, c'_1, is a maximal subcube of c_1 such that each signal undergoing a directed don't-care transition is set to its initial value (i.e., 0 for a rising transition and 1 for a falling transition). The *end subcube*, c'_2, is a maximal subcube of c_2 such that each signal undergoing a directed don't-care transition is set to its final value (i.e., 1 for a rising transition and 0 for a falling transition).

> **Example 5.5.9** Consider again the transition $[\bar{x} - \bar{y}-,x - y-]$ in Figure 5.31(a), where z is a rising directed don't care. The start subcube is $\bar{x} - \bar{y}\bar{z}$ and the end subcube is $x - yz$.

If we consider the transition cube $[c'_1, c'_2]$, the hazard considerations are the same as before. In other words, if this is a static $1 \rightarrow 1$ transition [i.e., $f(c'_1) = f(c'_2) = 1$], the entire transition cube must be included in some product term in the cover. If it is a dynamic $1 \rightarrow 0$ transition, then any product that intersects this transition cube must contain the start subcube, c'_1. Unlike burst mode, dynamic $0 \rightarrow 1$ transitions must be considered. Namely, any product that intersects the transition cube for a dynamic $0 \rightarrow 1$ transition must contain the end subcube, c'_2. These results are summarized in the following theorems.

Theorem 5.17 (Yun, 1999) *The output of a SOP implementation is hazard-free during a $1 \rightarrow 0$ extended burst-mode transition iff every product term intersecting the transition cube $[c_1, c_2]$ also contains the start subcube c'_1.*

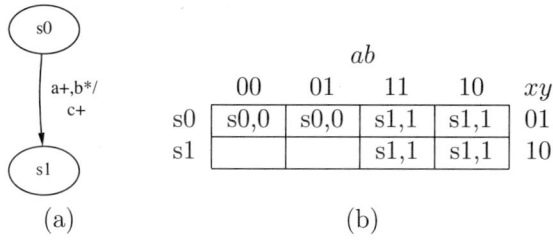

Fig. 5.32 (a) Simple extended burst-mode transition. (b) Corresponding flow table.

Theorem 5.18 (Yun, 1999) *The output of a SOP implementation is hazard-free during a $0 \rightarrow 1$ extended burst-mode transition iff every product term intersecting the transition cube $[c_1, c_2]$ also contains the end subcube c_2'.*

Example 5.5.10 Consider the Karnaugh map shown in Figure 5.31(b) and the dynamic $0 \rightarrow 1$ transitions $[\overline{x} - \overline{y}\,\overline{z}, x - \overline{y}z]$ and $[\overline{x}ly\overline{z}, xlyz]$. If we implement these two transitions with the logic $f = x\overline{y}z + xlz$, there is a dynamic $0 \rightarrow 1$ hazard. The problem is that the cube xlz illegally intersects the transition $[\overline{x} - \overline{y}\,\overline{z}, x - \overline{y}z]$ since it does not contain the entire end subcube. To eliminate this hazard, it is necessary to reduce xlz to $xlyz$.

Example 5.5.11 Consider the simple extended burst-mode transition and corresponding flow table shown in Figure 5.32. For this extended burst-mode transition, the transition cube $[\overline{a}\,\overline{x}\,y,\, a\,\overline{x}\,y]$ is a dynamic $0 \rightarrow 1$ transition for output c and next-state variable X and a dynamic $1 \rightarrow 0$ transition for next-state variable Y. This implies that $a\overline{x}y$ is a required cube for c and X, while $\overline{a}\,\overline{x}\,y$ is a required cube for Y. Assuming that b is a rising directed don't care, the start subcube for this transition is $\overline{a}\,\overline{b}\,\overline{x}\,y$ and the end subcube is $a\,b\,\overline{x}\,y$. The cube $\overline{x}y$ is a priveledged cube for c, X, and Y. Therefore, no cube in the cover for c or X can intersect this cube unless it includes $a\,b\,\overline{x}\,y$. No cube in the cover for Y can intersect this cube unless it includes $\overline{a}\,\overline{b}\,\overline{x}\,y$. The transition cube $[a\,\overline{x}\,y,\, a\,x\,\overline{y}]$ is a static $1 \rightarrow 1$ transition for output c and next-state variable X and a static $0 \rightarrow 0$ transition for next-state variable Y. This implies that $a\,b$ is a required cube for c and X.

Unfortunately, not every XBM transition can be implemented free of dynamic hazards. Consider the following example.

Example 5.5.12 The circuit shown in Figure 5.33(a) latches a conditional signal and converts it to a dual-rail signal. It can be described using the XBM machine in Figure 5.33(b). A circuit implementation for signal x derived from this XBM machine is shown in Figure 5.33(c) and its corresponding Karnaugh map is shown in Figure 5.33(d). In the circuit shown in Figure 5.33(c), if when c goes high, d is high, then x is set to high. This signal is fed-back as a state variable X. This feedback creates a static $1 \rightarrow 1$ transition $[dc\overline{X}\,\overline{Y}, dcX\overline{Y}]$, so the product $dc\overline{Y}$ must be completely covered by some product in the cover to

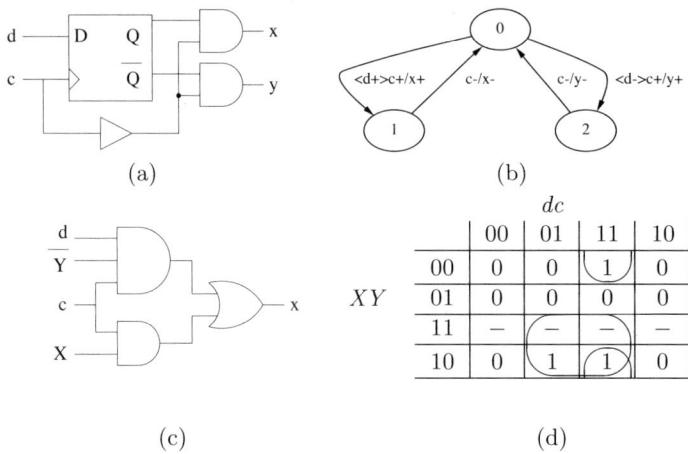

Fig. 5.33 (a) dff Circuit that latches a conditional signal and converts it to a dual-rail signal. (b) XBM machine to describe circuit from (a). (c) Asynchronous implementation with a dynamic hazard. (d) K-map showing the dynamic 1 → 0 hazard.

prevent a static 1 → 1 hazard. A hold time after x goes high, the conditional signal d is allowed to change. At some point in the future with d now at either value, c goes low. This starts a dynamic 1 → 0 transition $[-cX\overline{Y}, -\overline{c}X\overline{Y}]$ which can be represented by the product $X\overline{Y}$. This product must not be intersected unless the intersecting product term also includes the start subcube (i.e., $cX\overline{Y}$). However, the product $dc\overline{Y}$ must intersect this product, but it cannot contain the entire start subcube. The result is that there is no hazard-free solution.

To address this problem, we must modify the state assignment as well as the machine operation in these situations. This solution is discussed in Section 5.5.7 where we address state assignment.

5.5.6 State Minimization

Using the state minimization procedure described in Section 5.2, it is possible that no hazard-free cover exists for some variable in the design. This procedure will need to be modified for the MIC case. We illustrate this through an example.

Example 5.5.13 Consider the flow table fragment in Figure 5.34(a). Under our original definition of compatibility, states A and D can be merged to form the new state shown in Figure 5.34(b). There is a static 1 → 1 transition from input $a\overline{b}\overline{c}$ to $a\overline{b}c$ which has transition cube $a\overline{b}$. Recall that to be free of static 1-hazards, the product $a\overline{b}$ must be included in some product in the final cover. There is also a dynamic 1 → 0 transition from input $\overline{a}\overline{b}\overline{c}$ to $ab\overline{c}$ which has start point $\overline{a}\overline{b}\overline{c}$ and tran-

		Inputs $a\ b\ c$							
State		000	001	011	010	110	111	101	100
	A	A,1	C,0	–	A,1	B,0	–	–	A,1
		...							
	D	–	–	–	–	–	–	E,1	D,1

(a)

		Inputs $a\ b\ c$							
State		000	001	011	010	110	111	101	100
	AD	A,1	C,0	–	A,1	B,0	–	E,1	A,1

(b)

Fig. 5.34 (a) Fragment of a flow table. (b) Illegal state merging.

sition cube $\bar{c}$. Note that the product $a\bar{b}$ intersects this transition cube, but it does not include the start point $\bar{a}\bar{b}\bar{c}$. Therefore, this intersection is illegal and leads to a dynamic $1 \to 0$ hazard. However, if we restrict our product to $a\bar{b}c$ so that it no longer illegally intersects the dynamic transition, it no longer covers the entire static transition, so we have a static 1-hazard.

The use of conditional signals in XBM machines further complicates state minimization, which we again illustrate with an example.

Example 5.5.14 Consider the fragment of an XBM machine shown in Figure 5.35(a) and corresponding flow table shown in Figure 5.35(b). State A is not output compatible with either state B or C, due to the entry for input 10. States B and C are compatible, and combining them results in the new flow table shown in Figure 5.35(c). For input 11 in state A, the machine is transitioning to state B and there is a static $1 \to 1$ transition on output c which must be included completely by some product in the cover for it to be hazard-free. While in state BC, there is a dynamic $0 \to 1$ transition on output c, $[0-, 1-]$. Both the end cube and end subcube for this transition are $1-$, which must not be intersected without being included completely. However, in the minimized machine the required cube for the static transition during the state change cannot be expanded to include the entire end subcube, nor can it be reduced so as not to intersect the end cube. Therefore, there is no hazard-free solution to the flow table shown in Figure 5.35(c). Therefore, states B and C should not be combined.

To solve these problems, we must restrict the conditions under which two states are compatible. Under these new restrictions on state minimization, it can be proven that a hazard-free cover can always be found. Two states s_1 and s_2 are *dhf-compatible* (or dynamic hazard-free compatible) when they are compatible and for each output z and transition $[c_1, c_2]$ of s_1 and for each transition $[c_3, c_4]$ of s_2:

1. If z has a $1 \to 0$ transition in $[c_1, c_2]$ and a $1 \to 1$ transition in $[c_3, c_4]$, then $[c_1, c_2] \cap [c_3, c_4] = \emptyset$ or $c'_1 \in [c_3, c_4]$.

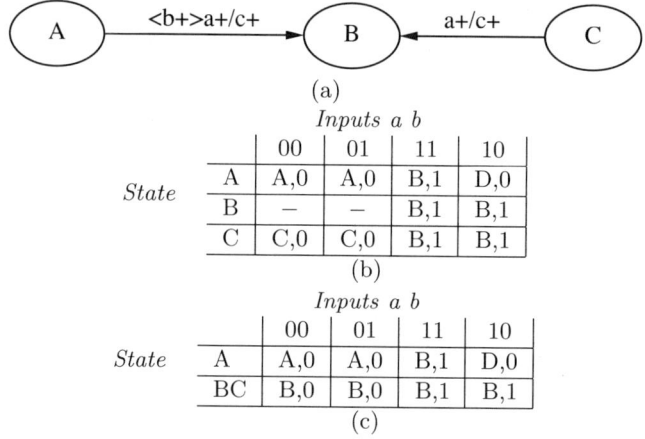

Fig. 5.35 (a) XBM machine fragment showing potential for dynamic hazard. (b) Initial flow table. (c) Illegal state merging.

2. If z has a $1 \to 0$ transition in $[c_1, c_2]$ and a $1 \to 0$ transition in $[c_3, c_4]$, then $[c_1, c_2] \cap [c_3, c_4] = \emptyset$, $c_1 = c_3$, $[c_1, c_2] \subseteq [c_3, c_4]$, or $[c_3, c_4] \subseteq [c_1, c_2]$.

3. If z has a $0 \to 1$ transition in $[c_1, c_2]$ and a $1 \to 1$ transition in $[c_3, c_4]$, then $[c_1, c_2] \cap [c_3, c_4] = \emptyset$ or $c'_2 \in [c_3, c_4]$.

4. It z has a $0 \to 1$ transition in $[c_1, c_2]$ and a $0 \to 1$ transition in $[c_3, c_4]$, then $[c_1, c_2] \cap [c_3, c_4] = \emptyset$, $c_2 = c_4$, $[c_1, c_2] \subseteq [c_3, c_4]$, or $[c_3, c_4] \subseteq [c_1, c_2]$.

The states s_1 and s_2 must also satisfy the following further restriction for each s_3, which can transition to s_1 in $[c_3, c_4]$ and another transition $[c_1, c_2]$ of s_2:

1. If z has a $1 \to 0$ transition in $[c_1, c_2]$ and a $1 \to 1$ transition in $[c_3, c_4]$, then $[c_1, c_2] \cap [c_3, c_4] = \emptyset$ or $c'_1 \in [c_3, m_4]$.

2. If z has a $0 \to 1$ transition in $[c_1, c_2]$ and a $1 \to 1$ transition in $[c_3, c_4]$, then $[c_1, c_2] \cap [c_3, c_4] = \emptyset$ or $c'_2 \in [c_3, c_4]$.

Similarly, for each s_3 which can transition to s_2 in $[c_3, c_4]$ and another transition $[c_1, c_2]$ of s_1:

1. If z has a $1 \to 0$ transition in $[c_1, c_2]$ and a $1 \to 1$ transition in $[c_3, c_4]$, then $[c_1, c_2] \cap [c_3, c_4] = \emptyset$ or $c'_1 \in [c_3, m_4]$.

2. If z has a $0 \to 1$ transition in $[c_1, c_2]$ and a $1 \to 1$ transition in $[c_3, c_4]$, then $[c_1, c_2] \cap [c_3, c_4] = \emptyset$ or $c'_2 \in [c_3, c_4]$.

	dc			
pxy	00	01	11	10
000	0	0	0	0
001	–	–	–	–
011	–	–	–	–
010	–	–	–	–
110	0	1	1	0
111	–	–	–	–
101	–	–	–	–
100	0	1	1	0

(a)

	dc			
pxy	00	01	11	10
000	0	0	1	0
001	–	–	–	–
011	–	–	–	–
010	–	–	–	–
110	0	1	1	0
111	–	–	–	–
101	–	–	–	–
100	0	1	1	0

(b)

Fig. 5.36 Partial Karnaugh maps for (a) X and (b) P.

5.5.7 State Assignment

The state assignment method described earlier can be used directly to find a critical race free state assignment under BM operation. Since state changes happen only after all inputs have changed, the state change can be thought to occur within the individual column, and the number of input changes does not matter.

For XBM machines, it may be necessary to add additional state variables to eliminate dynamic hazards when there are conditional signals. Consider again the example from Figure 5.33. In this example, there is an unavoidable dynamic hazard. To solve this problem, a new state variable is added for each conditional input burst in which the next input burst is unconditional and enables an output to fall. This state variable is set to high after the conditional input burst but before the output burst is allowed to begin. Intuitively, this state variable is storing the value of the conditional signal. It can be set low in the next output burst.

> **Example 5.5.15** Consider again the example from Figure 5.33. Partial Karnaugh maps (q is omitted) after adding a new state variable p are shown in Figure 5.36(a) for output X and Figure 5.36(b) for state variable P. The privileged cube is $px\bar{y}$, which cannot be intersected without including the start subcube, which is $cpx\bar{y}$. Now, the implementation for x is simply cp, which legally intersects this privileged cube. The new state variable, P, can be implemented with the function $cp + dc\bar{x}\bar{q}$ (note that q is not shown in the figure), which again does not illegally intersect the dynamic transition, so the resulting circuit shown in Figure 5.37 is now hazard-free.

5.5.8 Hazard-Free Two-Level Logic Synthesis

In this section we describe two-level logic synthesis in the presence of multiple input changes. In this case, the union of the required cubes forms the ON-

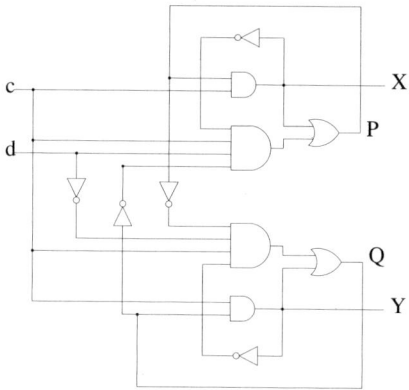

Fig. 5.37 Hazard-free dff circuit.

set for the function. Each of the required cubes must be contained in some product of the cover to ensure hazard freedom.

Example 5.5.16 Consider the Karnaugh map shown in Figure 5.38, which has the following transition cubes:

$$t_1 = [a\bar{b}\bar{c}d, ab\bar{c}\bar{d}]$$
$$t_2 = [a\bar{b}c\bar{d}, ab\bar{c}d]$$
$$t_3 = [\bar{a}b\bar{c}\bar{d}, \bar{a}b\bar{c}d]$$
$$t_4 = [\bar{a}bcd, a\bar{b}c\bar{d}]$$

Transition t_1 is a $1 \to 1$ transition, so it produces a required cube $a\bar{c}$. Transition t_2 is a $0 \to 0$ transition, so it does not produce any required cube. Transition t_3 is a $1 \to 0$ transition, so it produces a required cube for each transition subcube. First, the transition $[\bar{a}b\bar{c}\bar{d}, \bar{a}b\bar{c}\bar{d}]$ produces the required cube $\bar{a}\bar{c}\bar{d}$. Second, the transition $[\bar{a}b\bar{c}\bar{d}, \bar{a}b\bar{c}d]$ produces the required cube $\bar{a}b\bar{c}$. Transition t_4 is also a $1 \to 0$ transition, so again we produce a required cube for each transition subcube. The result are two additional required cubes bcd and $\bar{a}c$ (note that the required cubes $\bar{a}cd$ and $\bar{a}bc$ are both contained in $\bar{a}c$, so they can be discarded). This gives us a final required cube set of

$$\text{req-set} = \{a\bar{c}, \bar{a}\bar{c}\bar{d}, \bar{a}b\bar{c}, bcd, \bar{a}c\}$$

The transition cubes for each dynamic $1 \to 0$ transition are privileged cubes since they cannot be intersected unless the intersecting product includes its start subcube. Similarly, transition cubes for each dynamic $0 \to 1$ transition are also privileged cubes since they cannot be intersected unless the intersecting product includes its end subcube. If a cover includes a product that intersects a privileged cube without including its corresponding start subcube or end subcube, the cover is not hazard-free.

	ab			
	00	01	11	10
cd 00	1	1	1	1
01	0	1	1	1
11	1	1	1	0
10	1	1	0	0

Fig. 5.38 Karnaugh map for small two-level logic minimization example.

Example 5.5.17 Transitions t_3 and t_4 from Example 5.5.16 are dynamic $1 \to 0$ transitions, so their corresponding transition cubes are privileged cubes. This gives us the following set of privileged cubes:

$$\text{priv-set} = \{\bar{a}\bar{c}, c\}$$

Recall that we may not be able to produce a SOP cover that is free of dynamic hazards using only prime implicants, so we introduce the notion of a *dynamic-hazard-free implicant* (or *dhf-implicant*). A dhf-implicant is an implicant which does not illegally intersect any privileged cube. A *dhf-prime implicant* is a dhf-implicant that is contained in no other dhf-implicant. Note that a dhf-prime implicant may actually not be a prime implicant. A minimal hazard-free cover includes only dhf-prime implicants.

The first step to find all dhf-prime implicants is to find the ordinary prime implicants using the recursive procedure described earlier. We can start this procedure using a SOP that contains the required cubes and a set of implicants that represents the DC-set. After finding the prime implicants, we next check each prime to see if it illegally intersects a privileged cube. If it does, we attempt to shrink the cube to make it dhf-prime.

Example 5.5.18 The recursion to find the primes proceeds as follows:

$$\begin{aligned}
f(a,b,c,d) &= a\bar{c} + \bar{a}\bar{c}\bar{d} + \bar{a}b\bar{c} + bcd + \bar{a}c \\
f(a,b,0,d) &= a + \bar{a}\bar{d} + \bar{a}b \\
f(0,b,0,d) &= \bar{d} + b \\
f(1,b,0,d) &= 1 \\
\text{cs}(f(a,b,c,0)) &= \text{abs}((a+\bar{d}+b)(\bar{a}+1)) = a + \bar{d} + b \\
f(a,b,1,d) &= bd + \bar{a} \\
f(0,b,1,d) &= 1 \\
f(1,b,1,d) &= bd \\
\text{cs}(f(a,b,1,d)) &= \text{abs}((a+1)(\bar{a}+bd)) = \bar{a} + bd \\
\text{cs}(f(a,b,c,d)) &= \text{abs}((c+a+\bar{d}+b)(\bar{c}+\bar{a}+bd)) \\
&= \text{abs}(\bar{a}c + bcd + a\bar{c} + abd + \bar{c}\bar{d} + \bar{a}\bar{d} + b\bar{c} + \\
&\quad \bar{a}b + bd) \\
&= \bar{a}c + a\bar{c} + \bar{c}\bar{d} + \bar{a}\bar{d} + b\bar{c} + \bar{a}b + bd
\end{aligned}$$

The next step is to determine which primes illegally intersect a privileged cube. First, the prime $a\bar{c}$ does not intersect any privileged cube, so it is a dhf-prime. The primes $\bar{c}\bar{d}$ and $b\bar{c}$ intersect the privileged cube $\bar{a}\bar{c}$, but they include the start subcube $\bar{a}b\bar{c}\bar{d}$. Since they do not intersect the other privileged cube, c, they are also dhf-primes. The prime $\bar{a}c$ intersects the privileged cube c, but it also includes the start subcube $\bar{a}bcd$. Therefore, since it does not intersect the other privileged cube $\bar{a}\bar{c}$, it is also a dhf-prime. The prime $\bar{a}b$ intersects both privileged cubes, but it also includes both start subcubes. The last two primes, $\bar{a}\bar{d}$ and bd, also intersect both privileged cubes. However, the prime $\bar{a}\bar{d}$ does not include the start subcube of c (i.e., $\bar{a}bcd$), and the prime bd does not include the start subcube of $\bar{a}\bar{c}$ (i.e., $\bar{a}b\bar{c}\bar{d}$). Therefore, these last two primes are not dhf-prime implicants.

For these two remaining primes, we must shrink them until they no longer illegally intersect any privileged cubes. Let us consider first $\bar{a}\bar{d}$. We must add a literal that makes this prime disjoint from the privileged cube c. The only choice is to add $\bar{c}$ to make the new implicant, $\bar{a}\bar{c}\bar{d}$. This implicant no longer intersects the privileged cube c and it legally intersects the privileged cube $\bar{a}\bar{c}$ (i.e., it includes its start subcube). Note, though, that $\bar{a}\bar{c}\bar{d}$ is contained in the dhf-prime $\bar{c}\bar{d}$, so it is not a dhf-prime and can be discarded. Next, consider the prime bd. We must make it disjoint with the privileged cube $\bar{a}\bar{c}$. There are two possible choices, abd or bcd, but the implicant abd intersects the privileged cube c illegally, so it must be reduced further to $ab\bar{c}d$. This implicant, however, is contained in the dhf-prime $a\bar{c}$, so it can be discarded. The implicant bcd does not illegally intersect any privileged cubes and is not contained in any other dhf-prime, so it is a dhf-prime. Our final set of dhf-prime implicants is

$$\text{dhf-prime} = \{\bar{a}c, a\bar{c}, \bar{c}\bar{d}, b\bar{c}, \bar{a}b, bcd\}$$

The two-level hazard-free logic minimization problem for XBM operation is to find a minimum-cost cover which covers every required cube using only dhf-prime implicants.

Example 5.5.19 The constraint matrix is shown below.

	$\bar{a}c$	$a\bar{c}$	$\bar{c}\bar{d}$	$b\bar{c}$	$\bar{a}b$	bcd
$a\bar{c}$	—	1	—	—	—	—
$\bar{a}\bar{c}\bar{d}$	—	—	1	—	—	—
$\bar{a}b\bar{c}$	—	—	—	1	1	—
bcd	—	—	—	—	—	1
$\bar{a}c$	1	—	—	—	—	—

There are four essential rows which make $\bar{a}c$, $a\bar{c}$, $\bar{c}\bar{d}$, and bcd essential. To solve the remaining row, either the prime $b\bar{c}$ or $\bar{a}b$ can be selected. Therefore, there are two possible minimal hazard-free SOP covers which each include five product terms. Note that both covers require a non-

		abc							
		000	001	011	010	110	111	101	100
	00	0	0	0	0	0	0	0	0
de	01	0	0	0	0	0	0	0	0
	10	0	0	1	1	1	1	1	1
	11	0	0	1	1	0	0	1	1

Fig. 5.39 Karnaugh map for Example 5.5.20.

prime implicant, bcd. A minimal hazardous cover would only require four product terms, $a\bar{c}$, bd, $\bar{a}c$, and $\bar{c}\bar{d}$.

Our next example illustrates the hazard issues that must be addressed in the presence of directed don't cares and conditional signals.

Example 5.5.20 Consider the Karnaugh map shown in Figure 5.39 with the following generalized transition cubes:

$$t_1 = [\bar{a}\bar{b}\bar{d}, \bar{a}bd]$$
$$t_2 = [ab\bar{c}d\bar{e}, abcd\bar{e}]$$
$$t_3 = [a\bar{b}d, a\bar{b}\bar{d}]$$

The first transition is a dynamic $0 \to 1$ transition which is triggered by b and d going high. During this transition, c is a falling directed don't care and e is an unspecified conditional signal. This transition contributes the required cube $\bar{a}bd$ and the privileged cube $\bar{a}$. This privileged cube must not be intersected except by a product which includes the entire end subcube, which is $\bar{a}b\bar{c}d$. The second transition is a static $1 \to 1$ transition which contributes the required cube $abd\bar{e}$. Finally, the third transition is a dynamic $1 \to 0$ transition which is triggered by d going low. During this transition, c is a rising directed don't care and e is an unspecified conditional signal. The required cube for this transition is $a\bar{b}d$. This transition also makes $a\bar{b}$ a privileged cube which can only be intersected by products that include its start subcube, $a\bar{b}\bar{c}d$.

The primes for this example are:

$$\bar{a}bd, a\bar{b}d, ad\bar{e}, bd\bar{e}$$

The prime $ad\bar{e}$ intersects the privileged cube $a\bar{b}$, but it does not include its start subcube, $a\bar{b}\bar{c}d$. Therefore, this prime must be reduced to $abd\bar{e}$, which does not intersect the privileged cube. Similarly, the prime $bd\bar{e}$ intersects the privileged cube $\bar{a}$ and does not contain its end subcube $\bar{a}b\bar{c}d$. Again, this prime is reduced to $abd\bar{e}$ to eliminate the problem. Therefore, the final cover of the set of required cubes is

$$f = \bar{a}bd + a\bar{b}d + abd\bar{e}$$

Again, a nonprime implicant is needed to be hazard-free.

5.6 MULTILEVEL LOGIC SYNTHESIS

Typically, two-level SOP implementations cannot be realized directly for most technologies. The reason is that the AND or OR stages of the gate could have arbitrarily large fan-in (i.e., numbers of inputs). In CMOS, for example, gates with more than three or four inputs are considered to be too slow. Therefore, two-level SOP implementations must be decomposed into multilevel implementations using laws of Boolean algebra. Again, however, care must be taken not to introduce hazards. In this section we present a number of *hazard-preserving transformations*. Therefore, if we begin with a hazard-free SOP implementation and apply only hazard-preserving transformations, the resulting multilevel implementation is also hazard-free. The following theorem gives the laws of Boolean algebra which are hazard-preserving.

Theorem 5.19 (Unger, 1969) *Given any expression f_1, if we transform it into another expression, f_2, using the following laws:*

- $A + (B + C) \Leftrightarrow A + B + C$ *(associative law)*
- $A(BC) \Leftrightarrow ABC$ *(associative law)*
- $\overline{(A + B)} \Leftrightarrow \overline{A}\,\overline{B}$ *(DeMorgan's theorem)*
- $\overline{(AB)} \Leftrightarrow \overline{A} + \overline{B}$ *(DeMorgan's theorem)*
- $AB + AC \Rightarrow A(B + C)$ *(distributive law)*
- $A + AB \Rightarrow A$ *(absorptive law)*
- $A + \overline{A}B \Rightarrow A + B$

then a circuit corresponding to f_2 will have no combinational hazards not present in circuits corresponding to f_1.

In other words, if we transform a circuit into a new circuit using the laws listed above the two circuits have equivalent hazards. Therefore, if the original circuit is hazard-free, so is the new circuit. Note that the last three laws can be applied only in one direction. For example, the distributive law in the reverse direction [i.e., $A(B+C) \Rightarrow AB + AC$] preserves static hazards, but it may introduce new dynamic hazards. For example, the function $f = A(\overline{A}+B)$ is free of dynamic hazards. If we multiply it out, we get $A\overline{A} + AB$, which has a dynamic hazard when $B = 1$ and A changes.

We can also create new functions from others, with the following effects:

- Insertion or deletion of inverters at the output of a circuit only interchanges 0 and 1-hazards.

- Insertion or deletion of inverters at the inputs only relocates hazards to duals of the original transition.

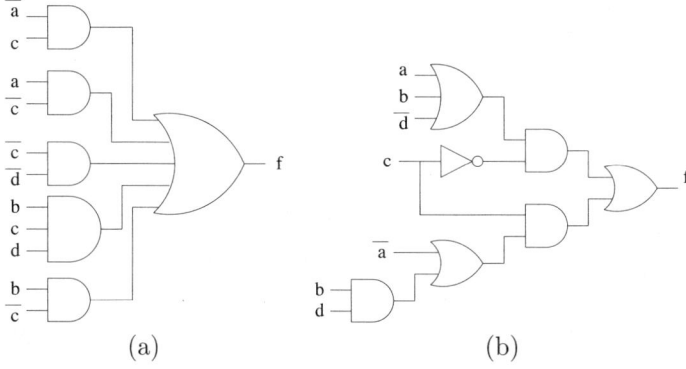

Fig. 5.40 (a) SOP implementation for small example. (b) Multilevel implementation for small example.

- The dual of a circuit (exchange AND and OR gates) produces a dual function with dual hazards.

Furthermore, there are many other hazard-preserving transformations not mentioned here.

In order to derive a hazard-free multilevel implementation, we first find a hazard-free SOP implementation. If we then convert it to a multilevel implementation using only the associative law, DeMorgan's theorem, factoring (not multiplying out), $A + AB \rightarrow A$, and $A + \overline{A}B \rightarrow A + B$, then it is also hazard-free. Similarly, to check a multilevel implementation for hazards, convert it to a SOP implementation using associative, distributive, and DeMorgan's laws and check for hazards (be sure not to perform any reductions using $A\overline{A} = 0$).

Example 5.6.1 Consider the two-level hazard-free implementation derived in Section 5.5.8, which has 11 literals as shown in Figure 5.40(a):

$$f = \overline{a}c + a\overline{c} + \overline{c}\overline{d} + bcd + b\overline{c}$$

Using factoring, we can obtain the following equation with eight literals, which is shown in Figure 5.40(b):

$$f = \overline{c}(a + b + \overline{d}) + c(\overline{a} + bd)$$

This is hazard-free since factoring is a hazard-preserving operation.

5.7 TECHNOLOGY MAPPING

After deriving optimized two-level or multilevel logic, the next step is to map these logic equations to some given gate library. This *technology mapping* step takes as input a set of technology-independent logic equations and a library

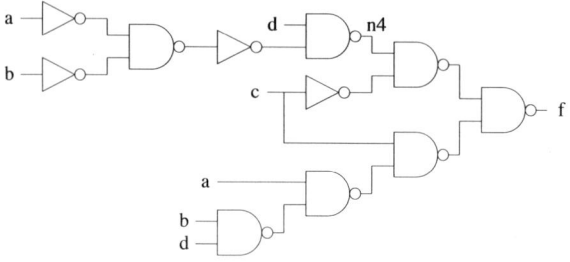

Fig. 5.41 NAND tree decomposition of the function $f = \bar{c}(a+b+\bar{d}) + c(\bar{a}+bd)$.

of cells implemented in some gate-array or standard-cell technology, and it produces a netlist of cells implementing the logic in the given technology. The technology mapping process is traditionally broken up into three major steps: *decomposition*, *partitioning*, and *matching/covering*. In this section we describe how these steps can be performed without introducing hazards.

The decomposition step transforms the network of logic equations into an equivalent network using only two-input/one-output *base functions*. A typical choice of base function is two-input NAND gates. Decomposition can be performed using recursive applications of DeMorgan's theorem and the associative law. As described in Section 5.6, these operations are hazard-preserving. This means that if you begin with a set of hazard-free logic equations, the equivalent network using only base functions is also hazard-free. Some technology mappers may perform simplification during the decomposition step. This process may remove redundant logic that has been added to eliminate hazards, so this simplification must be avoided.

Example 5.7.1 Consider the multilevel circuit from Section 5.6:

$$
\begin{aligned}
f &= \bar{c}(a+b+\bar{d}) + c(\bar{a}+bd) \\
f &= \bar{c}((a+b)+\bar{d}) + c(\bar{a}+bd) \text{ (associative law)} \\
f &= \bar{c}(\overline{(\bar{a}\,\bar{b})} + \bar{d}) + c(\bar{a}+b) \text{ (DeMorgan's theorem)} \\
f &= \bar{c}(\overline{(\bar{a}\,\bar{b})\,d}) + c(\overline{a\,(\overline{b\,d})}) \text{ (DeMorgan's theorem)} \\
f &= (\overline{\bar{c}(\overline{(\bar{a}\,\bar{b})\,d})})(\overline{c(\overline{a\,(\overline{b\,d})})}) \text{ (DeMorgan's theorem)}
\end{aligned}
$$

The NAND decomposition is depicted in Figure 5.41.

The partitioning step breaks up the decomposed network at points of multiple fanout into single output cones of logic which are to be individually mapped. Since the partitioning step does not change the topology of the network, it does not affect the hazard behavior of the network.

The matching and covering step examines each individual cone of logic and finds cells in the gate library to implement subnetworks within the cone. This matching step can be implemented either using *structural pattern-matching*

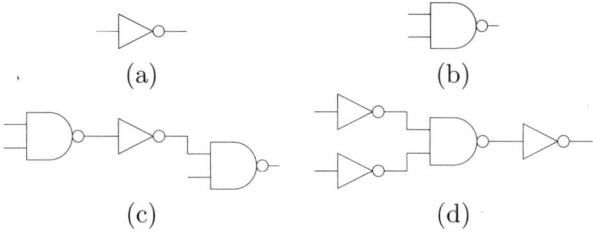

Fig. 5.42 (a) Inverter. (b) Two-input NAND gate. (c) Three-input NAND gate. (d) Two-input NOR gate.

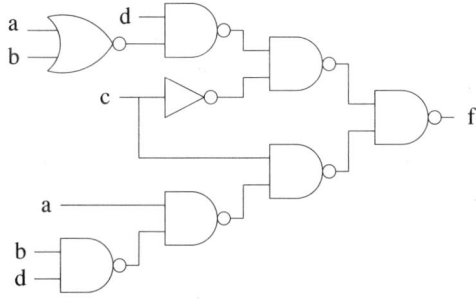

Fig. 5.43 Final mapped circuit for $f = \bar{c}(a+b+\bar{d}) + c(\bar{a}+bd)$.

techniques or *Boolean matching* techniques. In the structural techniques, each library element is also decomposed into base functions. Library elements are then compared against portions of the network to be mapped using pattern matching. Assuming that the decomposed logic and library gates are hazard-free, the resulting mapped logic is also hazard-free.

> **Example 5.7.2** Structural matching is applied to the network shown in Figure 5.41 with the library of gates shown in Figure 5.42. Assume that the inverter has cost 1, the two-input NAND gate has cost 3, the three-input NAND gate has cost 5, and the two-input NOR gate has cost 2. Consider the covering of the subtree starting at node $n4$. This subtree can either be covered with a three-input NAND gate and two inverters at a total cost of 7, or it can be covered with a two-input NAND gate and a two-input NOR gate at a cost of 5. The second choice would be made. The final mapped circuit is shown in Figure 5.43. Note that since the original logic and the library elements are hazard-free, the final mapped circuit is also hazard-free.

Boolean matching techniques attempt to find gates in the library which have equivalent Boolean functions. These techniques can perform better than structural methods, but they can also introduce hazards.

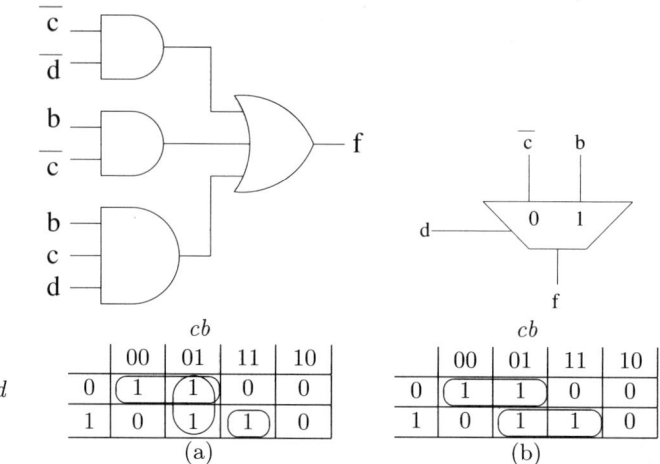

Fig. 5.44 (a) Original function mapped without hazards. (b) MUX implementation with a dynamic hazard.

Example 5.7.3 Consider part of the function from our earlier example shown in Figure 5.44(a). An equivalent function that may be found by a Boolean matching technique is the multiplexor shown in Figure 5.44(b). Karnaugh maps for each implementation are shown below the circuit diagrams. Recall that there is a dynamic $1 \to 0$ transition $[\bar{a}b\bar{c}\bar{d}, \bar{a}\bar{b}\bar{c}d]$. In the first implementation, there are no hazards since all products that intersect this transition contain the start cube. On the other hand, the multiplexor implementation includes a product term that illegally intersects the privileged cube. The result is the multiplexor has a dynamic $1 \to 0$ hazard.

To allow the use of Boolean matching and library gates which are not hazard-free in the structural methods, a different approach is needed. First, we must characterize the hazards found in the library gates. Next, during the technology mapping step, we check that the library gate being chosen has a subset of the hazards in the logic being replaced. If this is the case, this logic gate can be used safely.

Example 5.7.4 The problem with using the multiplexor implementation in Example 5.7.3 is the existence of the dynamic $1 \to 0$ transition $[\bar{a}b\bar{c}\bar{d}, \bar{a}\bar{b}\bar{c}d]$. If the original implementation had been $f = \bar{c}\bar{d} + bd$, this implementation would have a hazard for this dynamic transition also. Since the original implementation is derived to be hazard-free for all transitions of interest, this transition must not occur. Therefore, the multiplexor could be used in this situation.

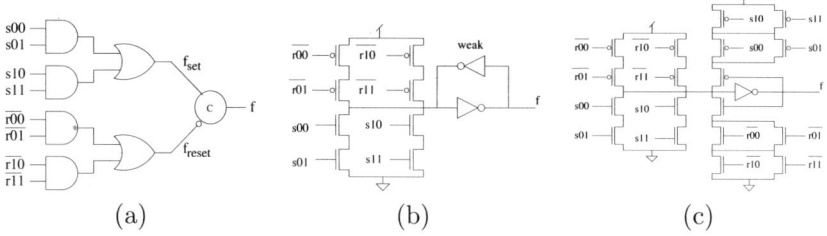

Fig. 5.45 (a) Generalized C-element logic structure with (b) weak-feedback and (c) fully static CMOS implementations.

5.8 GENERALIZED C-ELEMENT IMPLEMENTATION

Another interesting technology is to use *generalized C-elements* (gC) as the basic building blocks. In this technique, the implementation of the set and reset of a signal are decoupled. The basic structure is depicted in Figure 5.45(a), in which the upper sum-of-products represents the logic for the set, f_{set}, the lower sum-of-products represents the logic for the reset, f_{reset}, and the result is merged with a C-element. This can be implemented directly in CMOS as a single compact gate with weak feedback, as shown in Figure 5.45(b), or as a fully static gate, as shown in Figure 5.45(c).

A gC implementation reduces the potential for hazards. For example, static hazards cannot manifest on the output of a gC gate. Care has to be taken, though, during subsequent technology mapping to avoid introducing prolonged short-circuit current during decomposition. Prolonged short circuits are undesirable since they increase power dissipation as well as circuit switching time and should be avoided. Interestingly, avoiding such short circuits corresponds exactly to avoiding dynamic hazards caused by decomposing an N-stack (P-stack) in which its corresponding cube intersects a $1 \rightarrow 0$ ($0 \rightarrow 1$) transition without containing the start subcube (end subcube). By avoiding short circuits (i.e., not allowing decomposition of trigger signals which during a transition both enable and disable a P and N stack), the hazard constraints can be relaxed to no longer require that a product term intersecting a dynamic transition must include the start subcube. The problems with conditionals and dynamic hazards are also not present in gC implementations.

Given the hazard requirements discussed above, the hazard-free cover requirements for the set function, f_{set}, in an extended burst-mode gC become:

1. Each set cube of f_{set} must not include OFF-set minterms.

2. For every dynamic $0 \rightarrow 1$ transition $[c_1, c_2]$ in f_{set}, the end cube, c_2, must be completely covered by some product term.

3. Any product term of f_{set} intersecting the cube c_2 of a dynamic $0 \rightarrow 1$ transition $[c_1, c_2]$ must also contain the end subcube c'_2.

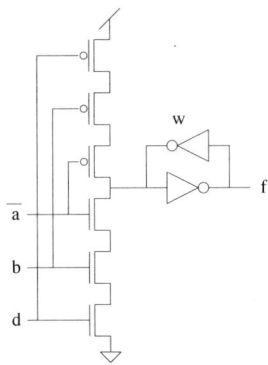

Fig. 5.46 Generalized C-element implementation for Example 5.5.20.

The second requirement describes the product terms that are required for the cover to turn on when it is supposed to. The first and third requirements describe the constraints that the required product terms must satisfy for the cover to be hazard-free. Hazard-freedom requirements for f_{reset} are analogous to f_{set}.

Example 5.8.1 Consider implementing the logic for Example 5.5.20 using a gC. For a gC, we need to consider only the two dynamic transitions. The logic derived for f_{set} and f_{reset} would be as follows:

$$f_{set} = \bar{a}bd$$
$$f_{reset} = a\bar{b}\bar{d}$$

The resulting circuit implementation is shown in Figure 5.46.

5.9 SEQUENTIAL HAZARDS

The correctness of Huffman circuits designed using the methods described in this chapter rely on the assumption that outputs and state variables stabilize before either new inputs or fed-back state variables arrive at the input to the logic. A violation of this assumption can result in a *sequential hazard*. The presence of a sequential hazard is dependent on the timing of the environment, circuit, and feedback delays.

To illustrate why delay is needed in the feedback, consider the partial flow table shown in Figure 5.47, starting in state 1 with x at 0 and changing x to 1. The result should be that the machine ends up in state 2 with the output staying at 0. Let us assume that part of the logic perceives the state change before it perceives the input change. The result is that the machine may appear to be in state 2 with input x still at 0. In this state, the next state is

	\multicolumn{2}{c}{x}
	0	1
1	①0	2,0
2	3,1	②0
3	③1	k,1

Fig. 5.47 Example illustrating the need for feedback delay.

3, so the machine may become excited to go to state 3. Let us assume that after reaching state 3 the logic now detects that x is 1. The result is that the machine now gets excited to go to state k. In other words, if the state change is fed back too quickly, it is possible that the final state is state k when it should be state 2. Another problem is that even if the input is perceived before the machine ends up in state 3, while it thinks it is in state 2 with input 0, it may start to set the output to 1. This means that the output may glitch. Regardless of the state assignment and the values of the outputs, there is no correct circuit realization for this flow table without delay in the feedback. A flow table with a configuration like this is said to contain an *essential hazard*. In general, a flow table has an essential hazard if after three changes of some input variable x, the resulting state is different than the one reached after a single change. In order to eliminate essential hazards, there is a *feedback delay requirement* which can be set conservatively to

$$D_f \geq d_{\max} - d_{\min}$$

where D_f is the feedback delay, $d_{\max}$ is the maximum delay in the combinational logic, and $d_{\min}$ is the minimum delay through the combinational logic.

Sequential hazards can also result if the environment reacts too quickly. Recall that Huffman circuits in this chapter are designed using the fundamental-mode environmental constraint, which says that inputs are not allowed to change until the circuit stabilizes. To satisfy this constraint, a conservative separation time needed between inputs can be expressed as follows:

$$d_i \geq 2d_{\max} + D_f$$

where d_i is the separation time needed between input bursts. This separation needs a $2d_{\max}$ term since the circuit must respond to the input change followed by the subsequent state change.

Finally, XBM machines require a *setup time* and *hold time* for conditional signals. In other words, conditional signals must stabilize a setup time before the compulsory signal transition which samples them, and it must remain stable a hold time after the output and state changes complete. Outside this window of time, the conditional signals are free to change arbitrarily.

5.10 SOURCES

The Huffman school of thought on the design of asynchronous circuits originated with his seminal paper [170] (later republished in [172]). This paper introduced flow tables and describes a complete methodology that includes all the topics addressed in this chapter: state minimization, state assignment, and hazard-free logic synthesis.

There are a substantial number of possible modes of operation for Huffman circuits [382]. Much of the early work is restricted to SIC operation, and the approaches to design are quite similar to those presented in Sections 5.2, 5.3, and 5.4. Quite different SIC synthesis methods have been proposed for transition (rather than level)-sensitive design styles [86, 358]. David proposed a direct synthesis method that uses a single universal cell [98] which avoids the need for state assignment and hazards. The original definition of MIC operation is that all inputs must change within some interval of time, and they are considered to have changed simultaneously [129]. Also, no further input changes are allowed for another period of time. Several methods for MIC design have been developed that either encode the inputs into a one-hot code with a spacer [129] or delay the inputs or outputs [129, 241]. Chuang and Das developed an MIC approach which uses edge-triggered flip-flops and a local clock signal to solve the hazard issues [85]. Other local clock approaches were developed by Rey and Vaucher [322], Unger [384], Hayes [160, 161], and most recently by Nowick [301]. A mixed approach that uses local clocking only when critical races exist was proposed by Yenersoy [411]. Stevens [365] extended the MIC model to allow multiple inputs to change at any time as long as the input changes are grouped together in bursts, and for this reason this mode of operation is often called *burst mode*. In the most general mode of operation called *unrestricted input change* (UIC), any input may change at any time as long as no input changes twice in a given time period. A UIC design method is described in [383] which relies on the use of inertial delay elements. Finally, when the period of time between input changes is set to the time it takes the circuit to stabilize, the circuit is said to be operating in fundamental mode [262].

Huffman circuit synthesis approaches that are quite different from those found in this chapter are found in [96, 97, 313, 395, 407]. These methods do not follow the same breakdown into the steps of state minimization, state assignment, and logic synthesis. For example, Vingron presents a method which through the construction of a coherency tree essentially performs state minimization and state assignment together.

The binate covering problem and its solution were first described by Grasselli and Luccio [155, 156]. In addition to the matrix reduction techniques found in Section 5.1, numerous others have been proposed [140, 155, 156, 179]. Recently, there have been some interesting new approaches proposed for solving binate covering problems [94, 314].

Techniques for state minimization of completely specified state machines were first described by Huffman [170]. The first work on incompletely specified state machines is due to Ginsburg [145, 146] and Paull and Unger [306]. The approach described in Section 5.2 is due to Grasselli and Luccio [155], and the running example is taken from their paper. Numerous other heuristic techniques for state minimization are described in [382]. Recent work in state minimization has been performed by Rho et al. [323] for synchronous design and Fuhrer and Nowick [132] for asynchronous design.

Again, the original work in critical race free state assignment originated with Huffman [170]. The initial work on state assignment allowed for state changes to take place through a series of steps, often changing only one state bit at a time. Such assignments have been called *shared row assignments*. Huffman determined that a universal state assignment (one that works for any arbitrary flow table) existed and requires $2S_0 - 1$ state variables, where $S_0 = \lceil \log_2 r \rceil$, where r is the number of rows in the flow table [172]. Saucier later showed that, in some cases, this bound could be broken [337]. Systematic methods for finding shared row assignments were developed by Maki and Tracey [242] and Saucier [338]. More recent work capable of handling large state machines automatically has been performed by Fisher and Wu [124] and Kang et al. [188].

To improve performance, Liu introduced the unicode single transition time (USTT) state assignment in which state transitions are accomplished in a single step [238]. Liu determined a bound on the number of state variables needed to be $2^{S_0} - 1$. Tracey developed an efficient procedure to find a minimal USTT state assignment [380]. This procedure is the one described in Section 5.3. Friedman et al. determined tighter bounds on the number of state variables needed for a universal USTT state assignment [128]. In particular, he showed that a universal USTT state assignment could be accomplished using no more than either $21S_0 - 15$ or $(S_0^3 + 5S_0)/6$ variables. A small error in one of their other tighter bounds was found by Nanya and Tohma [289]. While previous work concentrated on minimizing the number of state variables, Tan showed that this did not necessarily result in reduced logic [374]. Tan developed a state assignment technique which attempts to minimize literals instead of state variables [374]. Sawin and Maki present a variant of Tan's procedure which is capable of detecting faults [340]. Another approach which can yield simpler realizations was proposed by Mukai and Tohma [276]. Hollaar presents a design methodology based on the *one-hot* row assignment, which provides for a fairly direct circuit realization [168]. An exact method for minimizing output logic during state assignment has been developed by Fuhrer et al. [131]. An improved method has recently been proposed by Rutten and Berkelaar [333]. For large state machines, exact state assignment techniques can fail to produce a result. For this reason, Smith developed heuristic techniques to address large state machines [359]. To speed up the entire design process, Maki et al. developed methods to find the logic equations directly during state assignment [243].

While in USTT assignments a state is assigned a single state code, in multicode STT assignments a state may be assigned multiple codes. Kuhl and Reddy developed a multicode STT assignment technique [214] and Nanya and Tohma devised a universal multicode STT assignment [290]. A more recent multicode approach has been developed by Kantabutra and Andreou [189].

The recursive prime generation procedure described in Section 5.4.2 is derived from Blake's early work on Boolean reasoning [42]. An excellent description of this work is given in [49]. Setting up the prime implicant selection procedure as a covering problem as described in Section 5.4.3 was first done by McCluskey [261]. The original basic theory of combinational hazards described in Section 5.4.4 is due to Huffman [171]. The conditions to ensure hazard freedom under SIC are from McCluskey [263].

Static function and logic hazards under MIC operation were first considered by Eichelberger [122]. He demonstrated that all static logic hazards could be eliminated by using a cover that includes all the prime implicants. He also described a method using a ternary logic to detect logic hazards. Brzozowski et al. developed similar approaches to detecting races and hazards in asynchronous logic [59, 60, 55, 56]. Seger developed a ternary simulation method based on the almost-equal-delay model [342]. Methods for eliminating dynamic function and logic hazards were developed by Unger [382], Bredeson and Hulina [48], Bredeson [47], Beister [30], and Frackowiak [126]. Nowick [301] developed the restrictions on state minimization for BM machines which guarantee a hazard-free logic implementation, and Yun and Dill [420, 421] generalized these restrictions to handle XBM machines, as described in Section 5.5.6. Nowick and Dill developed the first complete algorithm and tool for two-level hazard-free logic minimization [298, 301]. The method described in Section 5.5 follows that used by Nowick and Dill [298, 301]. Recently, Theobald and Nowick developed implicit and heuristic methods to solve the two-level minimization problem [378]. The extensions needed to support extended burst-mode specifications were developed by Yun and Dill [418, 420, 421, 424]. Some recent efficient hazard-free logic minimizers have been developed by Rutten et al. [334, 332] and Jacobson et al. [180, 281].

The concept of hazard-preserving transformations for multilevel logic synthesis described in Section 5.6 is taken from Unger [382]. Kung developed further the idea of hazard-preserving transformations for logic optimization [215]. Lin and Devadas presented a hazard-free multilevel synthesis approach in which the structure of the circuit is based on its representation as a *binary decision diagram* (BDD) [234]. The technology mapping procedure described in Section 5.7 was developed by Siegel [352, 353]. Recently, Chou et al. developed technology mapping techniques that optimized for average-case performance [81]. Modifications of the design procedure to support generalized C-element implementations are due to Yun and Dill [420, 421]. A technology mapping procedure for gC circuits which optimizes for average-case performance appears in [181]. A combinational complex-gate approach was proposed by Kudva et al. [213].

Sequential hazards and the need for feedback delay were presented originally by Unger [382]. Langdon extended the definition of essential hazards to MIC operation [219]. Armstrong et al. showed that if all delay is concentrated at the gate outputs, it is possible to design circuits without feedback delay [11]. Langdon presented a similar result [220]. Brzozowski and Singh showed that in some cases the circuit could be redesigned without using any feedback at all [58]. Magó presented several alternative locations in which to put delay to eliminate sequential hazards and evaluated the amount of delay needed and its affect on the state assignment [241]. The setup and hold-time restriction on conditional signals in XBM machines is due to Yun and Dill [420]. Chakraborty et al. developed a timing analysis technique to check the timing assumptions needed to eliminate sequential hazards [71, 72].

Two excellent references on synthesis algorithms for synchronous design are the books by De Micheli [271] and Hachtel and Somenzi [158]. They were used as a reference for the material in this chapter on binate covering, state minimization, and logic minimization

Problems

5.1 Binate Covering
Solve the following constraint matrix using the bcp algorithm.

$$\mathbf{A} = \begin{bmatrix}
c_1 & c_2 & c_3 & c_4 & c_5 & c_6 & c_7 & c_8 & c_9 & c_{10} & c_{11} \\
- & - & - & - & - & - & - & - & - & 1 & - \\
1 & - & - & 1 & - & - & - & - & - & - & - \\
1 & 1 & - & 1 & 1 & 1 & 1 & - & - & - & - \\
1 & - & 1 & - & 1 & - & - & - & 1 & - & - \\
- & - & 1 & - & - & 1 & - & - & - & - & - \\
- & 1 & - & - & - & 1 & - & 1 & - & - & 1 \\
- & 1 & - & - & - & - & 1 & 1 & - & 1 & - \\
- & - & 1 & - & - & - & - & - & 1 & - & - \\
0 & - & - & - & - & - & - & 1 & - & - & - \\
0 & - & - & - & - & - & - & - & - & 1 & - \\
0 & - & 1 & - & - & - & - & - & 1 & - & - \\
1 & 0 & - & - & 1 & - & - & - & - & - & - \\
- & 0 & 1 & - & - & - & - & - & - & - & 1 \\
- & - & 0 & - & - & - & - & - & - & 1 & - \\
- & - & 0 & - & - & 1 & - & - & - & - & - \\
- & - & - & - & 0 & - & - & - & - & 1 & - \\
- & - & 1 & - & 0 & - & - & 1 & - & - & - \\
1 & - & - & - & 1 & 0 & - & - & - & - & - \\
1 & - & - & - & 1 & - & 0 & - & - & - & - \\
- & 1 & - & - & - & - & 0 & 1 & - & - & - \\
- & - & 1 & - & - & - & - & 0 & - & - & 1 \\
\end{bmatrix} \begin{matrix} 1 \\ 2 \\ 3 \\ 4 \\ 5 \\ 6 \\ 7 \\ 8 \\ 9 \\ 10 \\ 11 \\ 12 \\ 13 \\ 14 \\ 15 \\ 16 \\ 17 \\ 18 \\ 19 \\ 20 \\ 21 \end{matrix}$$

	00	01	11	10
1	–	2,0	–	5,1
2	–	1,0	3,–	–
3	4,1	3,–	1,1	5,–
4	–	–	2,–	1,–
5	6,–	3,–	2,1	–,0
6	5,–	6,1	1,–	2,–

Fig. 5.48 Flow table for Problem 5.3.

5.2 Binate Covering

Implement the BCP algorithm in your favorite programming language. It should read in a constraint matrix and output a list of columns used in the best solution found.

5.3 State Minimization

For the flow table shown in Figure 5.48:

5.3.1. Find compatible pairs using a pair chart.
5.3.2. Compute the maximal compatibles.
5.3.3. Set up and solve BCP using only the maximal compatibles.
5.3.4. Compute the prime compatibles.
5.3.5. Set up and solve BCP using the prime compatibles.
5.3.6. Form the reduced table.
5.3.7. Compare the results from 5.3.3 and 5.3.5.

5.4 State Minimization

For the flow table shown in Figure 5.49:

5.4.1. Find compatible pairs using a pair chart.
5.4.2. Compute the maximal compatibles.
5.4.3. Set up and solve BCP using only the maximal compatibles.
5.4.4. Compute the prime compatibles.
5.4.5. Set up and solve BCP using the prime compatibles.
5.4.6. Form the reduced table.
5.4.7. Compare the results from 5.4.3 and 5.4.5.

5.5 State Minimization

For the flow table shown in Figure 5.50:

5.5.1. Find compatible pairs using a pair chart.
5.5.2. Compute the maximal compatibles.
5.5.3. Compute the prime compatibles.
5.5.4. Set up and solve BCP using the prime compatibles.
5.5.5. Form the reduced table.

	00	01	11	10
1	3,–	2,–	1,1	1,–
2	6,–	–	4,1	1,–
3	–	3,1	–	–,0
4	2,1	–	–	–,0
5	–	3,0	1,–	–
6	4,0	–	5,–	–

Fig. 5.49 Flow table for Problem 5.4.

	00	01	11	10
1	3,0	1,–	–	–
2	6,–	2,0	1,–	–
3	–,1	–	4,0	–
4	1,0	–	–	5,1
5	–	5,–	2,1	1,1
6	–	2,1	6,–	4,1

Fig. 5.50 Flow table for Problem 5.5.

	00	01	11	10
1	1,0	2,0	3,0	1,0
2	1,0	2,0	3,0	2,1
3	3,1	5,0	3,0	4,0
4	1,0	–,–	4,0	4,0
5	3,1	5,0	5,1	4,0

Fig. 5.51 Flow table for Problem 5.6.

5.6 State Assignment
For the flow table shown in Figure 5.51:
  5.6.1. Find a state assignment without using the outputs.
  5.6.2. Find a state assignment using the outputs as state variables.

5.7 State Assignment
For the flow table shown in Figure 5.52:
  5.7.1. Find a state assignment without using the outputs.
  5.7.2. Find a state assignment using the outputs as state variables.

5.8 State Assignment
For the flow table shown in Figure 5.53:
  5.8.1. Find a state assignment without using the outputs.
  5.8.2. Find a state assignment using the outputs as state variables.

HUFFMAN CIRCUITS

	00	01	11	10
1	1,0	2,0	3,0	1,0
2	4,0	2,0	2,0	2,1
3	4,1	2,0	3,0	3,0
4	4,0	5,0	3,0	4,0
5	1,1	5,0	5,1	5,0

Fig. 5.52 Flow table for Problem 5.7.

	00	01	11	10
1	1,0	3,0	7,0	1,0
2	5,0	2,0	6,0	2,1
3	1,1	3,0	7,0	3,0
4	4,0	7,0	6,0	4,0
5	5,1	7,0	6,1	1,0
6	1,0	2,0	6,0	3,0
7	5,1	7,1	7,1	2,1

Fig. 5.53 Flow table for Problem 5.8.

5.9 Two-Level Logic Minimization

Do the following for the Karnaugh map shown in Figure 5.54:

5.9.1. Find all prime implicants using the recursive procedure.

5.9.2. Set up and solve a covering problem to pick the minimal number of prime implicants, ignoring hazards.

5.9.3. Set up and solve a covering problem to pick the minimal number of prime implicants for a hazard-free cover assuming SIC.

5.9.4. Assume that the only transitions possible are

$$t_1 = [\bar{a}b\bar{c}d, \bar{a}\bar{b}\bar{c}\bar{d}]$$
$$t_2 = [ab\bar{c}\bar{d}, a\bar{b}\bar{c}d]$$
$$t_3 = [\bar{a}\bar{b}c\bar{d}, \bar{a}bcd]$$
$$t_4 = [abc\bar{d}, a\bar{b}cd]$$

Identify the type of each transition and its transition cube.

5.9.5. Determine all required cubes for the transitions above.

5.9.6. Determine all privileged cubes for the transitions above.

5.9.7. Find the dhf-prime implicants.

5.9.8. Set up and solve a covering problem to pick the minimal number of prime implicants for a hazard-free cover assuming that the only transitions are those given above.

		ab			
		00	01	11	10
	00	1	1	1	1
cd	01	1	1	1	0
	11	0	1	1	0
	10	0	0	1	1

Fig. 5.54 Karnaugh map for Problem 5.9.

		ab			
		00	01	11	10
	00	0	0	0	1
cd	01	0	0	0	1
	11	1	0	0	1
	10	1	1	0	1

Fig. 5.55 Karnaugh map for Problem 5.10.

5.10 Two-Level Logic Minimization

Do the following for the Karnaugh map shown in Figure 5.55:

5.10.1. Find all prime implicants using the recursive procedure.

5.10.2. Set up and solve a covering problem to pick the minimal number of prime implicants, ignoring hazards.

5.10.3. Set up and solve a covering problem to pick the minimal number of prime implicants for a hazard-free cover assuming SIC.

5.10.4. Assume that the only transitions possible are:

$$t_1 = [\bar{a}\bar{b}\bar{c}\text{-}, \bar{a}\bar{b}c\text{-}]$$
$$t_2 = [\bar{a}bc\bar{d}, \bar{a}b\bar{c}\bar{d}]$$
$$t_3 = [ab\text{--}, a\bar{b}\text{--}]$$

Identify the type of each transition and its transition cube. Assume that d is a level signal and that c is a falling directed don't care in t_3.

5.10.5. Determine all required cubes for the transitions above.

5.10.6. Determine all privileged cubes for the transitions above.

5.10.7. Find the dhf-prime implicants.

5.10.8. Set up and solve a covering problem to pick the minimal number of prime implicants for a hazard-free cover assuming that the only transitions are those given above.

5.11 Burst-Mode Synthesis

Do the following for the BM machine shown in Figure 5.56:

5.11.1. Translate the BM machine into a flow table.

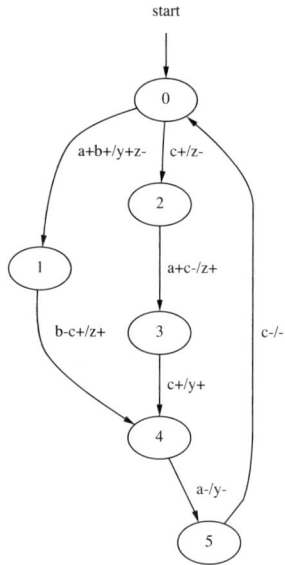

Fig. 5.56 BM machine for Problem 5.11. Note that $abc = 000$ and $yz = 01$ initially.

5.11.2. Perform state minimization on the flow table. Be sure to consider BM operation.

5.11.3. Perform state assignment on the reduced flow table.

5.11.4. Perform two-level logic minimization to find hazard-free logic to implement the output signals.

5.12 Extended Burst-Mode Synthesis

Do the following for the XBM machine shown in Figure 5.57:

5.12.1. Translate the XBM machine into a flow table.

5.12.2. Perform state minimization on the flow table. Be sure to consider XBM operation.

5.12.3. Perform state assignment on the reduced flow table.

5.12.4. Perform two-level logic minimization to find hazard-free logic to implement the output signals.

5.13 Multilevel Logic Synthesis

Apply hazard-preserving transformations to find a minimum literal factored form for your solution of Problem 5.9.

5.14 Multilevel Logic Synthesis

Apply hazard-preserving transformations to find a minimum literal factored form for your solution of Problem 5.10.

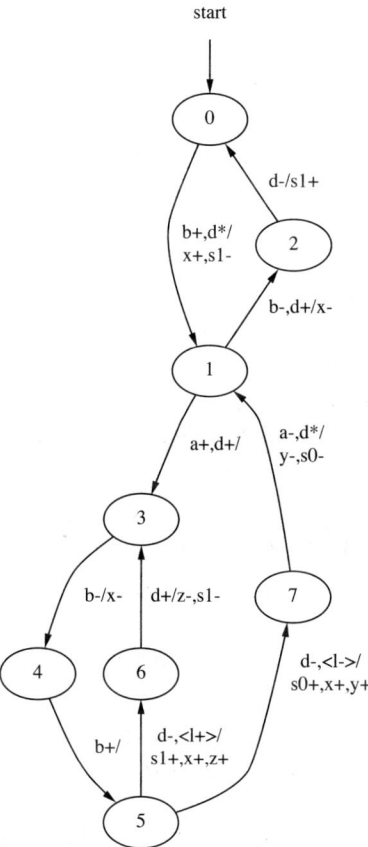

Fig. 5.57 XBM machine for Problems 5.12 and 5.21. Note that $abdl = 000-$ and $s0s1xyz = 01000$ initially.

	00	01	11	10
1	3,1	1,0	2,0	1,1
2	2,0	2,0	2,0	4,0
3	3,1	1,0	3,0	4,0
4	4,0	2,0	3,0	4,0

Fig. 5.58 Flow table for Problem 5.22.

5.15 Multilevel Logic Synthesis
Apply hazard-preserving transformations to find a minimum literal factored form for your solution of Problem 5.11.

5.16 Multilevel Logic Synthesis
Apply hazard-preserving transformations to find a minimum literal factored form for your solution of Problem 5.12.

5.17 Technology Mapping
Map your multilevel logic from Problem 5.13 using the library from Figure 5.41.

5.18 Technology Mapping
Map your multilevel logic from Problem 5.14 using the library from Figure 5.41.

5.19 Technology Mapping
Map your multilevel logic from Problem 5.15 using the library from Figure 5.41.

5.20 Technology Mapping
Map your multilevel logic from Problem 5.16 using the library from Figure 5.41.

5.21 Generalized C-Element Synthesis
Do the following for the XBM machine shown in Figure 5.57, targeting a gC implementation.

5.21.1. Translate the XBM machine into a flow table.

5.21.2. Perform state minimization on the flow table. Be sure to consider XBM operation.

5.21.3. Perform state assignment on the reduced flow table.

5.21.4. Perform two-level logic minimization to find hazard-free logic to implement the output signals using generalized C-elements.

5.22 Sequential Hazards
Find all essential hazards in the flow table shown in Figure 5.58.

6
Muller Circuits

The infinite is the finite of every instant.

—Zen Saying

When I can't handle events, I let them handle themselves.

—Henry Ford

Life is pleasant. Death is peaceful. It's the transition that's troublesome.

—Isaac Asimov

We are ready for any unforeseen event that may or may not occur.

—Vice President Dan Quayle, 9/22/90

In this chapter we introduce the Muller school of thought to the synthesis of asynchronous circuits. *Muller circuits* are designed under the *unbounded gate delay model*. Under this model, circuits are guaranteed to work regardless of gate delays, assuming that wire delays are negligible. Muller circuit design requires explicit knowledge of the behaviors allowed by the environment. It does not, however, put any restriction on the speed of the environment.

The design of Muller circuits requires a somewhat different approach as compared with traditional sequential state machine design. Most synthesis methods for Muller circuits translate the higher-level specification into a state graph. Next, the state graph is examined to determine if a circuit can be generated using only the specified input and output signals. If two states are found that have the same values of inputs and outputs but lead through an output transition to different next states, no circuit can be produced directly. In this case, either the protocol must be changed or new internal state signals must be added to the design. The method of determining the needed state variables is quite different from that used for Huffman circuits. Next, logic is derived using modified versions of the logic minimization procedures described earlier. The modifications needed are based upon the technology that is being used for implementation. Finally, the design must be mapped to gates in a given gate library. This last step requires a substantially modified technology mapping procedure as compared with traditional state machine synthesis methods.

We first give a formal definition of speed independence. Then we describe the "state assignment" method for Muller circuits. Finally, we describe logic minimization and technology mapping, respectively.

6.1 FORMAL DEFINITION OF SPEED INDEPENDENCE

In order to design a speed-independent circuit, it is necessary to have complete information about the behavior of both the circuit being designed and the environment. Therefore, we restrict our attention to *complete circuits*. A complete circuit C is defined by a finite set of *states*, S. At any time, C is said to be in one of these states.

The behavior of a complete circuit is defined by the set of *allowed sequences* of states. Each allowed sequence can be either finite or infinite, and the set of allowed sequences can also be finite or infinite. For example, the sequence $(s_1, s_2, s_3, \ldots)$ says that state s_1 is followed by state s_2, but it does not state the time at which this state transition takes place. Therefore, in order to determine when a state transition takes place, it is necessary that consecutive states be different. In other words, for an allowed sequence $(s_1, s_2, \ldots)$, any pair of consecutive states $s_i \neq s_{i+1}$. Another property is that each state $s \in S$ is the initial state of at least one allowed sequence. One can also derive additional allowed sequences from known ones. For example, if $(s_1, s_2, s_3, \ldots)$ is an allowed sequence, then so is $(s_2, s_3, \ldots)$. If $(s_1, s_2, \ldots)$ and $(t_1, t_2, \ldots)$ are allowed sequences and $s_2 = t_1$, then $(s_1, t_1, t_2, \ldots)$ is also an allowed sequence.

Example 6.1.1 Consider a complete circuit composed of four states, $S = \{a, b, c, d\}$, which has the following two allowed sequences:

1. $a, b, a, b, \ldots$
2. a, c, d

The sequences above imply that the following sequences are also allowed:

1. $b, a, b, a, \ldots$
2. c, d
3. d

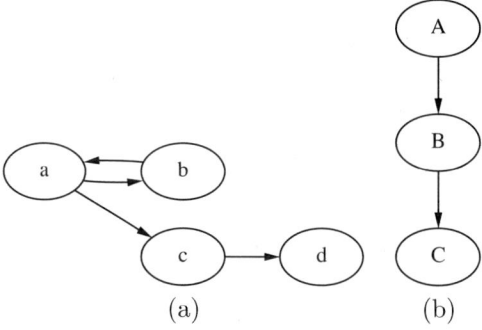

Fig. 6.1 (a) Simple state diagram. (b) Partial order of its equivalence classes.

4. a, b, a, c, d
5. a, b, a, b, a, c, d
6. b, a, c, d
7. etc.

A *state diagram* for this example is shown in Figure 6.1(a).

Two states $s_i, s_j \in S$ are *$\mathcal{R}$-related* (denoted $s_i \mathcal{R} s_j$) when:

1. $s_i = s_j$ or

2. s_i, s_j appear as a consecutive pair of states in some allowed sequence.

A sequence $(s_1, s_2, \ldots, s_m)$ is an *$\mathcal{R}$-sequence* if $s_i \mathcal{R} s_{i+1}$ for each $1 \le i \le m-1$.
A state s_i is *followed* by a state s_j (denoted $s_i \mathcal{F} s_j$) if there exists an $\mathcal{R}$-sequence $(s_i, \ldots, s_j)$. The $\mathcal{F}$-relation is reflexive and transitive, but not necessarily symmetric. If two states s_i and s_j are symmetric under the $\mathcal{F}$-relation (i.e., $s_i \mathcal{F} s_j$ and $s_j \mathcal{F} s_i$), they are said to be *equivalent* (denoted $s_i \mathcal{E} s_j$).

The equivalence relation, $\mathcal{E}$, partitions the finite set of states S of any circuit into equivalence classes of states. The $\mathcal{F}$-relation can be extended to these equivalence classes. If A and B are two equivalence classes, then $A \mathcal{F} B$ if there exists states $a \in A$ and $b \in B$ such that $a \mathcal{F} b$. Furthermore, if a is in the equivalence class A and b is in B and $A \mathcal{F} B$, then $a \mathcal{F} b$.

For any allowed sequence, there is a definite last class which is called the *terminal class*. A circuit C is *speed independent with respect to a state s* if all allowed sequences starting with s have the same terminal class.

Example 6.1.2 The circuit from Figure 6.1(a) is partitioned into three equivalence classes: $A = \{a, b\}$, $B = \{c\}$, and $C = \{d\}$. Applying the extended $\mathcal{F}$-relation to the equivalence classes, we get the following $A \mathcal{F} A$, $A \mathcal{F} B$, $A \mathcal{F} C$, $B \mathcal{F} B$, $B \mathcal{F} C$, and $C \mathcal{F} C$. This relation is depicted in Figure 6.1(b) without self-loops and transitive arcs. This circuit is not speed independent with respect to state a, since there exist allowed sequences starting in a that end in terminal class A and others that end in terminal class C. However, this circuit is speed independent with respect to states c and d since all allowed sequences starting in these states end in terminal class C.

Muller circuits are typically modeled using a state graph (SG) derived from a higher-level graphical model such as a STG or TEL structure as described in Chapter 4. We can reformulate the notion of allowed sequences on a state graph. An allowed sequence of states $(s_1, s_2, \ldots)$ is any sequence of states satisfying the following three conditions:

1. No two consecutive states s_i and s_{i+1} are equal.

2. For any state s_{j+1} and signal u_i, one of the following is true:
$$s_{j+1}(i) = s_j(i)$$
$$s_{j+1}(i) = s'_j(i)$$

3. If there exists a signal u_i and a state s_j such that $s_j(i) = s_r(i)$ and $s'_j(i) = s'_r(i)$ for all s_r in the sequence following s_j, then
$$s_j(i) = s'_j(i)$$

The second condition states that in an allowed sequence either a signal remains unchanged or changes to its implied value. The third condition states that if a signal is continuously excited to change, it eventually does change to its implied value.

Example 6.1.3 A simple speed-independent circuit and corresponding state graph is shown in Figure 6.2. Note that all states are contained in a single equivalence class, making it speed-independent.

6.1.1 Subclasses of Speed-Independent Circuits

There are several useful subclasses of speed-independent circuits. First, a circuit is *totally sequential with respect to a state s* if there is only one allowed sequence starting with s. Clearly, if there is only one allowed sequence starting with s, there can be only one terminal class which proves the following theorem.

Theorem 6.1 (Muller, 1959) *A circuit totally sequential with respect to s is also speed independent with respect to s.*

Example 6.1.4 A simple totally sequential circuit is shown in Figure 6.3(a), and its state graph is shown in Figure 6.3(b). In its initial state, $\langle 0R \rangle$, both x and y are 0, but y's implied value is 1. After y rises, the system moves to state $\langle R1 \rangle$, where x now has an implied value of 1. After x rises, the state is $\langle 1F \rangle$.

FORMAL DEFINITION OF SPEED INDEPENDENCE   211

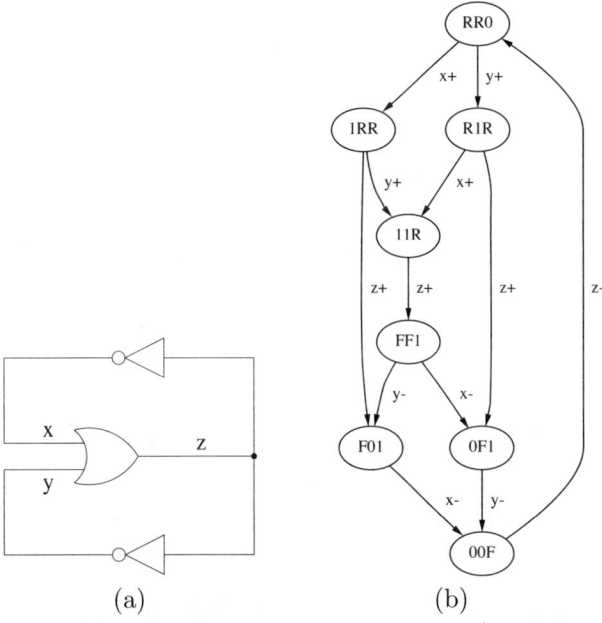

Fig. 6.2 (a) Speed-independent circuit. (b) Its state graph with state vector $\langle x, y, z \rangle$.

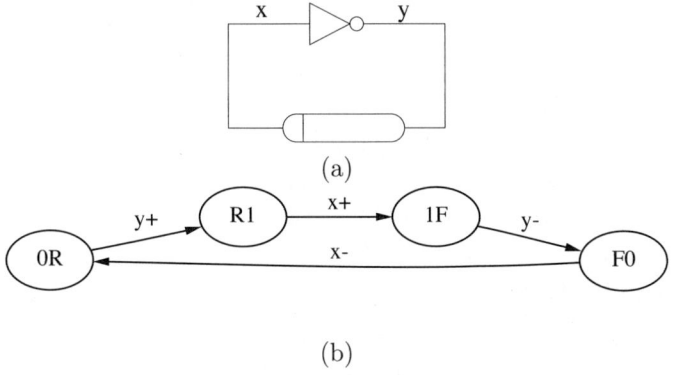

Fig. 6.3 (a) Totally sequential circuit. (b) Its state graph with state vector $\langle x, y \rangle$.

A circuit is *semi-modular* in a state s_i if in all states s_j reached after one signal has transitioned, any other signals excited in s_i are still excited in s_j. More formally:

$$\forall t_1, t_2 \in T \ . \ (s_i, t_1, s_j) \in \delta \wedge (s_i, t_2, s_k) \in \delta$$
$$\Rightarrow \exists s_l \in S \ . \ (s_j, t_2, s_l) \in \delta \wedge (s_k, t_1, s_l) \in \delta$$

A circuit is *semi-modular with respect to a state* s if all states reachable from s are semi-modular. A totally sequential circuit is semi-modular, but the converse is not necessarily true. A circuit that is semi-modular with respect to a state s is also speed independent with respect to s, but again the converse is not necessarily true. These results are summarized in the following theorems.

Theorem 6.2 (Muller, 1959) *A circuit totally sequential with respect to s is also semi-modular with respect to s.*

Theorem 6.3 (Muller, 1959) *A circuit semi-modular with respect to s is also speed independent with respect to s.*

> **Example 6.1.5** The circuit shown in Figure 6.2 is speed-independent, but it is not semi-modular. For example, in state $\langle 1RR \rangle$ signals y and z are both excited to rise, but after z rises, it goes to state $\langle F01 \rangle$, in which y is no longer excited. Another simple circuit and its corresponding state graph are shown in Figure 6.4. This circuit is semi-modular with respect to each state in the state graph.

Input transitions typically are allowed to be disabled by other input transitions, but output transitions are typically not allowed to be disabled. Therefore, another useful class of circuits are those which are *output semi-modular*. A SG is output semi-modular in a state s_i if only input signal transitions can disable other input signal transitions. More formally:

$$\forall t_1 \in T_O \ . \ \forall t_2 \in T \ . \ (s_i, t_1, s_j) \in \delta \wedge (s_i, t_2, s_k) \in \delta$$
$$\Rightarrow \exists s_l \in S \ . \ (s_j, t_2, s_l) \in \delta \wedge (s_k, t_1, s_l) \in \delta$$

where T_O is the set of output transitions (i.e., $T_O = \{u+, u- \mid u \in O\}$).

> **Example 6.1.6** If and only if all signals in the circuit shown in Figure 6.2 are inputs, then it is output semi-modular.

6.1.2 Some Useful Definitions

It is often useful to be able to determine in which states a signal is excited to rise or fall. The sets of *excitation states*, $ES(u+)$ and $ES(u-)$, provide this information and are defined as follows:

$$\begin{aligned} ES(u+) &= \{s \in S \mid s(u) = 0 \wedge u \in X(s)\} \\ ES(u-) &= \{s \in S \mid s(u) = 1 \wedge u \in X(s)\} \end{aligned}$$

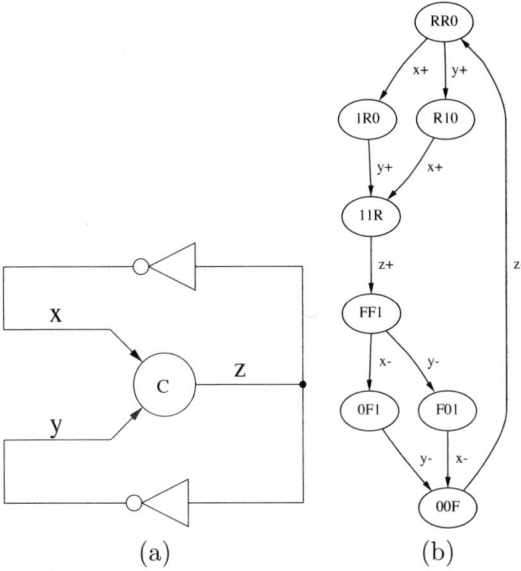

Fig. 6.4 (a) Semi-modular circuit. (b) State graph with state vector $\langle x, y, z \rangle$.

Recall that $X(s)$ is the set of signals that are excited in state s.

For each signal u, there are two sets of stable, or *quiescent*, states. The sets $QS(u+)$ and $QS(u-)$ are defined as follows:

$$QS(u+) = \{s \in S \mid s(u) = 1 \wedge u \notin X(s)\}$$
$$QS(u-) = \{s \in S \mid s(u) = 0 \wedge u \notin X(s)\}$$

Example 6.1.7 Consider the SG shown in Figure 6.4. The signal y has the following four sets:

$$ES(y+) = \{\langle RR0 \rangle, \langle 1R0 \rangle\}$$
$$ES(y-) = \{\langle FF1 \rangle, \langle 0F1 \rangle\}$$
$$QS(y+) = \{\langle R10 \rangle, \langle 11R \rangle\}$$
$$QS(y-) = \{\langle F01 \rangle, \langle 00F \rangle\}$$

An *excitation region* for signal u is a maximally connected subset of either $ES(u+)$ or $ES(u-)$. If it is a subset of $ES(u+)$, it is a *set region*, and it is denoted $ER(u+, k)$ where k indicates that it is the kth set region. Similarly, a *reset region* can be denoted $ER(u-, k)$.

The *switching region* for a transition $u*$, $SR(u*, k)$, is the set of states directly reachable through transition $u*$:

$$SR(u*, k) = \{s_j \in S \mid \exists s_i \in ER(u*, k).(s_i, u*, s_j) \in \delta\}$$

where "*" indicates either "+" for set regions or "−" for reset regions.

Example 6.1.8 Again consider the example SG shown in Figure 6.4. The signal y has the following excitation and switching regions:

$$\begin{aligned} ER(y+,1) &= \{\langle RR0 \rangle, \langle 1R0 \rangle\} \\ ER(y-,1) &= \{\langle FF1 \rangle, \langle 0F1 \rangle\} \\ SR(y+,1) &= \{\langle R10 \rangle, \langle 11R \rangle\} \\ SR(y-,1) &= \{\langle F01 \rangle, \langle 00F \rangle\} \end{aligned}$$

Another interesting subclass of speed-independent circuits are those which have *distributive* state graphs. A state graph is distributive if each excitation region has a unique *minimal state*. A minimal state for an excitation region, $ER(u*,k)$, is a state in $ER(u*,k)$, which cannot be directly reached by any other state in $ER(u*,k)$. More formally, a SG is distributive if

$$\forall ER(u*,k) \,.\, \exists \text{exactly one } s_j \in ER(u*,k) \,.\, \neg \exists s_i \in ER(u*,k) \,.\, (s_i, t, s_j) \in \delta$$

Example 6.1.9 The circuit in Figure 6.2 is not distributive, since for $ER(z+,1)$ there exists two minimal states, $\langle 1RR \rangle$ and $\langle R1R \rangle$. The circuit in Figure 6.4 is distributive.

Each cube in the implementation is composed of *trigger signals* and *context signals*. For an excitation region, a trigger signal is a signal whose firing can cause the circuit to enter the excitation region. The set of trigger signals for an excitation region $ER(u*,k)$ is

$$TS(u*,k) = \{v \subset N \mid \exists s_i, s_j \subset S.((s_i,t,s_j) \in \delta) \land (t = v+ \lor t = v-) \\ \land (s_i \notin ER(u*,k)) \land (s_j \in ER(u*,k))\}$$

Any nontrigger signal which is stable throughout an entire excitation region may be used as a context signal. The set of context signals for an excitation region $ER(u*,k)$ is

$$CS(u*,k) = \{v_i \in N \mid v_i \notin TS(u*,k) \land \forall s_j, s_l \in ER(u*,k).s_j(i) = s_l(i)\}$$

Example 6.1.10 Once again consider the example SG shown in Figure 6.4. The excitation regions for signal y have the following trigger and context signal sets:

$$\begin{aligned} TS(y+,1) &= \{z\} \\ TS(y-,1) &= \{z\} \\ CS(y+,1) &= \{y\} \\ CS(y-,1) &= \{y\} \end{aligned}$$

Note that although the signal y is rising in $ER(y+,1)$, it is considered stable since once y does rise the circuit has left the excitation region. Recall that a state labeled with an R is actually at the fixed logical value of 0, but it is excited to change.

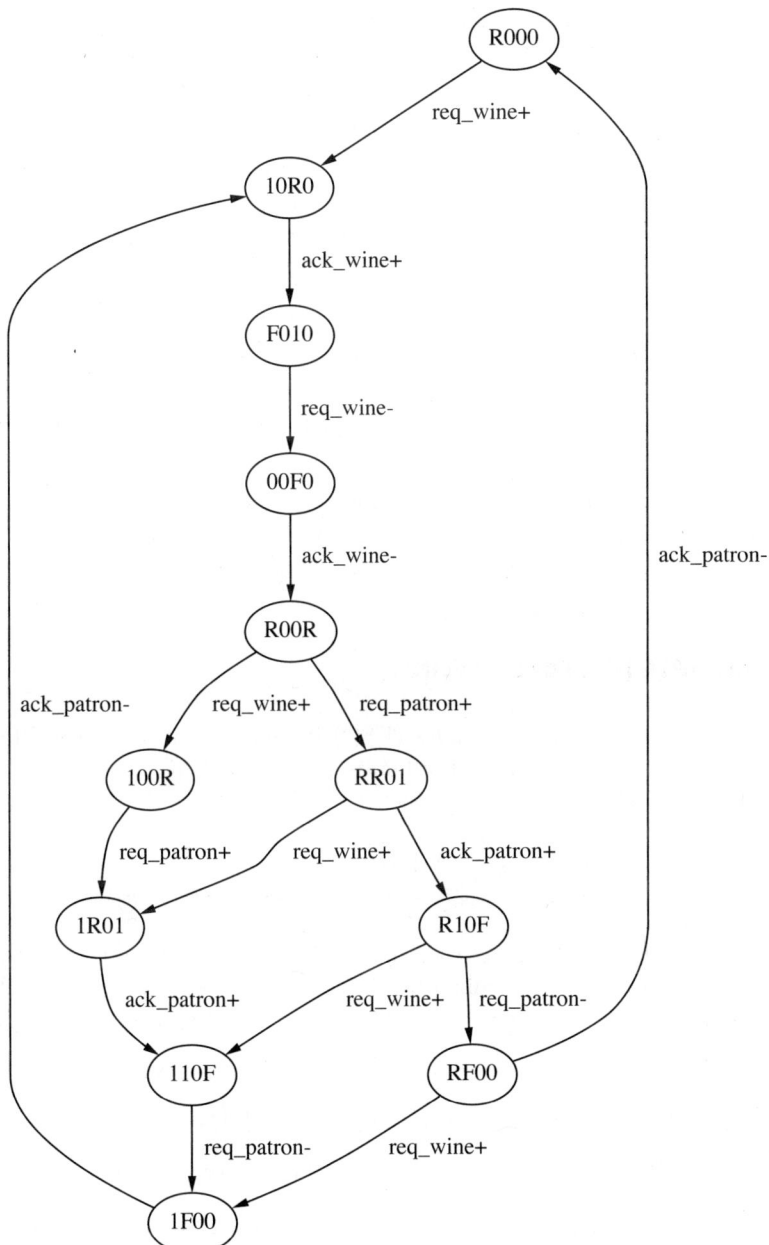

Fig. 6.5 State graph for the passive/active wine shop (statevector is ⟨*req_wine, ack_patron, ack_wine, req_patron*⟩).

Example 6.1.11 We conclude this section with a somewhat larger example. The SG for a passive/active wine shop is shown in Figure 6.5. The *ES*, *QS*, *ER*, *SR*, *TS*, and *CS* sets for the signal *req_patron* are shown below.

$$\begin{aligned}
ES(req\_patron+) &= \{\langle R00R\rangle, \langle 100R\rangle\} \\
ES(req\_patron-) &= \{\langle R10F\rangle, \langle 110F\rangle\} \\
QS(req\_patron+) &= \{\langle RR01\rangle, \langle 1R01\rangle\} \\
QS(req\_patron-) &= \{\langle RF00\rangle, \langle 1F00\rangle, \langle R000\rangle, \langle 10R0\rangle, \\
&\quad \langle F010\rangle, \langle 00F0\rangle\} \\
ER(req\_patron+,1) &= \{\langle R00R\rangle, \langle 100R\rangle\} \\
ER(req\_patron-,1) &= \{\langle R10F\rangle, \langle 110F\rangle\} \\
SR(req\_patron+,1) &= \{\langle RR01\rangle, \langle 1R01\rangle\} \\
SR(req\_patron-,1) &= \{\langle RF00\rangle, \langle 1F00\rangle\} \\
TS(req\_patron+,1) &= \{ack\_wine\} \\
TS(req\_patron-,1) &= \{ack\_patron\} \\
CS(req\_patron+,1) &= \{ack\_patron, req\_patron\} \\
CS(req\_patron-,1) &= \{ack\_wine, req\_patron\}
\end{aligned}$$

6.2 COMPLETE STATE CODING

Two states have *unique state codes* (USC) if they are labeled with different binary vectors. A SG has USC if all state pairs have USC. This is presented more formally below.

$$\begin{aligned}
USC(s_i, s_j) &\Leftrightarrow \lambda_S(s_i) \neq \lambda_S(s_j) \\
USC(S) &\Leftrightarrow \forall (s_i, s_j) \in S \times S \,.\, USC(s_i, s_j)
\end{aligned}$$

Two states have *complete state codes* (CSC) if they either have USC or do not have USC but do have the same output signals excited in each state. A SG has CSC if all state pairs have CSC. This is presented more formally below.

$$\begin{aligned}
CSC(s_i, s_j) &\Leftrightarrow USC(s_i, s_j) \lor X(s_i) \cap O = X(s_j) \cap O \\
CSC(S) &\Leftrightarrow \forall (s_i, s_j) \in S \times S \,.\, CSC(s_i, s_j)
\end{aligned}$$

A set of state pairs which violate the CSC property is defined as

$$CSCV(S) = \{(s_i, s_j) \in S \times S \mid \neg\, CSC(s_i, s_j)\}$$

Example 6.2.1 Consider the state graph shown in Figure 6.5. The states $\langle R000\rangle$ and $\langle R00R\rangle$ do not have USC since they share the same underlying state code $\langle 0000\rangle$. Similarly, the states $\langle 10R0\rangle$ and $\langle 100R\rangle$ do not have USC. Recall that the output signals for this circuit are *ack_wine* and *req_patron*. Therefore, states $\langle R000\rangle$ and $\langle R00R\rangle$ also do not have CSC, since *req_patron* is stable low in the first state and excited to rise in the second. Similarly, states $\langle 10R0 \rangle$ and $\langle 100R \rangle$ do not have CSC due to differences in output enablings. If *req_patron* had been an input, the first pair would have satisfied the CSC property. The second still would not, since *ack_wine* is only excited in the first state of the pair.

When a SG does not have CSC, the implied value for some output signals cannot be determined by simply considering the values of the signal wires. This ambiguity leads to a state of confusion for the circuit. Note that if a SG does not have USC but has CSC, there is no problem since a circuit only needs to be synthesized for the output signals. To synthesize a circuit from a SG that does not have CSC, the specification must be modified. One possibility is to reshuffle the protocol as described in Chapter 3. In this section we describe a method for inserting state variables to solve the CSC problem.

6.2.1 Transition Points and Insertion Points

The insertion of a state signal into a circuit involves the addition of a rising and falling transition on the new signal. *Transition points* are a useful way of specifying where these transitions occur. A transition point is an ordered pair of sets of transitions, $TP = (t_s, t_e)$, where t_s is a set of *start transitions* and t_e is a set of *end transitions*. The transition point represents the location in the protocol in which a transition on a new state signal is to be inserted. In a graphical model such as a STG, each transition point represents a transition with incoming arcs from each of the transitions in t_s and with outgoing arcs to the transitions in t_e. An *insertion point* consists of an ordered pair of transition points, $IP = (TP_R, TP_F)$, where TP_R is for the rising transition and TP_F is for the falling transition.

It is necessary to determine in which states a transition of a new state signal can occur when inserted into the circuit using a given TP. The signal transition becomes excited after all transitions in t_s have occurred. Assume that $t \in t_s$; then we know that t has occurred when we enter the switching region for t. We know that all $t \in t_s$ have occurred when we are in the switching region for each t. Therefore, the transition on the new state signal becomes excited when the circuit enters the intersection of all the switching regions for each transition in t_s [i.e., $\cap_{t \in t_s} SR(t)$]. The transition on this new signal is guaranteed to have completed before any transition in t_e can occur. Therefore, once this transition becomes excited, it may remain excited in any subsequent states until a state is reached through a transition in t_e. We can now recursively define the set of states in which a new transition inserted into TP is excited.

$$S(\text{TP}) = \{s_j \in S \mid s_j \in \cap_{t \in t_s} SR(t) \vee (\exists (s_i, t, s_j) \in \delta . s_i \in S(\text{TP}) \wedge t \notin t_e)\}$$

Example 6.2.2 Consider the state graph shown in Figure 6.5 with $TP = (\{req\_patron+\}, \{req\_patron-\})$. $SR(req\_patron+)$ is $\langle RR01 \rangle$ and $\langle 1R01 \rangle$, so $S(TP)$ is seeded with these states. To find the rest of the states in $S(TP)$, we follow state transitions from these states until $req\_patron-$ occurs. For example, $(\langle RR01 \rangle, ack\_patron+, \langle R10F \rangle) \in \delta$, so $\langle R10F \rangle$ is in $S(TP)$. The transition $(\langle 1R01 \rangle, ack\_patron+, \langle 110F \rangle) \in \delta$, so $\langle 110F \rangle$ is also in $S(TP)$. However, the transitions leaving these two states involve $req\_patron-$, so there are no more states in $S(TP)$. In summary, $S(\{req\_patron+\}, \{req\_patron-\})$ is

$$\{\langle RR01 \rangle, \langle 1R01 \rangle, \langle R10F \rangle, \langle 110F \rangle\}$$

Theoretically, the set of all possible insertion points includes all combinations of transitions in t_s and t_e for the rising transition and all combinations of transitions in t_s and t_e for the falling transition. Thus, the upper bound on the number of possible insertion points is $2^{|T|^4}$. Fortunately, many of these insertion points can be quickly eliminated because they either never lead to a satisfactory solution of the CSC problem or the same solution is found using a different insertion point.

A transition point must satisfy the following three restrictions:

1. The start and end sets should be disjoint (i.e., $t_s \cap t_e = \emptyset$).

2. The end set should not include input transitions (i.e., $\forall t \in t_e \,.\, t \notin T_I$).

3. The start and end sets should include only concurrent transitions (i.e., $\forall t_1, t_2 \in t_s \,.\, t_1 \parallel t_2$ and $\forall t_1, t_2 \in t_e \,.\, t_1 \parallel t_2$).

The first requirement, that t_s and t_e be disjoint, simply eliminates unnecessary loops. The second requirement is necessary since the interface behavior is assumed to be fixed. In other words, we are not allowed to change the way the environment reacts to changes in output signals. In particular, if we allowed input transitions in t_e, this would force the environment to delay an input transition until a state signal changes. The third requirement is that all transitions in t_s and t_e be mutually concurrent. A transition point which contains transitions that are not concurrent describes the same behavior (i.e., the same set of states where the new signal transition is excited) as a transition point that satisfies these requirements.

Example 6.2.3 Again, consider the state graph shown in Figure 6.5. The transition point $(\{req\_patron+\}, \{req\_patron+\})$ violates the first requirement, and it is clearly not useful as it implies that the new signal transition is excited in all states. The transition point $(\{req\_patron+\}, \{ack\_patron+\})$ violates the second requirement, and it would force ack_patron to wait to see the state signal change before it could rise. This would require the interface behavior to be changed, which is not allowed. The transition point $(\{ack\_wine-, req\_patron+\}, \{req\_patron-\})$ violates the third requirement since ack_wine- is not concurrent with but rather precedes $req\_patron+$. This transition point implies the same set of states as ($\{req\_patron+\}$, $\{req\_patron-\}$). The transition point ($\{req\_patron+\}$, $\{ack\_wine+, req\_patron-\}$) also violates the third requirement since $req\_patron-$ is not concurrent with ack_wine+, but rather, precedes it. This transition point also implies the same states as ($\{req\_patron+\}$, $\{req\_patron-\}$).

Some of the legal transition points are given below.

$$
\begin{array}{rcl}
(\{ack\_wine+\} &,& \{ack\_wine-\}) \\
(\{ack\_wine-\} &,& \{ack\_wine+\}) \\
(\{req\_wine-\} &,& \{ack\_wine-\}) \\
(\{req\_patron+\} &,& \{req\_patron-\}) \\
(\{ack\_patron+\} &,& \{req\_patron-\}) \\
(\{req\_wine+, req\_patron-\} &,& \{ack\_wine+\}) \\
(\{req\_wine+, req\_patron-\} &,& \{ack\_wine-\}) \\
(\{req\_wine+, ack\_patron-\} &,& \{ack\_wine+\}) \\
(\{req\_wine+, ack\_patron-\} &,& \{ack\_wine-\})
\end{array}
$$

Once two legal and useful TPs have been found, they are combined into an insertion point $IP = (TP_R, TP_F)$ and checked for compatibility. Two transition points are incompatible when either of the following is true:

$$TP_R(t_s) \cap TP_F(t_s) \neq \emptyset$$
$$TP_R(t_e) \cap TP_F(t_e) \neq \emptyset$$

For a state graph to have consistent state assignment, a transition on a signal must be followed by an opposite transition before another transition of the same type can occur. An incompatible insertion point always creates an inconsistent state assignment.

Example 6.2.4 The transition points ($\{ack\_wine+\}$, $\{ack\_wine-\}$) and ($\{req\_wine+, req\_patron-\}$, $\{ack\_wine-\}$) would not form a compatible insertion point since ack_wine- is in both $TP_R(t_e)$ and $TP_F(t_e)$.

6.2.2 State Graph Coloring

After finding all compatible insertion points, the next step is to determine the effect of inserting a state variable into each insertion point. This could be determined simply by inserting the state signal and rederiving the SG. This approach is unnecessarily time consuming and may produce a SG with an inconsistent state assignment. To address both of these problems, the original SG is partitioned into four subpartitions, corresponding to the states in which the new signal is rising, falling, stable high, and stable low.

Partitioning is accomplished by coloring each state in the original SG. First, all states in $S(TP_R)$ are colored as *rising*, which indicates that the state signal would be excited to rise in these states. The states in $S(TP_F)$ are colored as *falling*. If during the process of coloring *falling* states, a state is found that has already been colored as *rising*, this insertion point leads to an inconsistent state assignment and must be discarded. Once both the rising and falling states have been colored, all states following those colored rising before reaching any colored falling are colored as *high*. Similarly, all states between those colored as falling and those colored as rising are colored as *low*. While coloring *high* or *low*, if a state to be colored is found to already have a color, then again the insertion point leads to an inconsistent state assignment.

Example 6.2.5 Consider the insertion point

$$IP((\{req\_patron+\},\{req\_patron-\}),(\{ack\_wine-\},\{ack\_wine+\}))$$

We would first color all states in $S(\{req\_patron+\},\{req\_patron-\})$ which we found previously to be $\{\langle RR01\rangle, \langle 1R01\rangle, \langle R10F\rangle, \langle 110F\rangle\}$ to be rising. Next, we would color all states in $S(\{ack\_wine-\},\{ack\_wine+\})$ to be falling. However, this would result in each of the following states $\{\langle RR01\rangle, \langle 1R01\rangle, \langle R10F\rangle, \langle 110F\rangle\}$ to be colored both rising and falling. Therefore, this insertion point results in an inconsistent state assignment and is discarded.

Example 6.2.6 Consider the insertion point

$$IP((\{ack\_wine+\},\{ack\_wine-\}),(\{req\_patron+\},\{req\_patron-\}))$$

Coloring the SG would result in the following partition:

$$\begin{aligned} rising &= \{\langle F010\rangle, \langle 00F0\rangle\} \\ falling &= \{\langle RR01\rangle, \langle 1R01\rangle, \langle R10F\rangle, \langle 110F\rangle\} \\ high &= \{\langle R00R\rangle, \langle 100R\rangle\} \\ low &= \{\langle RF00\rangle, \langle 1F00\rangle, \langle R000\rangle, \langle 10R0\rangle\} \end{aligned}$$

This coloring is also shown in Figure 6.6. Notice that for this insertion point, the states $\langle R000\rangle$ and $\langle 10R0\rangle$ are colored as low and states $\langle R00R\rangle$ and $\langle 100R\rangle$ are colored as high. This means that in the first two states the new state signal is stable low and in the last two states the new state signal is stable high. The result is that these states, which previously had a CSC violation, no longer have a CSC violation since the new state signal disambiguates them.

6.2.3 Insertion Point Cost Function

After partitioning the SG determines that an insertion point leads to a consistent state assignment, the next step is to determine if the insertion point found is better than one found previously. The primary component of the cost function is the number of CSC violations which would remain after a state signal is inserted into a given IP. The number of remaining CSC violations for a given IP is determined by eliminating from CSCV any pair of violations in which one state is colored *high* while the other is colored *low*. Next, states that previously had a USC violation may now have a CSC violation due to the

COMPLETE STATE CODING 221

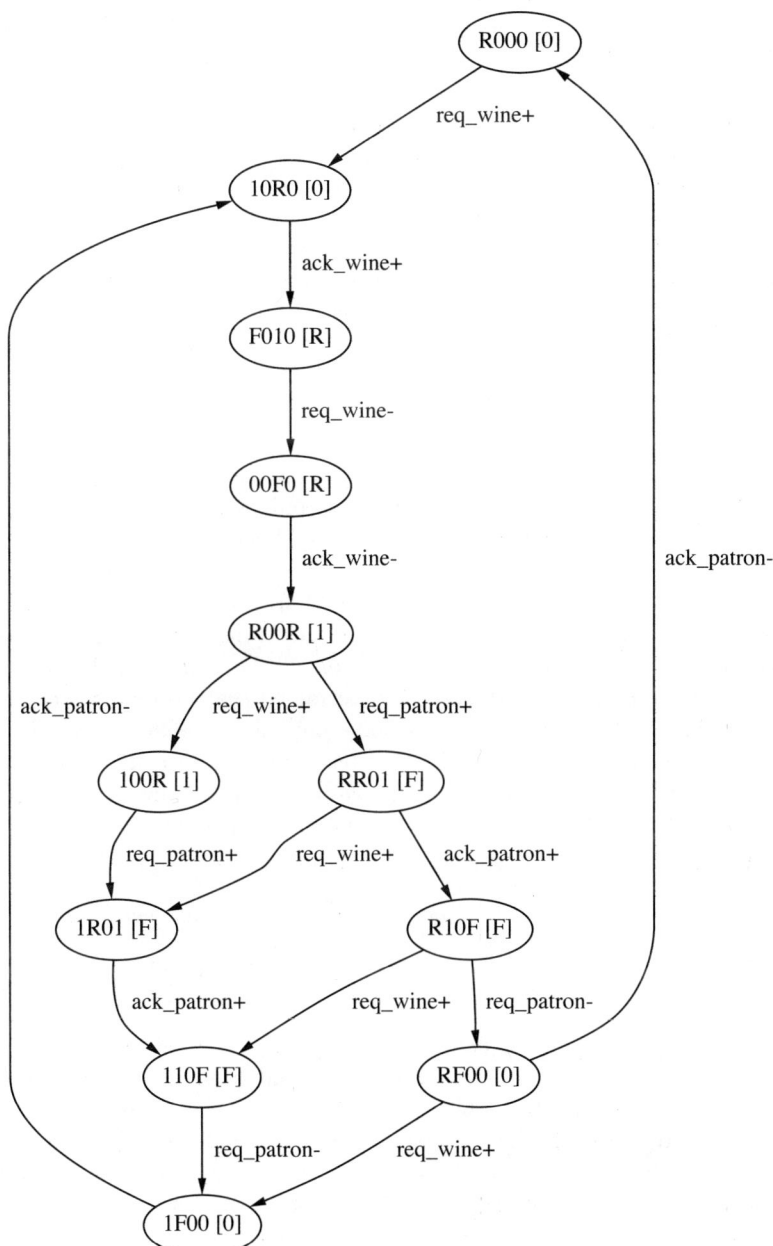

Fig. 6.6 Colored state graph for the passive/active wine shop (statevector is ⟨*req_wine, ack_patron, ack_wine, req_patron*⟩).

insertion of the state signal. In particular, for each pair of states with a USC violation (but not a CSC violation), if one is colored *rising* while the other is colored *low*, there is now a CSC violation since these states have different output enablings. Similarly, if one is colored *falling* and the other is colored *high*, there is also a new CSC violation. Each new CSC violation of the type just described must be added to the total remaining.

When two IPs leave the same number of CSC violations, a secondary cost function must be used. Each additional criterion is considered in order until the tie is broken. One possible additional cost function is to select the IP with the smallest sum, $|TP_R(t_e)| + |TP_F(t_e)|$, since it delays a smaller number of other transitions. Another is to select the IP with the smallest sum, $|TP_R(t_s)| + |TP_F(t_s)|$, since fewer enabling signals leads to simpler circuits.

6.2.4 State Signal Insertion

Once a good insertion point has been selected, the next step is to insert the state signal into the SG. This can be accomplished either by adding the transition to the higher-level representation, such as a STG or TEL structure, or by expanding the SG. If the STG is to be modified, arcs are added from each transition in t_s to the new state signal transition. Similarly, arcs are added from the new transition to each of the transitions in t_e. The same steps are followed for the reverse transition of the state signal. An initial marking for the new arcs must also be determined that preserves liveness and safety of the STG. After both transitions have been added to the STG, the state signal is assigned an initial value based on the coloring of the initial state. If the initial state is colored as *high* or *falling*, the initial value is high. Otherwise, the initial value is low. At this point, a new SG can be found.

> **Example 6.2.7** Again consider the insertion point
>
> IP(({*ack_wine+*}, {*ack_wine−*}), ({*req_patron+*}, {*req_patron−*}))
>
> The STG for this insertion point is shown in Figure 6.7. The state graph for this STG is shown in Figure 6.8, and it has CSC.

Alternatively, the new SG can be found directly. Each state in the original state graph is extended to include one new signal value. If a state is colored *low*, the new signal is '0' in that state. If a state is colored *high*, the new signal is '1'. If a state is colored *rising*, it must be split into two new states, one in which the new signal is 'R' and another in which the new signal is '1'. Similarly, if a state is colored *falling*, it must be split into two new states, where one has the new signal as 'F' and the other has it as '0'.

> **Example 6.2.8** Compare the state graphs shown in Figures 6.6 and 6.8. Notice that all states colored with a stable value, 0 or 1, appear in the new state graph simply extended with that new variable. In the case of states colored with an unstable value, R or F, must be split to show the change in the state variable. For example, the state $\langle F010 \rangle$ is

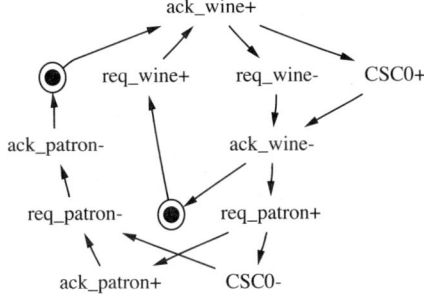

Fig. 6.7 STG for the passive/active wine shop which has CSC.

extended to $\langle F010R \rangle$ and a new state must be added in which the state signal changes, $\langle F0101 \rangle$. For the state $\langle 00F0 \rangle$, the enabling of ack_wine must be removed in the extended state since it is in the end set and does not become enabled until after $CSC0$ changes. Therefore, the new extended state becomes $\langle 0010R \rangle$. A new state must also be added for after $CSC0$ rises in which ack_wine is now enabled (i.e., $\langle 00F01 \rangle$). The states for the falling of the state signal are expanded similarly.

6.2.5 Algorithm for Solving CSC Violations

The algorithm for solving CSC violations is shown in Figure 6.9. It first checks if there are any CSC violations. If there are, it finds all legal transition points. Next, it considers each pair of transition points as a potential insertion point. For each legal insertion point, it colors the state graph. If the colored state graph is consistent and the cost of this insertion point is better than the best found so far, it records it. Finally, it inserts the new signal into the best insertion point found. Often, the insertion of a single state signal is not sufficient to eliminate all CSC violations. Therefore, after deriving the new SG, it may be necessary to solve CSC violations in the new SG. It does this by calling the CSC solver recursively and adds an additional state signal.

6.3 HAZARD-FREE LOGIC SYNTHESIS

After generating a SG with CSC, we apply a modified logic minimization procedure to obtain a hazard-free logic implementation. The modifications necessary are dependent upon the assumed technology. In this section we describe the necessary modifications for three potential technologies: complex gates, generalized C-elements, and basic gates. We conclude this section with a description of an extremely efficient logic minimization algorithm which can be applied to an important subclass of specifications.

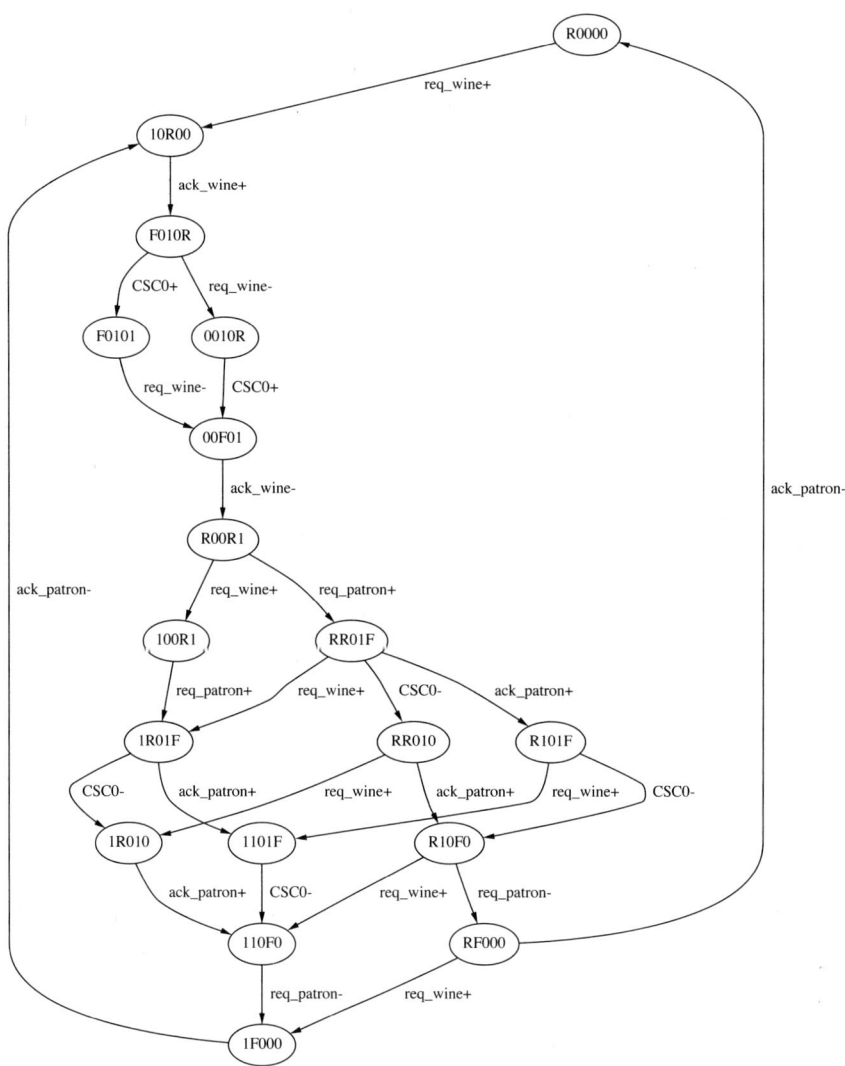

Fig. 6.8 State graph for the passive/active wine shop with CSC (state vector is ⟨*req_wine, ack_patron, ack_wine, req_patron, CSC0*⟩).

```
csc_solver(SG) {
  CSCV = find_csc_violations(SG);
  if (|CSCV| = 0) return SG;         /* No CSC violations so return.*/
  best = |CSCV|;
  best_IP = (∅, ∅);                  /* Initialize best insertion point.*/
  TP = find_all_transition_points(SG);
  foreach TP_R ∈ TP
    foreach TP_F ∈ TP
      if IP = (TP_R, TP_F) is legal then {
        CSG = color_state_graph(SG, TP_R, TP_F);
        CSCV = find_csc_violations(CSG);
        if (CSG is consistent) and ((|CSCV| < best) or
           ((|CSCV| = best) and (cost(IP) < cost(best_IP))))) then {
          best = |CSCV|;
          best_IP = (TP_R, TP_F);    /* Record new best IP.*/
        }
      }
  SG = insert_state_signal(SG, best_IP);
  SG = csc_solver(SG);               /* Add more signals, if needed.*/
  return SG;
}
```

Fig. 6.9 Algorithm for solving CSC violations.

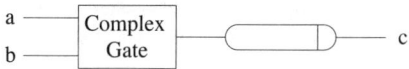

Fig. 6.10 Atomic gate model.

6.3.1 Atomic Gate Implementation

In the first synthesis method, we assume that each output is implemented using a single complex *atomic gate*. A gate is atomic when its delay is modeled by a single delay element connected to its output as depicted in Figure 6.10.

Using this model, after obtaining a SG with CSC, we can apply traditional logic minimization to find a logic implementation. The ON-set for a signal u is the set of all states in which u is either excited to rise [i.e., $ES(u+)$] or stable high [i.e., $QS(u+)$]. The OFF-set is the set of all states in which u is either excited to fall [i.e., $ES(u-)$] or stable low [i.e., $QS(u-)$]. The DC-set is the set of all unreachable states, or equivalently, those states not included in either the ON-set or OFF-set. In other words, the state space is partitioned for a signal u as follows:

$$\begin{aligned} \text{ON-set} &= \{\lambda_S(s) \mid s \in (ES(u+) \cup QS(u+))\} \\ \text{OFF-set} &= \{\lambda_S(s) \mid s \in (ES(u-) \cup QS(u-))\} \\ \text{DC-set} &= \{0,1\}^{|N|} - (\text{ON-set} \cup \text{OFF-set}) \end{aligned}$$

We apply the recursive prime generation procedure described earlier to find all prime implicants. Finally, we set up and solve a covering problem to find the minimum number of primes that covers the minterms in the *ON-set*.

Example 6.3.1 Consider the SG with CSC shown in Figure 6.8. For the signal *ack_wine*, we would find the following sets:

$$
\begin{aligned}
\textit{ON-set} &= \{10000, 10100, 00100, 10101\} \\
\textit{OFF-set} &= \{00101, 00001, 10001, 00011, 10011, 01011, 00010, \\
&\quad 10010, 01010, 11010, 01000, 11000, 11011, 00000\} \\
\textit{DC-set} &= \{00110, 00111, 01001, 01100, 01101, 01110, 01111, \\
&\quad 10110, 10111, 11001, 11100, 11101, 11110, 11111\}
\end{aligned}
$$

The primes found for *ack_wine* are

$$P = \{\texttt{1-1--}, \texttt{-11--}, \texttt{--11-}, \texttt{--1-0}, \texttt{-1-01}, \texttt{10-00}\}$$

The constraint matrix for *ack_wine* is shown below.

	1-1--	-11--	--11-	--1-0	-1-01	10-00
10000	–	–	–	–	–	1
10100	1	–	–	1	–	1
00100	–	–	–	1	–	–
10101	1	–	–	–	–	–

The primes `1-1--`, `--1-0`, and `10-00` are essential and cover the entire ON-set. In a similar fashion, we can find the logic implementations for *req_patron* and *CSC0*. The final circuit is shown in Figure 6.11(a).

The circuit in Figure 6.11(a) is hazard-free if all the delay is modeled as being at the output of the final gate in each signal network (i.e., the output of the OR gates). Unfortunately, if this circuit is mapped to basic gates and the delays of these gates are considered individually, the implementation may be hazardous. Consider the sequence of states shown in Figure 6.11(b). After *req_wine* goes high, *u2* goes high, causing *ack_wine* to go high. Assume that the gate feeding *u3* is slow to rise. At this point, if *req_wine* goes low, *u2* could go low, causing *ack_wine* to become excited to fall. Now, if *u3* finally rises, *ack_wine* would become excited to rise. The result is a $1 \rightarrow 0 \rightarrow 1$ glitch on the signal *ack_wine*.

6.3.2 Generalized C-Element Implementation

Another implementation strategy is to use generalized C-elements (gC) (see Figure 5.45). Using the gC approach, two minimization problems must now be solved for each signal u. The first implements the set of the function [i.e., $set(u)$], and the second implements the reset [i.e., $reset(u)$]. To implement $set(u)$, the ON-set is only the states in which u is excited to rise. The OFF-set is again the states in which u is excited to fall or is stable low. The DC-set now includes the states in which u is stable high as well as the unreachable

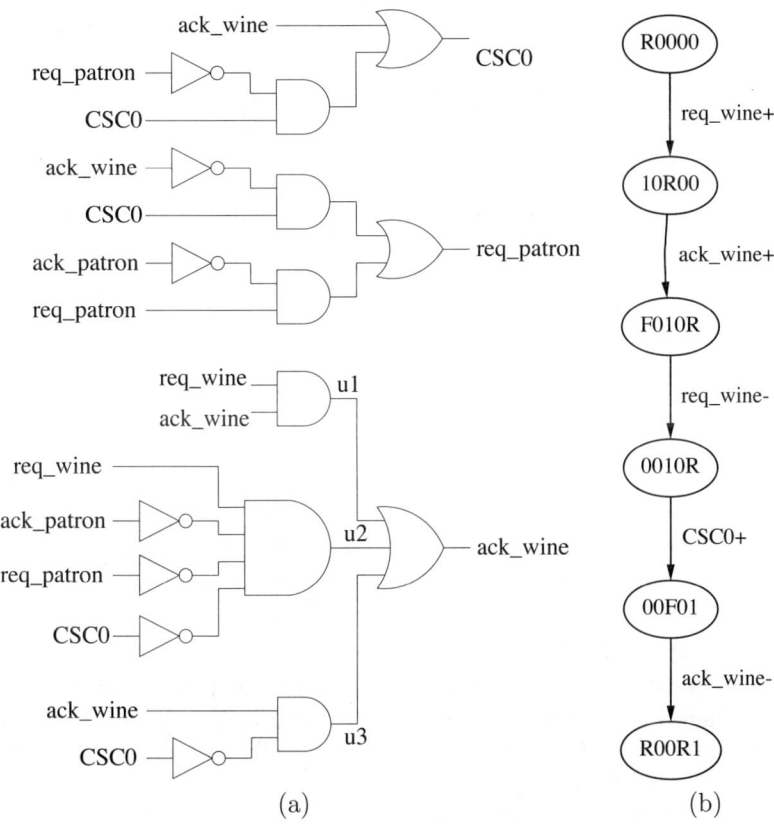

Fig. 6.11 (a) Atomic gate implementation of the passive/active shop. (b) Sequence of states leading to a gate-level hazard.

states. The stable high states are don't cares because once a gC is set, its feedback holds its state. In other words, we partition the state space as follows:

$$\begin{aligned} ON\text{-}set &= \{\lambda_S(s) \mid s \in (ES(u+))\} \\ OFF\text{-}set &= \{\lambda_S(s) \mid s \in (ES(u-) \cup QS(u-))\} \\ DC\text{-}set &= \{0,1\}^{|N|} - (ON\text{-}set \cup OFF\text{-}set) \end{aligned}$$

To implement $reset(u)$, the ON-set is the set of states in which u is excited to fall, and the OFF-set is the set of states in which u is either rising or high. The DC-set includes both the unreachable states and the states in which u is low. In other words, the partition becomes

$$\begin{aligned} ON\text{-}set &= \{\lambda_S(s) \mid s \in (ES(u-))\} \\ OFF\text{-}set &= \{\lambda_S(s) \mid s \in (ES(u+) \cup QS(u+))\} \\ DC\text{-}set &= \{0,1\}^{|N|} - (ON\text{-}set \cup OFF\text{-}set) \end{aligned}$$

We can now apply standard methods to find a minimum number of primes to implement the set and reset functions.

Example 6.3.2 Again consider the SG with CSC shown in Figure 6.8. For $set(ack\_wine)$, we would find the following sets:

$$
\begin{aligned}
ON\text{-}set &= \{10000\} \\
OFF\text{-}set &= \{00101, 00001, 10001, 00011, 10011, 01011, 00010, \\
&\quad 10010, 01010, 11010, 01000, 11000, 11011, 00000\} \\
DC\text{-}set &= \{00110, 00111, 01001, 01100, 01101, 01110, 01111, \\
&\quad 10110, 10111, 11001, 11100, 11101, 11110, 11111, \\
&\quad 10100, 00100, 10101\}
\end{aligned}
$$

The primes for ack_wine are again found to be

$$P \;=\; \{\texttt{1-1--}, \texttt{-11--}, \texttt{--11-}, \texttt{--1-0}, \texttt{-1-01}, \texttt{10-00}\}$$

Only one prime, however, is needed to cover the ON-set: `10-00`. For the reset function, we also only need one prime `0---1`. In a similar fashion, we can find the logic implementations for the set and reset functions for req_patron and $CSC0$. The final gC circuit is shown in Figure 6.12.

Consider the sequence of states shown again in Figure 6.12(b). Again, req_wine rising causes ack_wine to rise. After ack_wine rises, req_wine is allowed to fall. There is no longer a potential for a hazard since the feedback in the gC implementation holds ack_wine stable high until $CSC0$ rises at which point ack_wine is supposed to fall.

When the set function for a signal u, $set(u)$, is on in all states in which u should be rising or high, the state holding element can be removed. The implementation for u is simply equal to the logic for $set(u)$. Similarly, if $reset(u)$ is on in all states in which u should be falling or low, the signal u can be implemented with $\overline{reset(u)}$. This process is called *combinational optimization*.

Example 6.3.3 Consider the SG shown in Figure 6.13(a). Using the generalized C-element implementation approach, we derive the circuit shown in Figure 6.13(b). The set of states in which signal d is rising or stable high is $110R$ and $11R1$ (note that in state $111F$, signal d is excited to fall). The logic for the set function, $ab\overline{c}$, evaluates to 1 in both of these states. Therefore, there is no need for a state-holding device, so we can use the circuit shown in Figure 6.13(c) for signal d instead.

If we use AND gates (with inverted inputs) and C-elements instead of gCs to implement the circuit shown in Figure 6.12, it would still be hazard-free. In general, however, this is not the case. Consider the SG and circuit shown in Figure 6.13. If we implement the circuit for signal c, using a two-input AND (with complemented first terminal), two-input OR, an inverter, and a C-element, this circuit is not hazard-free. Consider a sequence of states

HAZARD-FREE LOGIC SYNTHESIS   229

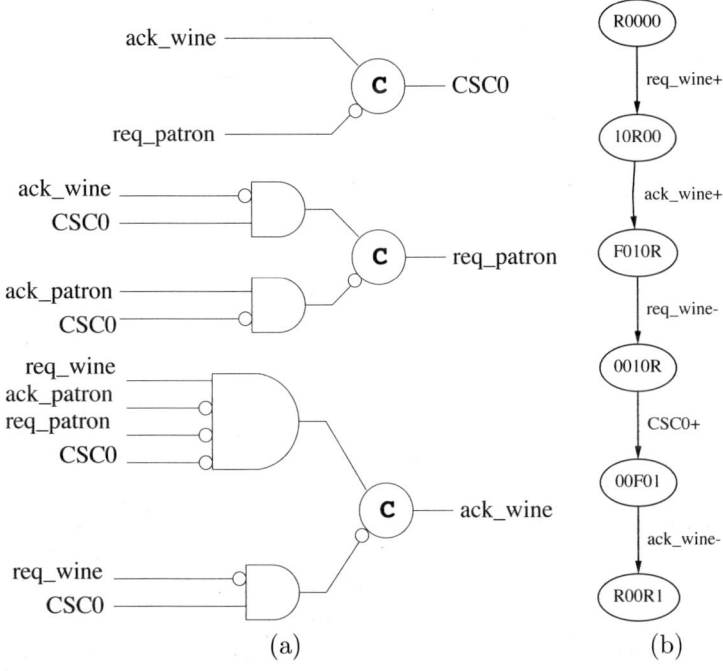

Fig. 6.12 (a) Generalized C-element implementation of passive/active shop. (b) Sequence of states leading to a gate-level hazard for the atomic gate circuit.

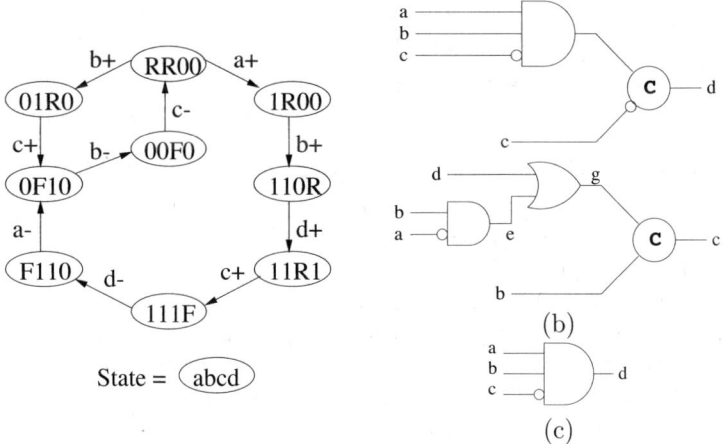

Fig. 6.13 (a) SG for small example. (b) Generalized C-element implementation. (c) Optimized combinational implementation of signal d.

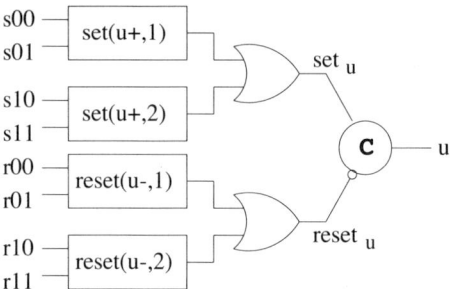

Fig. 6.14 Standard C-implementation.

beginning in state $\langle F110 \rangle$. Initially, signal e (the output of the two-input AND gate) is low, but after a goes low it becomes excited to rise. Let us assume that this gate is slow and e does not rise immediately. Next, b goes low, which disables e causing a hazard. The result can manifest in many ways. It may not cause any problem. It may cause c to fall slower. If c starts to fall and the glitch on e propagates to g causing g to rise, then c may stop falling and actually be restored back to its high value by the feedback in the C-element. Finally, if c falls and a and b rise before the glitch on e propagates to g, the glitch could cause c to turn on prematurely (i.e., before d has risen).

6.3.3 Standard C-Implementation

To avoid the hazard concerns discussed above, we could modify our logic minimization procedure to produce a gate-level hazard-free implementation called a *standard C-implementation*. The general structure of the standard C-implementation is shown in Figure 6.14. While the structure is similar to the gC-implementation, the method in which it is designed is quite different. First, each *region function* [i.e., $set(u+, k)$ or $reset(u-, k)$] implements a single (or possibly a set of) excitation region(s) for the signal u. In the gC-implementation, an excitation region can be implemented by multiple product terms. Second, each region function turns on only when it enters a state in its excitation region, turns off monotonically sometime after the signal u changes, and must stay off until the excitation region is entered again. To guarantee this behavior, each region function must satisfy certain correctness constraints, leading to a modified logic minimization procedure.

Each region function is implemented using a single atomic gate, corresponding to a *cover* of an excitation region. The cover of a set region $C(u+, k)$ [or a reset region $C(u-, k)$] is a set of states for which the corresponding region function in the implementation evaluates to one. While a single region function can be used to implement multiple excitation regions, we first present a method in which each region function only implements a single excitation region. Later, we extend the method to allow gate sharing.

For a cover to produce a gate-level hazard-free implementation, it must satisfy certain *correctness constraints*. The idea behind these constraints is that each region function can only change when it is needed to actively drive the output signal to change. Consider a region function for a set region. This gate turns on when the circuit enters a state in the set region. When the region function changes to 1, it excites the OR gate, which in turn excites the C-element (assuming that the reset network is low) to set u to 1. Only after u has risen can the region function be excited to fall. The region function then must fall monotonically. The signal u will not be able to fall until the region function has fallen and the OR gate for the set network has fallen. Once the region function falls, it is not allowed to be excited again until the circuit again enters a state in the corresponding set region. To guarantee this behavior, a correct cover must satisfy a *covering* and an *entrance constraint*.

First, a correct cover needs to satisfy a covering constraint which states that the reachable states in the cover must include the entire excitation region but must not include any states outside the union of the excitation region and associated quiescent states, that is,

$$ER(u*, k) \subseteq [C(u*, k) \cap S] \subseteq [ER(u*, k) \cup QS(u*)]$$

Second, the cover of each excitation region must also satisfy an entrance constraint which states that the cover must only be entered through excitation region states:

$$[(s_i, t, s_j) \in \delta \wedge s_i \notin C(u*, k) \wedge s_j \in C(u*, k)] \Rightarrow s_j \in ER(u*, k)$$

As the following theorem states, if all covers satisfy these two constraints, the resulting standard C-implementation is correct.

Theorem 6.4 (Beerel, 1998) *If for all outputs $u \in O$ all region function covers $C(u*, k)$ satisfy the covering and entrance constraints the standard C-implementation is correct (i.e., complex-gate equivalent and hazard-free).*

The goal of logic minimization is now to find an optimal sum-of-products function for each region function that satisfies the definition of a correct cover given above. An implicant of an excitation region is a product that may be part of a correct cover. In other words, a product c is an implicant of an excitation region $ER(u*, k)$ if the set of reachable states covered by c is a subset of the states in the union of the excitation region and associated quiescent states, that is,

$$[c \cap S] \subseteq [ER(u*, k) \cup QS(u*)]$$

For each set region $ER(u+, k)$, the ON-set is those states in $ER(u+, k)$. The OFF-set includes not only the states in which u is falling or low, but also the states outside this excitation region where u is rising. This additional restriction is necessary to make sure that a region function can only turn on in its excitation region, and it will not glitch on in another excitation region for the same signal. More formally, we partition the state space as follows:

$$\begin{aligned}
\textit{ON-set} &= \{\lambda_S(s) \mid s \in (ER(u+,k))\} \\
\textit{OFF-set} &= \{\lambda_S(s) \mid s \in (ES(u-) \cup QS(u-)) \cup (ES(u+) - ER(u+,k))\} \\
\textit{DC-set} &= \{0,1\}^{|N|} - (\textit{ON-set} \cup \textit{OFF-set})
\end{aligned}$$

The ON-set, OFF-set, and DC-set for a reset region $ER(u-,k)$ can be defined similarly:

$$\begin{aligned}
\textit{ON-set} &= \{\lambda_S(s) \mid s \in (ER(u-,k))\} \\
\textit{OFF-set} &= \{\lambda_S(s) \mid s \in (ES(u+) \cup QS(u+)) \cup (ES(u-) - ER(u-,k))\} \\
\textit{DC-set} &= \{0,1\}^{|N|} - (\textit{ON-set} \cup \textit{OFF-set})
\end{aligned}$$

The prime implicants can again be found using standard techniques.

> **Example 6.3.4** Consider again the SG shown in Figure 6.13(a). There are two set regions for c: $ER(c+,1) = 01R0$ and $ER(c+,2) = 11R1$. Let's examine the implementation of $ER(c+,1)$. For this excitation region, we find the following partition of the state space:
>
> $$\begin{aligned}
\textit{ON-set} &= \{0100\} \\
\textit{OFF-set} &= \{0000, 1000, 0010, 1100, 1101\} \\
\textit{DC-set} &= \{0001, 0011, 0101, 0110, 0111, \\
&\quad\quad 1001, 1010, 1011, 1110, 1111\}
\end{aligned}$$
>
> The primes found are as follows:
>
> $$P = \{\texttt{01--}, \texttt{1-1-}, \texttt{-11-}, \texttt{0--1}, \texttt{-0-1}, \texttt{--11}\}$$

The entrance constraint creates a set of *implied states* for each implicant c [denoted $IS(c)$]. An implied state of an implicant c is a state that is not covered by c but due to the entrance constraint must be covered if the implicant is to be part of the cover. In other words, a state s is an implied state of an implicant c for the excitation region $ER(u*,k)$ if it is not covered by c, and s is a predecessor of a state that is both covered by c and not in the excitation region. This means that the product c becomes excited in a quiescent state instead of an excitation region state. If there does not exist some other product in the cover which contains this implied state, the cover violates the entrance constraint. More formally, the set of implied states for an implicant c is defined as follows:

$$IS(c) = \{s_i \mid s_i \notin c \land \exists s_j . (s_i, t, s_j) \in \delta \land (s_j \in c) \land (s_j \notin ER(u*,k))\}$$

An implicant may have implied states that are outside the excitation region and the corresponding quiescent states. Therefore, these implied states may not be covered by any other implicant. If this implicant is the only prime implicant which covers some excitation region state, the covering problem cannot be solved using only prime implicants.

> **Example 6.3.5** Consider the prime implicant 01-- which is the only prime that covers the ON-set (i.e., the state $01R0$). This implicant can be entered through the transition $(F110, a-, 0F10)$. However, the state $0F10$ is not in $ER(c+, 1)$. Therefore, this would be an entrance violation, so the state $F110$ is an implied state for this implicant. This state must be covered by some other prime in the cover to satisfy the entrance constraint. There are two primes that cover this state: 1-1- and -11-. If we include either of these primes, the cover can be entered through the transition $(11R1, c+, 111F)$, so $11R1$ is an implied state for these two primes. However, the state $11R1$ is in the OFF-set since it is part of another excitation region for c [i.e., $ER(c+, 2)$]. Therefore, we cannot include either of these primes in the cover, since we cannot cover their implied states. Therefore, no correct cover exists using only prime implicants.

To address this problem, we must introduce the notion of *candidate implicants*. An implicant is a candidate implicant if there exists no other implicant which properly contains it and has a subset of the implied states. In other words, c_i is a candidate implicant if there *does not exist* an implicant c_j that satisfies the following two conditions:

$$c_j \supset c_i$$
$$IS(c_j) \subseteq IS(c_i)$$

Prime implicants are always candidate implicants, but not all candidate implicants are prime. An optimal cover can always be found using only candidate implicants.

Theorem 6.5 (Beerel, 1998) *An optimal cover of a region function always exists and consists of only candidate implicants.*

We find all candidate implicants using the algorithm in Figure 6.15. This algorithm is similar to the *prime_compatibles* algorithm from Chapter 5. The only real differences are that it checks implied states instead of class sets and uses the function *lit_extend* (instead of *max_subsets*) to find all implicants with one more literal than the given prime.

> **Example 6.3.6** Returning to our example, we would seed the list of candidate implicants with the primes found earlier. They are all of size 2. Let us first consider 01--. As stated earlier, the implied state for this prime is $F110$. Since there is an implied state, we must consider extending this prime with an additional literal. The implicant 010- has no implied states, and it is a subset of no other candidate implicant with no implied states, so it is added to the list of candidate implicants. The implicants 011- and $0F10$ have $F110$ as an implied state, so they are not candidate implicants. Finally, 0111 has no implied states, but

```
candidate_implicants(SG, P) {
    done = ∅;                           /* Initialize already computed set.*/
    for (k = |largest(P)|; k ≥ 1; k − −) {  /* Loop largest to smallest.*/
        foreach (q ∈ P; |q| = k) enqueue(C, q)  /* Queue all of size k.*/
        foreach (c ∈ C; |c| = k) {      /* Consider candidates of size k.*/
            if (IS(SG, c) = ∅) then continue    /* If empty, skip.*/
            foreach (s ∈ lit_extend(c)) {       /* Check extensions by 1 lit.*/
                if (s ∈ done) then continue     /* If computed, skip.*/
                Γ_s = IS(SG, s)         /* Find extension's implied states.*/
                prime = true            /* Initialize prime as true.*/
                foreach (q ∈ C; |q| ≥ k) {      /* Check larger candidates.*/
                    if (s ⊂ q) then {   /* If contained in prime, check it.*/
                        Γ_q = IS(SG, q)         /* Compute implied states.*/
                        if (Γ_s ⊇ Γ_q) then {   /* If smaller, not candidate.*/
                            prime = false;
                            break
                        }
                    }
                }
                if (prime = 1) then enqueue(C, s)   /* If prime, queue it.*/
                done = done ∪ {s}               /* Mark as computed.*/
            }
        }
    }
    return(C);                          /* Return candidate implicants.*/
}
```

Fig. 6.15 Algorithm to find candidate implicants.

it is a subset of 0--1 which also has no implied states, so it is also not a candidate implicant. Next, we consider extending the prime 1-1- since it has implied state 11R1. The implicant 101- has no implied states, so it is a candidate implicant. This is the last candidate implicant found, and the complete set of candidate implicants is:

$$\{01\text{--}, 010\text{-}, 1\text{-}1\text{-}, 101\text{-}, \text{-}11\text{-}, 0\text{--}1, \text{-}0\text{-}1, \text{--}11\}$$

We can now formulate a covering problem by introducing a Boolean variable x_i for each candidate implicant c_i. The variable $x_i = 1$ when the candidate implicant is included in the cover and 0 otherwise. Using these variables, we can construct a product-of-sums representation of the covering and entrance constraints.

First, a *covering clause* is constructed for each state s in the excitation region. Each clause consists of a disjunction of candidate implicants that cover s. More formally,

$$\bigvee_{i : s \in c_i} x_i$$

To satisfy the covering clause for each state s in $ER(u*, k)$, at least one x_i must be set to 1. This means that for each excitation region state s, there must be an implicant chosen that includes it in the cover. The set of covering clauses for an excitation region guarantees that all excitation region states are covered. Since candidate implicants are not allowed to include states outside the excitation region and corresponding quiescent states, the cover is guaranteed to satisfy the covering constraint.

Example 6.3.7 For our example there is only one excitation region state, 01R0, which is included only in candidate implicants c_1 (01--) and c_2 (010-), so we get the following covering clause:

$$(x_1 + x_2)$$

For each candidate implicant c_i, a *closure clause* is constructed for each of its implied states $s \in IS(c_i)$. Each closure clause represents an implication which states that if a candidate implicant is included in the cover, its implied states must also be included in some other implicant in the cover. This can be expressed formally as follows:

$$\overline{x_i} \vee \bigvee_{j: s \in c_j} x_j$$

The closure clauses ensure that the cover satisfies the entrance constraint.

Example 6.3.8 The candidate implicant c_1 (01--) has implied state 0F10. This state is included in the implicants c_3 (1-1-) and c_5 (-11-). Therefore, we get the following closure clause:

$$(\overline{x_1} + x_3 + x_5)$$

The complete formulation is

$$(x_1 + x_2)(\overline{x_1} + x_3 + x_5)\overline{x_3}\,\overline{x_5}\,\overline{x_8}$$

Our goal now is to find an assignment of the x_i variables that satisfies the function with the minimum cost where the cost is the number of implicants. Since there are negated variables, the covering problem is binate. To solve the binate covering problem, we again construct a constraint matrix to represent the product-of-sums described above. The matrix has one row for each clause and one column for each candidate implicant. The rows can be divided into a *covering section* and a *closure section*, corresponding to the covering and closure clauses. In the covering section, for each excitation region state s, a row exists containing a 1 in every column, corresponding to a candidate implicant that includes s. In the closure section, for each implied state s of each candidate implicant c_i, a row exists containing a 0 in the column corresponding to c_i and a 1 in each column corresponding to a candidate implicant c_j that covers the implied state s.

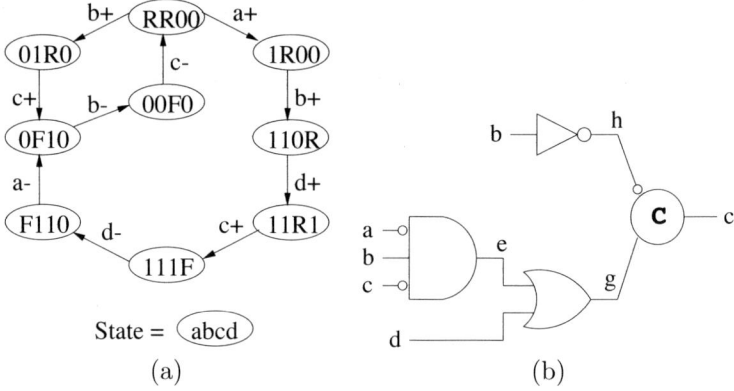

Fig. 6.16 (a) SG for small example. (b) Circuit found using basic gate approach.

Example 6.3.9 The constraint matrix for $ER(c+, 1)$ is depicted below.

	01--	010-	1-1-	101-	-11-	0--1	-0-1	--11
1	1	1	1	–	–	–	–	–
2	0	–	1	–	1	–	–	–
3	–	–	0	–	–	–	–	–
4	–	–	–	–	0	–	–	–
5	–	–	–	–	–	–	–	0

Rows 3, 4, and 5 are essential making the candidate implicants 1-1-, -11-, and --11 unacceptable, so these rows and the corresponding columns must be removed to produce the following matrix:

	01--	010-	101-	0--1	-0-1
1	1	1	1	–	–
2	0	–	–	–	–

Now, row 2 is essential making the implicant 01-- unacceptable, resulting in the following matrix:

	010-	101-	0--1	-0-1
1	1	1	–	–

Now, row 1 is essential, making the implicant 010- essential. Therefore, the cover includes only the implicant 010- (i.e., $\bar{a}b\bar{c}$). Note that this matrix can only be solved by selecting an implicant that is not prime. The resulting circuit implementation is shown in Figure 6.16. Consider again the sequence of states that led to a hazard beginning in state $F110$. When a falls, the signal e is not excited since c is high. Therefore, the hazard described earlier has been removed.

Again, we can apply a combinational optimization to the result. For standard C-implementations, we can remove the state-holding element when either the set of covers for the set function for a signal u include all states where u is rising or high [i.e., $\bigcup_k C(u+, k) \supseteq ES(u+) \cup QS(u+)$], or the covers for the reset function include all states where u is falling or low [i.e., $\bigcup_k C(u-, k) \supseteq ES(u-) \cup QS(u-)$].

Another optimization is to allow a single gate to implement multiple excitation regions. The procedure finds a gate that covers each excitation region using modified correctness constraints. The covering constraint is modified to allow the cover to include states from other excitation regions, that is,

$$ER(u*, k) \subseteq [C(u*, k) \cap S] \subseteq \left[\bigcup_l ER(u*, l) \cup QS(u*) \right]$$

The entrance constraint must also be modified to allow the cover to be entered from any corresponding excitation region state:

$$[(s_i, t, s_j) \in \delta \wedge s_i \notin C(u*, k) \wedge s_j \in C(u*, k)] \Rightarrow s' \in \bigcup_l ER(u*, l)$$

An additional constraint is also now necessary to guarantee that a cover either includes an entire excitation region or none of it:

$$ER(u*, l) \not\subseteq C(u*, k) \Rightarrow ER(u*, l) \cap C(u*, k) = \emptyset$$

Example 6.3.10 Consider the state graph shown in Figure 6.17(a). The signal c has two set regions: $ER(c+, 1) = 10R$ and $ER(c+, 2) = 11R$. Using the earlier constraints, the primes are found to be

$$P(c+, 1) = \{\texttt{10-}, \texttt{1-1}, \texttt{-11}\}$$
$$P(c+, 2) = \{\texttt{11-}, \texttt{1-1}, \texttt{-11}\}$$

The cover for $ER(c+, 1)$ is the prime `10-`, since it covers the excitation region state $10R$ and has no implied states. For $ER(c+, 2)$, the prime `11-` has implied state $FR1$. This implied state can be covered by `1-1`, but this prime has implied state $10R$ which is in the offset. Therefore, the prime `11-` must be expanded to `110` which is the final solution. Therefore, the set function for c is $a\bar{b} + ab\bar{c}$ [see Figure 6.17(b)].

Using the new constraints that allow gate sharing, the primes are now found to be

$$P(c+, 1) = \{\texttt{1--}, \texttt{-11}\}$$
$$P(c+, 2) = \{\texttt{1--}, \texttt{-11}\}$$

With the new entrance constraint, a transition is allowed to enter a prime through any excitation region state. Therefore, the prime `1--` has no implied states and is a minimal cover. Therefore, using the new constraints, the set function for c is simply a [see Figure 6.17(c)].

238 MULLER CIRCUITS

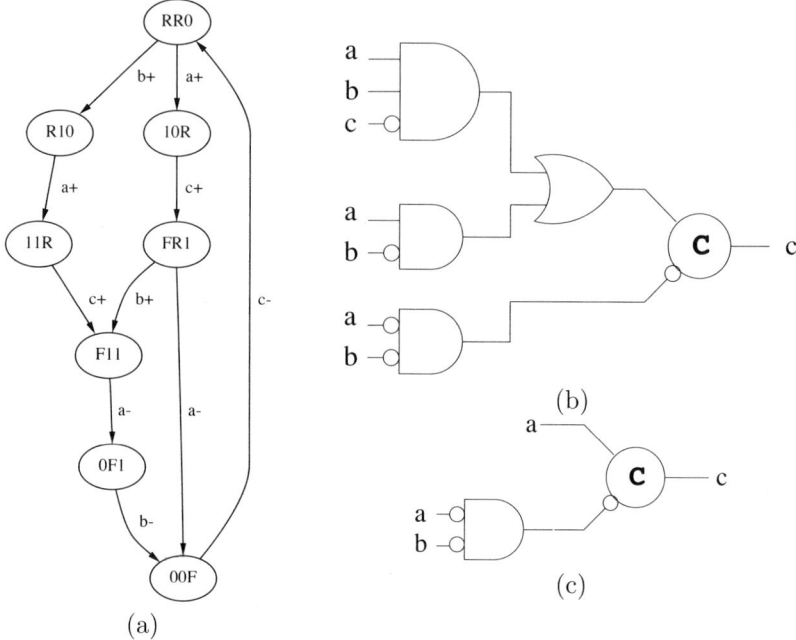

Fig. 6.17 (a) SG for gate-sharing example. (b) Original circuit without gate sharing. (c) Final circuit with gate sharing.

6.3.4 The Single-Cube Algorithm

The foregoing algorithms for logic minimization are often more general than necessary since many region functions can be implemented with a single product, or cube. This subsection presents a more efficient algorithm, shown in Figure 6.18, that finds an optimal single-cube cover, if one exists.

For a single-cube cover to hazard-freely implement a region function, all literals in the cube must correspond to signals that are *stable* throughout the excitation region. Otherwise, the single-cube cover would not cover all excitation region states. When a single-cube cover exists, an excitation region $ER(u*, k)$ can be sufficiently approximated using an *excitation cube* [denoted $EC(u*, k)$] which is the supercube of the states in the excitation region and is defined on each signal v as follows:

$$EC(u*, k)(v) \equiv \begin{cases} 0 & \text{if } \forall s \in ER(u*, k) \,.\, s(v) = 0 \\ 1 & \text{if } \forall s \in ER(u*, k) \,.\, s(v) = 1 \\ - & \text{otherwise} \end{cases}$$

If a signal has a value of 0 or 1 in the excitation cube, the signal can be used in the cube implementing the region. The set of states implicitly represented by

```
single_cube(SG,technology) {
    foreach u ∈ O {                        /* Consider each output signal.*/
        EC =find_excitation_cubes(SG);
        foreach EC(u*,k) ∈ EC {            /* Find cover for each EC.*/
            TC(u*,k) =find_trigger_cube(SG,EC(u*,k));
            CS(u*,k) =find_context_signals(SG,EC(u*,k),TC(u*,k));
            V(u*,k) =find_violations(SG,EC(u*,k),TC(u*,k),technology);
            CC =build_cover_table(CS(u*,k),V(u*,k));
            C(u*,k) =solve_cover_table(CC,TC(u*,k));
        }
        solution(u) = optimize_logic(C);   /* Combo opt, gate sharing.*/
    }
    return solution;
}
```

Fig. 6.18 Single-cube algorithm.

the excitation cube is always a superset of the set of excitation region states [i.e., $EC(u*,k) \supseteq ER(u*,k)$].

The set of trigger signals for an excitation region $ER(u*,k)$ can also be represented with a cube called a *trigger cube* $TC(u*,v)$, defined as follows for each signal v:

$$TC(u*,k)(v) \equiv \begin{cases} s_j(v) & \text{If } \exists (s_i,t,s_j) \in \delta \cdot (t = v+ \vee t = v-) \wedge \\ & (s_i \notin EC(u*,k)) \wedge (s_j \in EC(u*,k)) \\ - & \text{otherwise} \end{cases}$$

The single-cube algorithm requires the cover of each excitation region to contain all its trigger signals [i.e., $C(u*,k) \subseteq TC(u*,k)$]. Since only stable signals can be included, a necessary condition for our algorithm to produce an implementation is that all trigger signals be stable [i.e., $EC(u*,k) \subseteq TC(u*,k)$].

The excitation cubes and trigger cubes are easily found with a single pass through the SG. For each excitation region, an excitation cube is built by forming the supercube over all states in the excitation region. The trigger cube is built by finding each signal that takes the circuit into the excitation cube.

Example 6.3.11 The excitation cubes and trigger cubes corresponding to all the excitation regions in the example SG in Figure 6.13(a) are shown in Table 6.1. Notice that every trigger signal is stable and our algorithm proceeds to find the optimal single-cube cover.

The goal of the single-cube algorithm is to find a cube $C(u*,k)$ where $EC(u*,k) \subseteq C(u*,k) \subseteq TC(u*,k)$ such that it satisfies the required correctness constraints for the given technology. The cube should also be maximal (i.e., cover as many states as possible). The single-cube algorithm begins with a cube consisting only of the trigger signals [i.e., $C(u*,k) = TC(u*,k)$]. If this cover contains no states that violate the required correctness constraints,

Table 6.1 Excitation and trigger cubes with cube vector $\langle a, b, c, d \rangle$.

$u*, k$	$EC(u*, k)$	$TC(u*, k)$
$c+, 1$	0100	-1--
$c+, 2$	1101	---1
$c-, 1$	0010	-0--
$d+, 1$	1100	-1--
$d-, 1$	1111	--1-

we are done. This, however, is often not the case, and context signals must be added to the cube to remove any *violating states*. For each violation detected, the procedure determines the choices of context signals which would exclude the violating state. Finding the smallest set of context signals to resolve all violations is a covering problem.

If the implementation technology is generalized C-elements, then for a set region a state is a violating state when the trigger cube intersects a set of states where the signal is falling or stable low. Similarly, for a reset region, a state is a violating state when the trigger cube intersects the set of states where it is rising or stable high. More formally, the sets of violating states are defined as follows:

$$V(u+, k) = \{s \in S \mid s \in TC(u+, k) \land s \subset ES(u-) \cup QS(u-)\}$$
$$V(u-, k) = \{s \in S \mid s \in TC(u-, k) \land s \in ES(u+) \cup QS(u+)\}$$

Example 6.3.12 Returning again to our example, the trigger cube for $ER(c+, 1)$ is -1--, which includes the reachable states 01R0, 0F10, F110, 111F, 11R1, and 110R. The only violating state is 110R since it is in $QS(c-)$. For $ER(c+, 2)$, $ER(c-, 1)$, and $ER(d-, 1)$, there are no violating states. The trigger cube for $ER(d+, 1)$ is -1--, which illegally intersects the states 111F, F110, 0F10, and 01R0.

The next thing we need to do is determine which context signals remove these violating states. A signal is allowed to be a context signal if it is stable in the excitation cube [i.e., $EC(u*, k)(v) = 0$ or $EC(u*, k)(v) = 1$]. A context signal removes a violating state when it has a different value in the excitation cube and the violating state. In other words, a context signal v removes a violating state s when $EC(u*, k)(v) = \overline{s(v)}$.

Example 6.3.13 Consider first the violating state 110R for $ER(c+, 1)$. The only possible context signal is a, which is 0 in the excitation cube and 1 in the violating state. Therefore, we must reduce the cover to 01-- to remove this violating state. For $ER(d+, 1)$, the violating state 111F can be removed using either context signals c or d, $F110$ can be removed with c, $0F10$ can be removed with a or c, and finally, 01R0 can be removed with a.

HAZARD-FREE LOGIC SYNTHESIS   241

In order to select the minimum number of context signals, to remove all the violating states, we need to set up a covering problem. The constraint matrix for the covering problem has a row for each violating state and a column for each context signal.

Example 6.3.14 The constraint matrix for $ER(d+, 1)$ is shown below.

	a	c	d
111F	–	1	1
F110	–	1	–
0F10	1	1	–
01R0	1	–	–

The context signals a and c are essential and solve the entire matrix. This means that the implementation for the set function for d is $a\bar{b}\bar{c}$. The final circuit implementation is shown in Figure 6.13(b). Again, the combinational optimization can be applied to obtain the circuit shown in Figure 6.13(c) for signal d.

To produce a standard C-implementation, we must use the covering and entrance constraints. First, for each excitation cube, $EC(u*, k)$, the procedure finds all states in the initial cover [i.e., $TC(u*, k)$] which violate the covering constraint. In other words, a state s in $TC(u*, k)$ is a violating state if the signal u is excited in the opposite direction, is stable at the opposite value, or is excited in the same direction but the state is not in the current excitation region. The set of covering violations, $CV(u*, k)$, can be defined more formally as follows:

$$CV(u+, k) = \{s \in S \mid s \in TC(u+, k) \land s \in ES(u-) \cup QS(u-) \cup (ES(u+) - EC(u+, k))\}$$

$$CV(u-, k) = \{s \in S \mid s \in TC(u-, k) \land s \in ES(u+) \cup QS(u+) \cup (ES(u-) - EC(u-, k))\}$$

Example 6.3.15 With this modification, for $EC(c+, 1)$, in addition to the state $110R$, the state $11R1$ is also a violating state since it is in the other set region. The sets of violating states for $EC(c+, 2)$, $EC(c-, 1)$, and $EC(d-, 1)$ are still empty, and the set of violating states for $EC(d+, 1)$ is unchanged.

Next, all state transitions which either violate or may violate the entrance constraint must be found. For each state transition $(s_i, v*, s_j)$, this is possible when s_j is a quiescent state, s_j is in the initial cover, and v excludes s_i. The set of entrance violations, $EV(u*, k)$, can be defined more formally as follows:

$$EV(u+, k) = \{s_j \in S \mid (s_i, v*, s_j) \in \delta \land s_j \in QS(u+) \land s_j \in TC(u+, k) \\ \land EC(u+, k)(v) = \overline{s_i(v)}\}$$

$$EV(u-, k) = \{s_j \in S \mid (s_i, v*, s_j) \in \delta \land s_j \in QS(u-) \land s_j \in TC(u-, k) \\ \land EC(u-, k)(v) = \overline{s_i(v)}\}$$

When a potential entrance violation is detected, a context signal must be added which excludes s_j from the cover when v is included in the cover. Therefore, if v is a trigger signal, the state s_j is a violating state. If v is a possible context signal choice, s_j becomes a violating state when v is included in the cover.

Example 6.3.16 Let's consider $EC(c+, 1)$ again. The state transition $(F110, a-, 0F10)$ is a potential entrance violation since $0F10$ is in $QS(c+)$, $0F10$ is in $TC(c+, 1)$, and the signal a excludes the state $F110$ from the cover. Therefore, the state $0F10$ is a violating state when a is included in the cover. Similarly, the state transition $(111F, d-, F110)$ becomes an entrance violation when d is included as a context signal. For $EC(d-, 1)$, the state transition $(01R0, c+, 0F10)$ is a potential entrance violation, and since c is a trigger signal, $0F10$ is a violating state.

Again, to select the minimum number of context signals, we have a covering problem. Since inclusion of certain context signals causes some states to have entrance violations, the covering problem is binate. To solve this binate covering problem, we create a constraint matrix for each region. There is a row in the constraint matrix for each violation and each violation that could potentially arise from a context signal choice, and there is a column for each context signal. The entry in the matrix contains a 1 if the context signal excludes the violating state. An entry in the matrix contains a 0 if the inclusion of the context signal would require a new violation to be resolved.

Example 6.3.17 The constraint matrix for $EC(c+, 1)$ is shown below.

	a	c	d
$110R$	1	–	–
$11R1$	1	–	1
$0F10$	0	1	–
$F110$	1	1	0

The context signal a is essential, so it must be in the cover. Selecting a, however, causes the state $0F10$ to become a violating state. To exclude this state, the signal c must be added as an additional context signal. Therefore, the implementation of $EC(c+, 1)$ is $\bar{a}b\bar{c}$. The standard C-implementation is shown in Figure 6.16(b).

If a violation is detected for which there is no context signal to resolve it, the constraint matrix construction fails. In this case, or if a trigger signal is not stable, we must either constrain concurrency, add state variables, or use the more general algorithm described earlier to find a circuit. We conclude this section with a few examples to show how these situations appear.

Example 6.3.18 First, consider $ER(z+, 1)$ for the nondistributive SG shown in Figure 6.19. The excitation cube for this region would be `--0` and the trigger cube would be `11-`. Clearly, the trigger signals are not stable in the excitation region, so no single-cube cover exists.

In general, the single-cube algorithm cannot be applied to nondistributive state graphs.

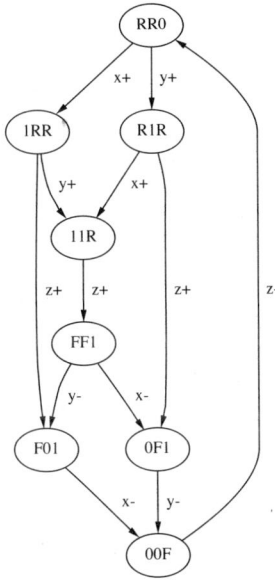

Fig. 6.19 Nondistributive SG [state vector $\langle x, y, z \rangle$].

Example 6.3.19 As a second example, consider $ER(w-, 1)$ for the SG in Figure 6.20(a). The excitation cube is 10-- and the trigger cube is --0-. Again, we find that the trigger signal is not stable in the excitation region, so no single-cube cover exists. If we remove the offending state $F010$, we obtain the SG in Figure 6.20(b). The signal w can now be implemented with a single cube, but there is a problem with x. Consider $ER(x+, 1)$ which has excitation cube 00-- and trigger cube 0---. There is no problem with the trigger signal. The state $R011$ is a violating state since it is included in the trigger cube and x is stable low. There is, however, no possible context signal which can be added to remove this violating state. Therefore, again there is no single-cube cover.

6.4 HAZARD-FREE DECOMPOSITION

The synthesis method described in Section 6.3.4 puts no restrictions on the size of the gates needed to implement the region functions. In all technologies, however, there is some limitation on the number of inputs that a gate can have. For example, in CMOS, it is typically not prudent to put more than four transistors in series, as it significantly degrades performance. Furthermore, large transistor stacks can have charge-sharing problems. These problems are especially dangerous in generalized C-elements, where excess charge can lead to the gate latching an incorrect value. Therefore, it is often necessary to

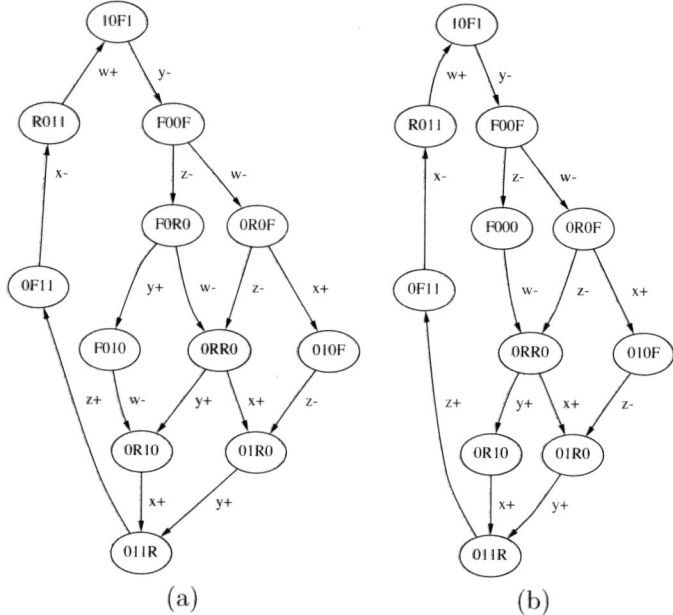

Fig. 6.20 (a) SG with an unstable trigger signal (state vector $\langle w, x, y, z \rangle$). (b) SG with an unresolvable violation (state vector $\langle w, x, y, z \rangle$).

decompose high-fanin gates into limited-fanin gates found in the given gate library. For Huffman circuits, decomposition of high-fanin gates can be done in an arbitrary fashion, preserving hazard freedom (if it existed in the original circuit). Unfortunately, this problem is much more difficult for Muller circuits.

Example 6.4.1 Let us assume that we have a library of gates which allows no more than two inputs per gate. Therefore, for the circuit shown in Figure 6.13(c), we would need to decompose the three-input AND gate into two two-input AND gates. Two possible ways of decomposing this gate are shown in Figure 6.21(b) and (c). Consider first the circuit shown in Figure 6.21(b). In state $\langle F110 \rangle$, inputs a, b, c, and internal signal e are high while d is low. After a falls, we move to state $\langle 0F10 \rangle$, and e becomes excited to go low. However, let us assume that the AND-gate generating signal e is slow. Next, b falls, moving us to state $\langle 00F0 \rangle$. If at this point c falls before e falls, d can become excited to rise prematurely. The result is that there is a hazard on the signal d, and there is a potential for a circuit failure.

Next, consider the circuit shown in Figure 6.21(c), beginning in state $\langle F110 \rangle$. This time e begins low, and it does not become excited to change until after a falls, b falls, c falls, and a rises again. At this point, however, b is already low, which maintains d in its low state until b rises again. In fact, there is no sequence of transitions that can cause this circuit to experience a hazard.

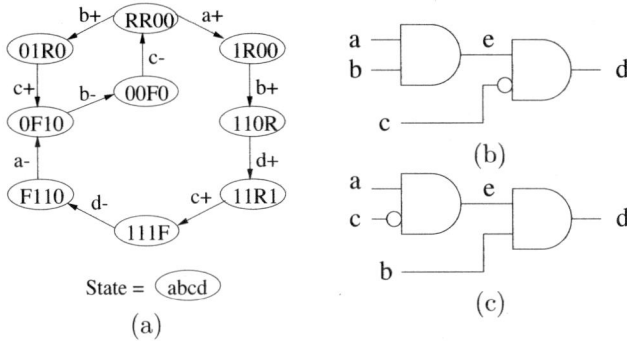

Fig. 6.21 (a) SG for example. (b) Hazardous and (c) hazard-free decomposition.

This example illustrates the need for special care during decomposition to guarantee a hazard-free implementation. Essentially, we need to find a new internal signal which can be added to produce a simpler circuit. Therefore, in this section we present a simple technique for finding hazard-free decompositions which is similar to the insertion point procedure described earlier to solve CSC violations.

6.4.1 Insertion Points Revisited

Let us analyze the circuits in Figure 6.21 using the idea of insertion points. For the hazard-free circuit shown in Figure 6.21(c), the transition point to set the new signal e is $(\{a+\}; \{d+\})$. This means that e is rising in the states $\langle 1000 \rangle$ and $\langle 110R \rangle$, but it is guaranteed to have risen before d can rise. In other words, d rising acknowledges that e has risen. The transition point to reset the new signal is $(\{c+\}, \{d-\})$. This means that e is falling only in state $\langle 111F \rangle$, and it is guaranteed to have fallen before d can fall. Therefore, again d changing acknowledges the change on e. The fact that changes on this internal signal are acknowledged by primary output changes is what guarantees hazard freedom.

Now consider the hazardous circuit shown in Figure 6.21(b). The transition point to set e is $(\{b+\}, \{d+\})$, so it is changing only in state $\langle 110R \rangle$. This means that again d rising acknowledges that e has risen. The transition point for e to fall has a start set of $\{a-\}$, but it has no end set. There is no transition that is prevented from occurring until e has fallen. It is this lack of acknowledgment that leads to the hazard described previously. One could consider using $b-$ in an end set, but this is an input signal, and we are not allowed to change the interface behavior. If, instead, we use $c-$ as an end set, we just move the problem, as e now requires a three-input gate.

Again, we have a large number of potential insertion points which need to be filtered. The filters discussed for CSC violations still apply. Due to the nature of the decomposition problem, however, there are some additional restrictions that can be used.

Consider the decomposition of a cover, $C(u*, k)$, which is composed of a single cube, or AND gate. For the new signal to be useful to decompose this gate, either its set or reset function must be composed of signals in the original AND gate. Therefore, we can restrict the start set for one of the transition points to transitions on just those signals in the gate being decomposed. We also know that this original gate is composed of trigger and context signals. By their definition, context signals always change before trigger signals, so they are not concurrent with the trigger signals. Therefore, we only need to consider start sets which include either trigger or context signals from the original gate, but not both. We need to consider all possible combinations of the trigger signals as potential start sets, but we only need to consider concurrent subsets of the context signals as potential start sets. Finally, since this new signal transition must complete before $u*$ can be excited, we only need to consider transitions that occur after those in the start set and before $u*$ as potential candidates to be in the end set.

If both the cover of a set region and a reset region of u must be decomposed, the same restrictions on a transition point can be used for the reverse transition on the new signal. If, however, the new signal is not needed to decompose both a set and reset region, there is a little more flexibility in the start and end sets. First, the start set should include concurrent transitions which occur after $u*$ and before any transitions in the first start set. Including the reverse transition of $u*$ in the end set is often useful, but any transition after $u*$ could potentially be used in the end set.

6.4.2 Algorithm for Hazard-Free Decomposition

The complete algorithm for hazard-free decomposition is shown in Figure 6.22. This algorithm takes a SG and an initial design. It then finds all gates which have a fanin larger than the maximum allowed size. If there are none, the algorithm returns the current design. Otherwise, it records the current design as the best found so far, and it finds all transition points which can be used to decompose the high-fanin gates. It then considers each insertion point in turn. If the state graph colored using a legal insertion point is consistent, it inserts the new signal and resynthesizes the design. If the resulting design has less high-fanin gates or has a lower cost, it is recorded. Again, a single state signal may not decompose all high-fanin gates, so this algorithm recursively calls itself to add more signals until all high-fanin gates have been decomposed. This algorithm is illustrated using the following example.

> **Example 6.4.2** Consider the SG shown in Figure 6.8 and its circuit implementation shown in Figure 6.12(a). Let us assume that we are only allowed gates with at most three inputs. Therefore, we need to decompose the four-input AND gate used to implement the set region for *ack_wine*. For this cover, the trigger signals are *req_wine* and *ack_patron*,

```
decomposition(SG, design, maxsize) {
   HF = find_high_fanin_gates(design, maxsize);
   if (|HF| = 0) return design;         /* No high-fanin gates, return.*/
   best = |HF|;
   best_IP = design;                    /* Initialize best found.*/
   TP = find_all_transition_points(SG, design, HF);
   foreach TP_R ∈ TP
     foreach TP_F ∈ TP
       if IP = (TP_R, TP_F) is legal then {
         CSG = color_state_graph(SG, TP_R, TP_F);
         if (CSG is consistent) then {
           SG' = insert_state_signal(SG, IP);
           design = synthesis(SG');             /* Find new circuit.*/
           HF = find_high_fanin_gates(design, maxsize);
           if (((|HF| < best) or ((|HF| = best) and
               (cost(design) < cost(best_IP))))) then {
             best = |HF|;
             best_IP = design;
           }
         }
       }
   design = decomposition(SG, design);   /* Add more signals.*/
   return design;
}
```

Fig. 6.22 Algorithm for decomposing high-fanin gates.

and the context signals are *req_patron* and *CSC0*. We would like the new gate to be off the critical path, so we first attempt to implement it using context signals. The transition *CSC0−* always precedes *req_patron−*, so these two transitions cannot be included in the same start set. Therefore, there are only two possible transition points using context signals:

$$(\{CSC0-\}, \{ack\_wine+\})$$
$$(\{req\_patron-\}, \{ack\_wine+\})$$

If we need to use the trigger signals instead, there are three possible transition points to consider:

$$(\{req\_wine+\}, \{ack\_wine+\})$$
$$(\{ack\_patron-\}, \{ack\_wine+\})$$
$$(\{req\_wine+, ack\_patron-\}, \{ack\_wine+\})$$

Next, we need to consider the transition points for the reverse transition of the new signal. We would like to insert the reverse transition somewhere between *ack_wine+* and *ack_wine−*, so we restrict our attention to transitions between them. In other words, the start and end sets are only made up of the following transitions: *ack_wine+*, *CSC0+*, *req_wine−*, and *ack_wine−*. Consider first using *ack_wine+* in the start set. If we do this, no other transition can be in the start set, since all of the other choices occur after ack_wine+. Of the remaining transitions, the end set cannot include req_wine-, since it is an input transition. Therefore, there are two possible transition points:

$$(\{ack\_wine+\}, \{CSC0+\})$$
$$(\{ack\_wine+\}, \{ack\_wine-\})$$

If we use $CSC0+$ in the start set, its end set must be ack_wine-, since req_wine is an input. If we use req_wine- in the start set, either $CSC0+$ or ack_wine- can be in the end set. Finally, since $CSC0+$ and req_wine- change concurrently, they can appear in the start set together. In this case, the end set must be ack_wine-. Therefore, we have the following transition points:

$$(\{CSC0+\}, \{ack\_wine-\})$$
$$(\{req\_wine-\}, \{CSC0+\})$$
$$(\{req\_wine-\}, \{ack\_wine-\})$$
$$(\{CSC0+, req\_wine-\}, \{ack\_wine-\})$$

Now that we have enumerated all the transition points, the next step is to form insertion points out of combinations. We again color the graph to determine if the insertion point leads to a consistent state assignment. We should also check if any USC violations become CSC violations as a result of the signal insertion. If neither of these problems arise, we need to further check the insertion point by deriving a new state graph and synthesizing the circuit. If the new circuit meets the fanin constraints, the insertion point is accepted. If not, we try the next insertion point.

For our example, we must select the following transition point for one transition of the new signal:

$$(\{req\_patron-\}, \{ack\_wine+\})$$

However, we are free to select any of the reverse transition points to meet the gate size of 3 constraint. An example circuit is shown in Figure 6.23 using the following transition point for the reverse transition:

$$(\{ack\_wine+\}, \{ack\_wine-\})$$

6.5 LIMITATIONS OF SPEED-INDEPENDENT DESIGN

The circuit shown in Figure 6.23 requires several gates with inverted inputs. If the bubble on the ack_patron input to the set AND gate for ack_wine is removed and replaced with an inverter, the circuit is no longer hazard-free. Consider the state of the circuit just after ack_wine has gone low and before

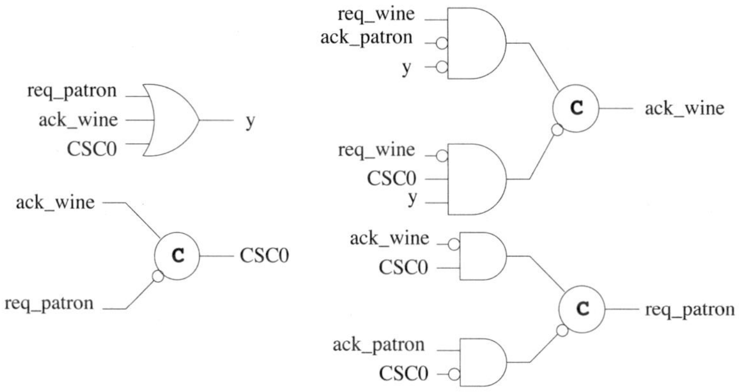

Fig. 6.23 Passive/active wine shop using gates with a maximum stack size of 3.

req_patron has gone high. In this state *CSC0* is high, so *req_patron* can go high followed by *ack_patron* going high. This enables this new inverter to go low. In the meantime, *CSC0* can go low, followed by *req_patron* going low, and finally, *ack_patron* can go low, resulting in the inverter being disabled. In other words, this sequence would result in a hazard at the output of this new inverter. Clearly, this long sequence of events is highly unlikely to be faster than the switching time of an inverter. However, under the speed-independent delay model, this sequence must be considered to be possible. If timing information is known, however, it may be possible to determine that such a hazard is impossible. This is the subject of the next chapter.

6.6 SOURCES

Speed-independent switching circuit theory originated with Muller and Bartky [277, 278, 279]. Perhaps, the best description of Muller's work is in Volume II of Miller's textbook on switching theory [272]. The definition of speed independence given in Section 6.1 follows that given in [272, 279]. Several of the concepts and examples in Section 6.1 came from this early work [272, 279]. Many of the definitions, however, have been tuned to fit the speed-independent design methods described in the following sections. A method for efficient identification of speed-independent circuits is given by Kishinevsky et al. [202].

Similar to the speed-independent model is the *self-timed* model proposed by Seitz [345, 347]. In this model, wires in *equipotential regions* are assumed to have zero delay. The length of these wires must be short enough that a change on one end can be perceived at the other end in less than the *transit time* of an electron through a transistor. If delay elements are added to all wires which are not in an equipotential regions, this model is reduced to essentially the speed-independent model.

The complete state coding problem and a method to solve it was first described by Bartky [24]. The method for solving the CSC problem described in Section 6.2 follows the work of Krieger [211]. The idea of coloring the state graph originated with Vanbekbergen et al. [391]. This work often produced inefficient circuits since it allowed the state graph to be colored arbitrarily to solve the CSC problem. To improve upon the logic generated, Ykman-Couvreur and Lin restricted the coloring such that the new state signals are only inserted into excitation and switching regions [412, 413]. The insertion points described here are a generalization of this idea. Cortadella et al. generalized the idea further, expanding the set of possible insertion points [93]. A quite different approach taken to solving the CSC problem is taken by Lavagno et al. [223]. In this work, the state graph is converted to an FSM, and traditional FSM critical race free state assignment approach is taken. Another approach proposed by Gu and Puri first decomposes the STG specification of the circuit into manageable piece, and it then applies Boolean satisfiability to find a solution for each of the smaller subgraphs [157].

Most early speed-independent design approaches required complex atomic gates as well as several of the more recent design methodologies proposed by Chu [83], Meng et al. [268], and Vanbekbergen [389, 390]. A speed-independent design method which targets three-valued logic can be found in [405]. Martin's speed-independent design methodology was the first to utilize generalized C-elements [249, 254]. The first work to synthesize speed-independent circuits using only basic gates (NANDs and NORs) was done by Varshavsky and his students [393]. This work, however, produced rather inefficient circuits and was limited to circuits without choice. Adding the C-element to the list of basic gates, Beerel and Meng developed conditions for correct standard C-implementations [25]. Similar conditions with some generalizations were later presented by Kondratyev et al. [210]. Quite a different approach is taken by Sawasaki et al. which uses hazardous set and reset logic but uses a special flip-flop rather than a C-element to filter the hazards, keeping the outputs hazard-free [339]. The single-cube algorithm described in Section 6.3.4 was developed by Myers [283]. A comparison of this algorithm to the more general algorithm appears in [29]. The theory in Section 6.3.3 comes from this paper. A method that uses BDDs to find all possible correct covers is presented in [377]. A synthesis technique that assigned don't cares in such a way as to ensure initializability is given in [73]. Several researchers have developed methods to synthesize gate-level hazard-free speed-independent circuits directly from a STG [185, 187, 235, 305, 350]. This avoids the state explosion inherent in SG-based synthesis methods.

The decomposition problem was first discussed by Kimura, who called these decompositions, *extensions* [197, 198]. An extension is called a *good extension* if its behavior ignoring the new signals produces exactly the same allowed sequences as the original. In particular, Kimura investigated the effect of adding buffers into a wire and stated that a circuit suffers the *delay problem of the first kind* when adding buffers to wires that have delay results in a bad extension [197]. In [198] he introduced delay problems of the second kind, in which the circuit is no longer partially semi-modular with respect to the original signals, and the third kind, in which the circuit is not totally semi-modular. The work by Varshavsky and his students developed methods to decompose logic, but as mentioned earlier, they were limited in their utility and often produced inefficient circuits [393]. Lavagno et al. leveraged synchronous technology mapping methods by producing a circuit that is hazard-free using the atomic gate assumption, but then added delay elements to remove hazards introduced by decomposition [221, 222, 225]. Siegel and De Micheli developed conditions in which high-fanin AND gates in a standard C-implementation can be decomposed [351, 353]. Myers et al. introduced an approach to breaking up high-fanin gC gates using a method of decomposition and resynthesis [284]. Perhaps one of the most important works in this area is Burns's method which utilizes implicit methods to explore large families of potential decompositions of a generalized C-element implementation [66]. Recent work by Cortadella et al. and Kondratyev et al. has built upon this approach to allow for a greater range of potential decompositions [92, 209]. The discussion in Section 6.4 was inspired by Krieger's insertion point idea [211] and recent work by Burns and others.

Problems

6.1 Speed Independence
For the SG shown in Figure 6.24(a), find its corresponding partial order of equivalence classes. A property is said to hold in a state if it holds in all states that follow it. For each state in the SG, determine which of the following properties hold:

1. Speed independence
2. Semi-modularity
3. Distributivity
4. Totally sequential

6.2 Complete State Coding
For the STG in Figure 6.24(b), find the state graph and then find all state pairs which violate complete state coding. Solve the CSC violations by adding a new state variable. Show the new state graph.

6.3 Hazard-Free Logic Synthesis
From the SG in Figure 6.25, find a hazard-free logic implementation for the output signals x, d, and c using the:

6.3.1. Atomic gate approach.

252   MULLER CIRCUITS

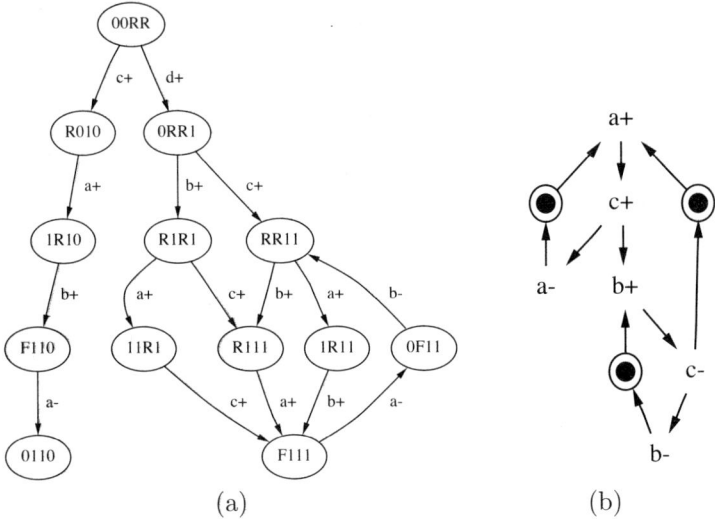

Fig. 6.24   (a) SG for Problem 6.1. (b) STG for Problem 6.2.

6.3.2. Generalized C-element approach (use the single-cube algorithm).

6.3.3. Standard C-element approach (use the single-cube algorithm).

6.4   Exceptions
From the SG shown in Figure 6.26 for the output signals $r2$, $a1$, and x.

6.4.1. Find the excitation cubes and trigger cubes.

6.4.2. Use the single-cube algorithm to find a standard C-implementation, if possible. If not possible, explain why not.

6.4.3. Use the more general algorithm to find a multicube cover for the unimplemented signals from 6.4.2.

6.5   Speed-Independent Design
Perform the following on the VHDL in Figure 6.27.

6.5.1. Assuming that all signals are initially low, find the state graph.

6.5.2. Find all state pairs which violate complete state coding.

6.5.3. Solve the CSC violations by using reshuffling. Show the new state graph.

6.5.4. Solve the CSC violations by adding a new state variable q. Show the new state graph.

6.5.5. Use Boolean minimization to find the circuit from either of the solutions above.

6.5.6. Find the circuit from your reshuffled and state variable solutions using a generalized C-element implementation technique. Comment on which is best and why.

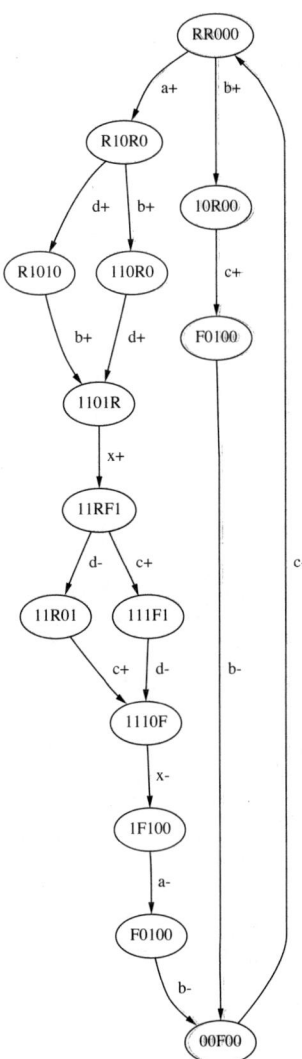

Fig. 6.25 SG for Problem 6.3.

254   MULLER CIRCUITS

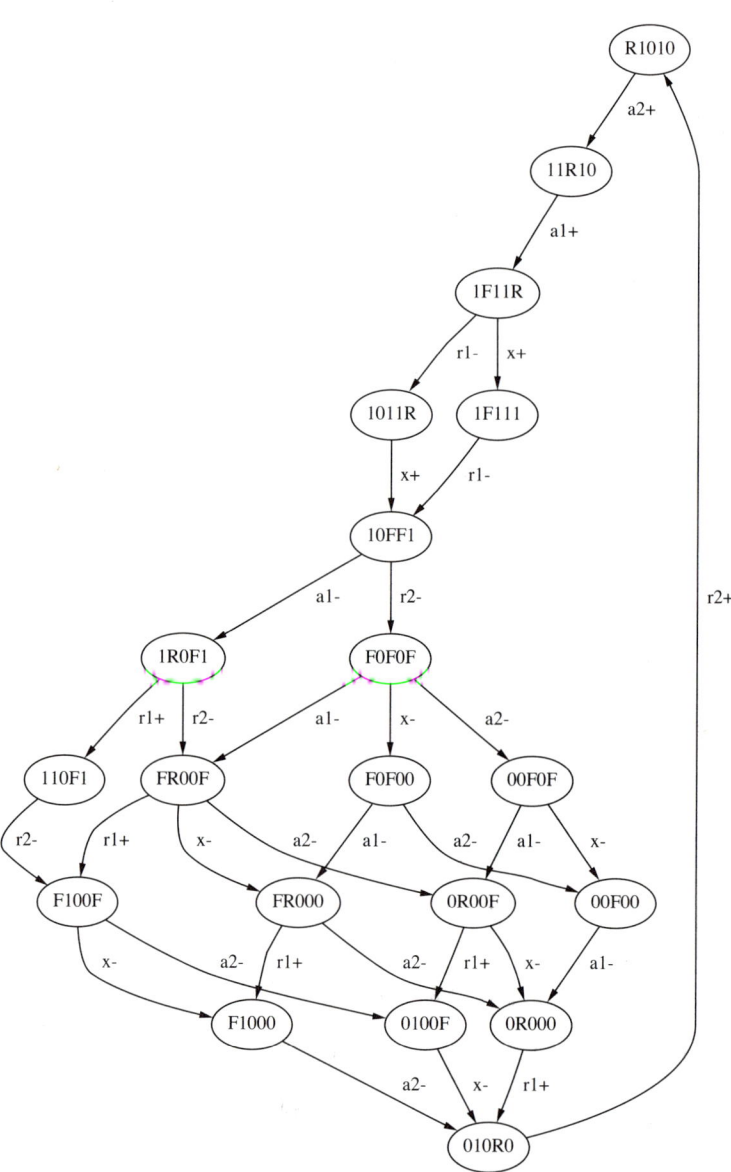

Fig. 6.26 SG for Problem 6.4.

```
library ieee;
use ieee.std_logic_1164.all;
use work.nondeterminism.all;
use work.handshake.all;
entity p6 is
end entity;
architecture hse of p6 is
  signal ai,bi,x:std_logic;  --@ in
  signal ao,bo:std_logic;
begin
main:process
begin
  guard(ai,'1');
  if (x = '1') then
    assign(ao,'1',1,2);
    guard(ai,'0');
    assign(ao,'0',1,2);
  else
    assign(bo,'1',1,2);
    guard(bi,'1');
    assign(bo,'0',1,2);
    guard(bi,'0');
    assign(ao,'1',1,2);
    guard(ai,'0');
    assign(ao,'0',1,2);
  end if;
end process;
ai:process
  variable z:integer;
begin
  z:=selection(2);
  if (z=1) then
    assign(x,'1',1,2);
    assign(ai,'1',1,2);
    guard(ao,'1');
    assign(ai,'0',1,2);
    assign(x,'0',1,2);
    guard(ao,'0');
  else
    assign(ai,'1',1,2);
    guard(ao,'1');
    assign(ai,'0',1,2);
    guard(ao,'0');
  end if;
end process;  bi:process
begin
  guard(bo,'1');
  assign(bi,'1',1,2);
  guard(bo,'0');
  assign(bi,'0',1,2);
end process;
end hse;
```

Fig. 6.27 VHDL for Problem 6.5.

6.6 Standard C-Implementation

Find a standard C-implementation for the circuit specified in Figure 6.28.

6.7 Hazard-Free Decomposition

For the state graph shown in Figure 6.29 and output signals *lo*, *ro*, and *x*:

6.7.1. Find a gC implementation using the single-cube algorithm.

6.7.2. Use the insertion point method to decompose any gates which have a fanin greater than two.

```vhdl
library ieee;
use ieee.std_logic_1164.all;
use work.nondeterminism.all;
use work.handshake.all;
entity p6 is
end entity;
architecture hse of p6 is
  signal ai:std_logic; --@ in
  signal bi:std_logic; --@ in
  signal x:std_logic; --@ in
  signal ao:std_logic;
  signal bo:std_logic;
begin
main:process
begin
  guard(ai,'1');
  if (x = '1') then
    assign(ao,'1',1,2);
    guard(ai,'0');
    assign(ao,'0',1,2);
  else
    assign(bo,'1',1,2);
    guard(bi,'1');
    assign(ao,'1',1,2);
    assign(bo,'0',1,2);
    guard_and(bi,'0',ai,'0');
    assign(ao,'0',1,2);
  end if;
end process;
ai:process
  variable z:integer;
begin
  z:=selection(2);
  if (z=1) then
    assign(x,'1',1,2);
    assign(ai,'1',1,2);
    guard(ao,'1');
    assign(x,'0',1,2);
    assign(ai,'0',1,2);
    guard(ao,'0');
  else
    assign(ai,'1',1,2);
    guard(ao,'1');
    assign(ai,'0',1,2);
    guard(ao,'0');
  end if;
end process;   bi:process
begin
  guard(bo,'1');
  assign(bi,'1',1,2);
  guard(bo,'0');
  assign(bi,'0',1,2);
end process;
end hse;
```

Fig. 6.28 VHDL for Problem 6.6.

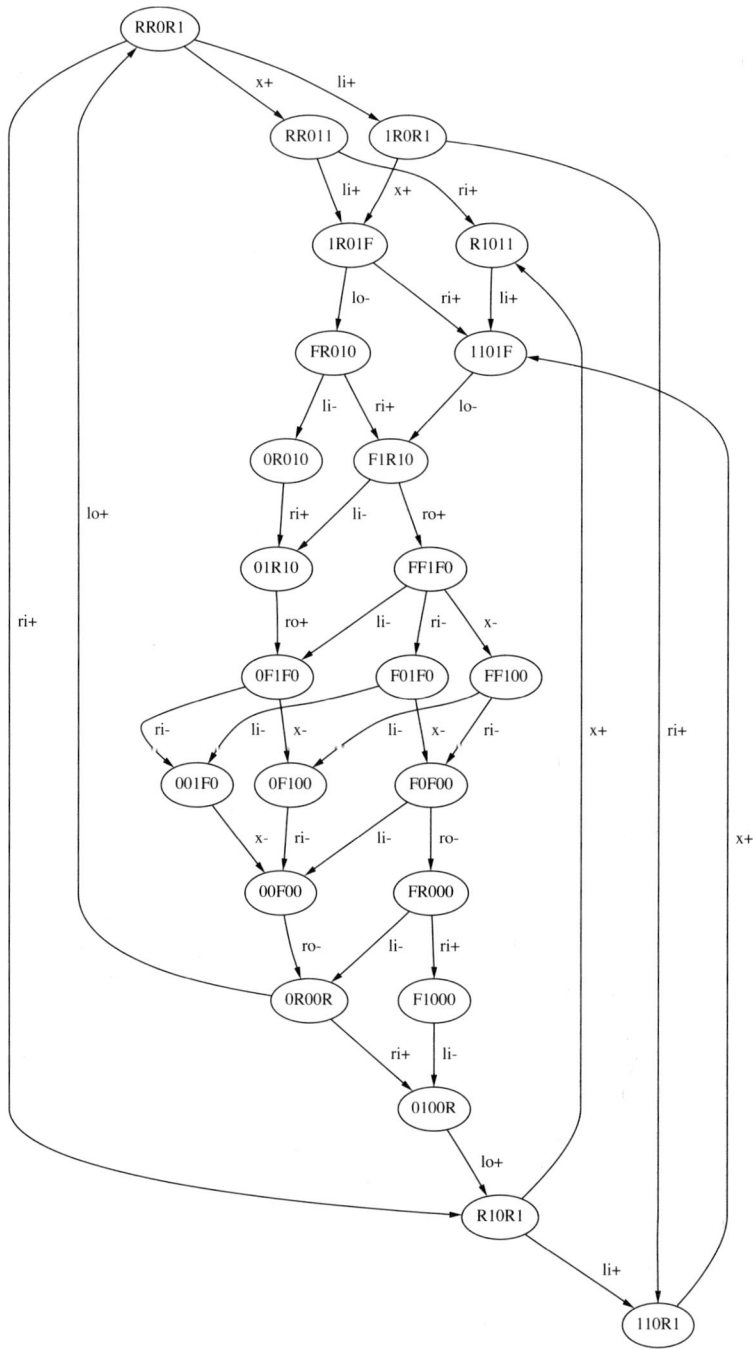

Fig. 6.29 SG for Problem 6.7.

7
Timed Circuits

Time is what prevents everything from happening at once.
—John Archibald Wheeler (1911–)

Dost thou love life?Then don't squander time, for that is the stuff life is made of.
—Benjamin Franklin

We must use time as a tool, not as a couch.
—John F. Kennedy

Time is a great teacher, but unfortunately it kills all its pupils.
—Hector Berlioz

In previous chapters, synthesis of asynchronous circuits has been performed using very limited knowledge about the delays in the circuit being designed. Although this makes for very robust systems, a range of delay from 0 to infinity (as in the speed-independent case) is extremely conservative. It is quite unlikely that large functional units could respond after no delay. It is equally unlikely that gates and wires would take an infinite amount of time to respond. When timing information is known, this information can be utilized to identify portions of the state space which are unreachable. These unreachable states introduce additional don't cares in the logic synthesis problem, and they can be used to optimize the implementation that is produced. In this chapter we present a design methodology that utilizes known timing information to produce *timed circuit* implementations.

7.1 MODELING TIMING

In this section we introduce semantics to support timing information during the synthesis process. This is done using a simple example.

Example 7.1.1 Consider the case where the shopkeeper actively calls the winery when he needs wine and the patron when he has wine to sell. To save time, he decides to call the patron immediately after calling the winery, without waiting for the wine to arrive. Although the patron gets quite irate if the wine is not there, the shopkeeper simply locks the door until the wine arrives. So the process goes like this: The shopkeeper calls the winery, calls the patron, peers out the window until he sees both the wine delivery boy and the patron, lets them in, and completes the sale.

After a while, he discovers that the winery always delivers its wine between 2 and 3 minutes after being called. The patron, on the other hand, takes at least 5 minutes to arrive, even longer when he is sleeping off the previous bottle. Using this timing information, he has determined that he does not need to keep the door locked. Furthermore, he can take a short nap behind the counter, and he only needs to wake when he hears the loud voice of his devoted patron. For he knows that when the patron arrives, the wine must already have been delivered, making the arrival of the wine redundant.

The timing relationships described in the example are depicted in Figure 7.1 using a TEL structure. Recall that the vertices of the graph are events and the edges are rules. Each rule is labeled with a bounded timing constraint of the form $[l, u]$, where l is the lower bound and u is the upper bound on the firing of the rule. In this example, all events are simple sequencing events and all level expressions are assumed to be *true*.

In order to analyze timed circuits and systems, it is necessary to determine the reachable *timed states*. An *untimed state* is a set of marked rules. A timed state is an untimed state from the original machine and the current value of all the active *timers* in that state. There is a timer t_i associated with each arc in the graph. A timer is allowed to advance by any amount less than its upper bound, resulting in a new timed state.

Example 7.1.2 In the example in Figure 7.1, the rule $r_6 = \langle$ *Wine is Purchased, Call Winery* $\rangle$ is initially marked (indicated with a dotted rule), so the initial untimed state is $\{r_6\}$. In the initial state, timer t_6 has value 0. Therefore, the initial timed state is $(\{r_6\}, t_6 = 0)$. In the initial state, the timer t_6 is allowed to advance by any amount less than or equal to 3.

Recall that when a timer reaches its lower bound, the rule becomes satisfied. When a timer reaches its upper bound, the rule is expired. An event enabled by a single rule must happen sometime after its rule becomes satisfied and before it becomes expired. When an event is enabled by multiple rules, it must happen after all of its rules are satisfied but before all of its rules are

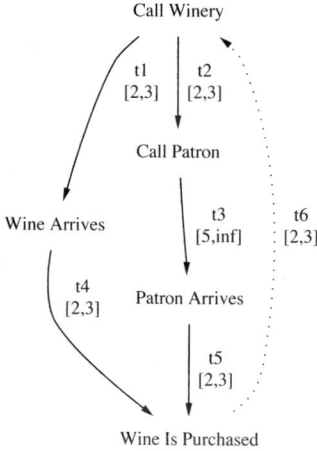

Fig. 7.1 Simple timing problem.

expired. To model the set of all possible behaviors, we extend the notion of allowed sequences to timed states and include the time at which a state transition occurs. These *timed sequences* are composed of transitions which can be either time advancement or a change in the untimed state.

Example 7.1.3 Let us consider one possible sequence of events. If $t_6 \geq 2$, the initial rule becomes satisfied. If t_6 reaches the value of 3, the rule is expired. The *Call Winery* event must happen sometime between 2 and 3 time units after the last bottle of wine is purchased. After the *Call Winery* event, timers t_1 and t_2 are initialized to 0. They must then advance in lockstep. When t_1 and t_2 reach a value of 2, the events *Wine Arrives* and *Call Patron* become *enabled*. At this point, time can continue to advance or one of these transitions can occur. Let us assume that time advances to time 2.1, and then the event *Call Patron* happens. After this event, the timer t_2 can be discarded, and a new timer t_3 is introduced with an initial value of 0. Time can be allowed to continue to advance until timer t_1 equals 3, at which point the *Wine Arrives* event will be forced to occur. Let us assume that the *Wine Arrives* when t_1 reaches 2.32. The result is that t_1 is discarded and t_4 is introduced with value 0 (note that t_3 currently has value 0.22). Time is again allowed to advance. When t_4 reaches a value of 2, the rule between *Wine Arrives* and *Wine Is Purchased* becomes satisfied, but the wine cannot be purchased because the patron has not yet arrived. Time continues until t_4 reaches a value of 3 when the same rule becomes expired, but the patron has still not arrived. At this point, we can discard t_4, as it no longer is needed to keep track of time. We replace it with a marker denoting that that rule has expired. Currently, t_3 is at a value of 3.22, so we must wait at least another 2.78 time units before the patron will arrive. When t_3 reaches a value of 5, the rule $\langle$ Call Patron, Patron Arrives, 5, inf $\rangle$ becomes satisfied, and the patron can arrive at any time. Again, we can discard the timer t_3, since there is an infinite upper bound. After the patron arrives, we introduce t_5, and between 2 and 3 time units, the wine is purchased, and we repeat. The corresponding timed sequence for this trace is as follows: $(([\{r_6\}, t_6 = 0], 0), ([\{r_6\}, t_6 = 2.22], 2.22), ([\{r_1, r_2\}, t_1 = t_2 = 0], 2.22), ([\{r_1, r_2\}, t_1 = t_2 = 2.1], 4.32), ([\{r_1, r_3\}, t_1 = 2.1, t_3 = 0], 4.32), ([\{r_1, r_3\}, t_1 = 2.32, t_3 = 0.22], 4.54), ([\{r_3, r_4\}, t_3 = 0.22, t_4 = 0], 4.54), ([\{r_3, r_4\}, t_3 = 3.22, t_4 = 3], 7.54), ([\{r_3, r_4\}, t_3 = 3.22], 7.54), ([\{r_3, r_4\}], 9.32), ([\{r_4, r_5\}, t_5 = 0], 9.32), ([\{r_4, r_5\}, t_5 = 2.2], 11.52), ([\{r_6\}, t_6 = 0], 11.52))$.

Since time can take on any real value, there is an uncountably infinite number of timed states and timed sequences. In order to perform timed state space exploration, it is necessary either to group the timed states into a finite number of equivalence classes or restrict the values that the timers can attain. In the rest of this chapter we address this problem through the development of an efficient method for *timed state space exploration*.

7.2 REGIONS

The first technique divides the timed state space for each untimed state into equivalence classes called *regions*. A region is described by the integer component of each timer and the relationship between the fractional components. As an example, consider a state with two timers t_1 and t_2, which are allowed to take any value between 0 and 5. The set of all possible equivalence classes is depicted in Figure 7.2(a). When the two timers have zero fractional components, the region is a point in space. When timer t_1 has a zero fractional component and timer t_2 has a nonzero fractional component, the region is a vertical line segment. When timer t_1 has a nonzero fractional component and t_2 has a zero fractional component, the region is a horizontal line segment. When both timers have nonzero but equal fractional components, the region is a diagonal line segment. Finally, when both timers have nonzero fractional components and one timer has a larger fractional component, the region is a triangle. Figure 7.2(a) depicts 171 distinct timed states.

Example 7.2.1 Let us consider a timed sequence from the example in Figure 7.1. Assume that the winery has just been called. This would put us in the timed state shown at the top of Figure 7.3(a). In this timed state, timers t_1 and t_2 would be initialized to 0 (i.e., $t_1 = t_2 = 0$), and their fractional components would both be equal to 0 [i.e., $f(t_1) = f(t_2) = 0$]. The geometric representation of this timed state is shown at the top of Figure 7.3(b), and it is simply a point at the origin.

In this timed state, the only possible next timed state is reached through time advancement. In other words, the fractional components

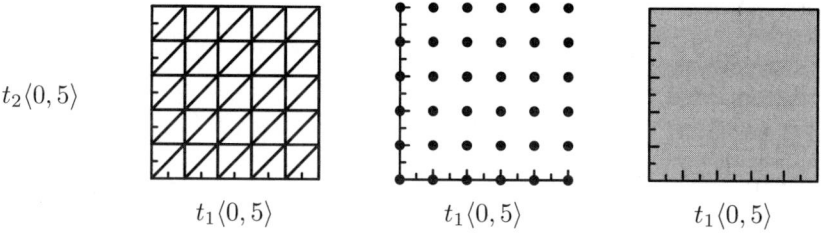

Fig. 7.2 (a) Possible timed states using regions. (b) Possible timed states using discrete time. (c) Timed state represented using a zone.

for the two timers are allowed to advance in lockstep, and they can take on any value greater than 0 and less than 1. The result is the second timed state shown in Figure 7.3(a). This timed state can be represented using a diagonal line segment as shown in the second graph in Figure 7.3(b).

Once the fractional components of these timers reach the value 1, we must move to a new timed state, increase the integer component of these timers to 1, and reset the fractional components. The result is the third timed state shown in Figure 7.3(a), which is again a point in space.

Advancing time is accomplished by allowing the fractional components to take on any value greater than 0 and less than 1, producing another diagonal line segment. When the fractional components reach 1, we again move to a new timed state where the integer components are increased to 2 and the fractional components are reset. In this timed state, there are three possible next timed states, depending on whether time is advanced, the wine arrives, or the patron is called. Let us assume that time is again advanced, leading to the timed state $t_1 = t_2 = 2, f(t_1) = f(t_2) > 0$.

In this timed state, the patron is called. In the resulting timed state, we eliminate the timer t_2 and introduce the timer t_3 with value 0. The fractional component of t_1 is greater than the fractional component of t_3, since we know that $t_1 > 0$ and $t_3 = 0$ [i.e., $f(t_1) > f(t_3) = 0$]. Pictorially, the timed state is a horizontal line segment, as shown in the seventh graph in Figure 7.3(b).

In this timed state, either the wine can arrive or time can be advanced. Let us assume that time advances. As time advances, we allow both fractional components to increase between 0 and 1. However, since we know that t_1's fractional component is greater than t_3's, the resulting region in space is the triangle shown in the eighth graph in Figure 7.3(b).

Once the fractional component for t_1 reaches 1, we must increase the integer component for timer t_1 to 3. Note that in this timed state, the fractional component can be anywhere between 0 and 1. Also note

264   TIMED CIRCUITS

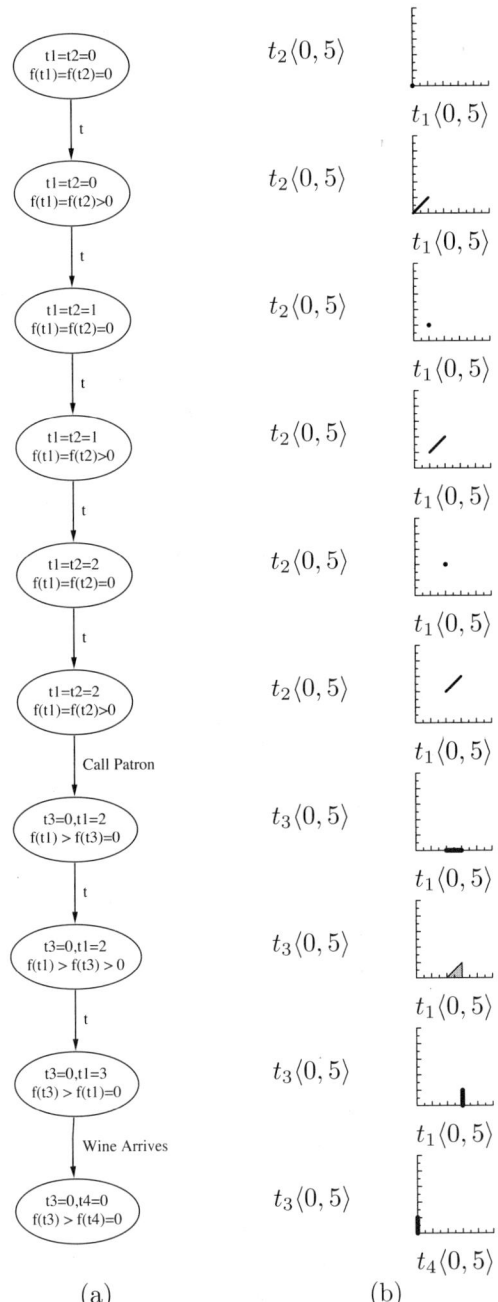

Fig. 7.3 (a) Sequence of timed states using regions. (b) Geometric representation of the sequence of timed states.

that since we reset the fractional component for t_1 to 0, the fractional component for t_3 is now greater than that for t_1.

In this timed state, the timer t_1 has reached its upper bound, so the event *Wine Arrives* must occur. To produce a new timed state, we eliminate the timer for t_1 and introduce the timer for t_4. We know that the fractional component for t_3 is greater than that of t_4 since $f(t_3) > 0$ and $f(t_4) = 0$.

This timed sequence represents only one of the many possible sequences of timed states from the point that the winery is called until both the patron is called and the wine arrives. The portion of the timed state space represented using regions for all possible timed sequences involving these three events is shown in Figure 7.4. Using this region-based technique, it requires 26 timed states to represent all the timing relationships for only four untimed states.

Indeed, the timed state explosion can be quite severe since the worst-case complexity is

$$|S|\frac{n!}{\ln 2}\left(\frac{k}{\ln 2}\right)^n 4^{1/k}$$

where S is the number of untimed states, n is the number of rules that can be enabled concurrently, and k is the maximum value of any timing constraint.

7.3 DISCRETE TIME

For timed Petri nets and TEL structures, all timing constraints are of the form $\leq$ or $\geq$, since timing bounds are inclusive. In other words, we never need to check if a timer is strictly less than or greater than a bound. It has been shown that in this case, the fractional components are not necessary. Therefore, we only need to keep track of the *discrete-time* states. If we consider again two timers that can take values from 0 to 5, the possible timed states are shown in Figure 7.2(b). The number of timed states in this figure is now only 36. The worst-case complexity is now

$$|S|(k+1)^n$$

This represents a reduction in worst-case complexity compared to region-based techniques by a factor of more than $n!$.

Example 7.3.1 Using discrete time, each time advancement step simply increments all timers by 1 time unit. For example, after the winery has been called, we are in a state with timers t_1 and t_2 initially 0 as shown at the top of Figure 7.5. In this timed state, time is advanced by 1 time unit, bringing us to a new timed state where the timers are each equal to 1. In this timed state, time must again be advanced bringing us to a new timed state where the timers equal 2. In this timed state, time can again advance, the wine can arrive, or the patron can be called. Exploring each possibility, we derive 13 possible timed states from the

266 TIMED CIRCUITS

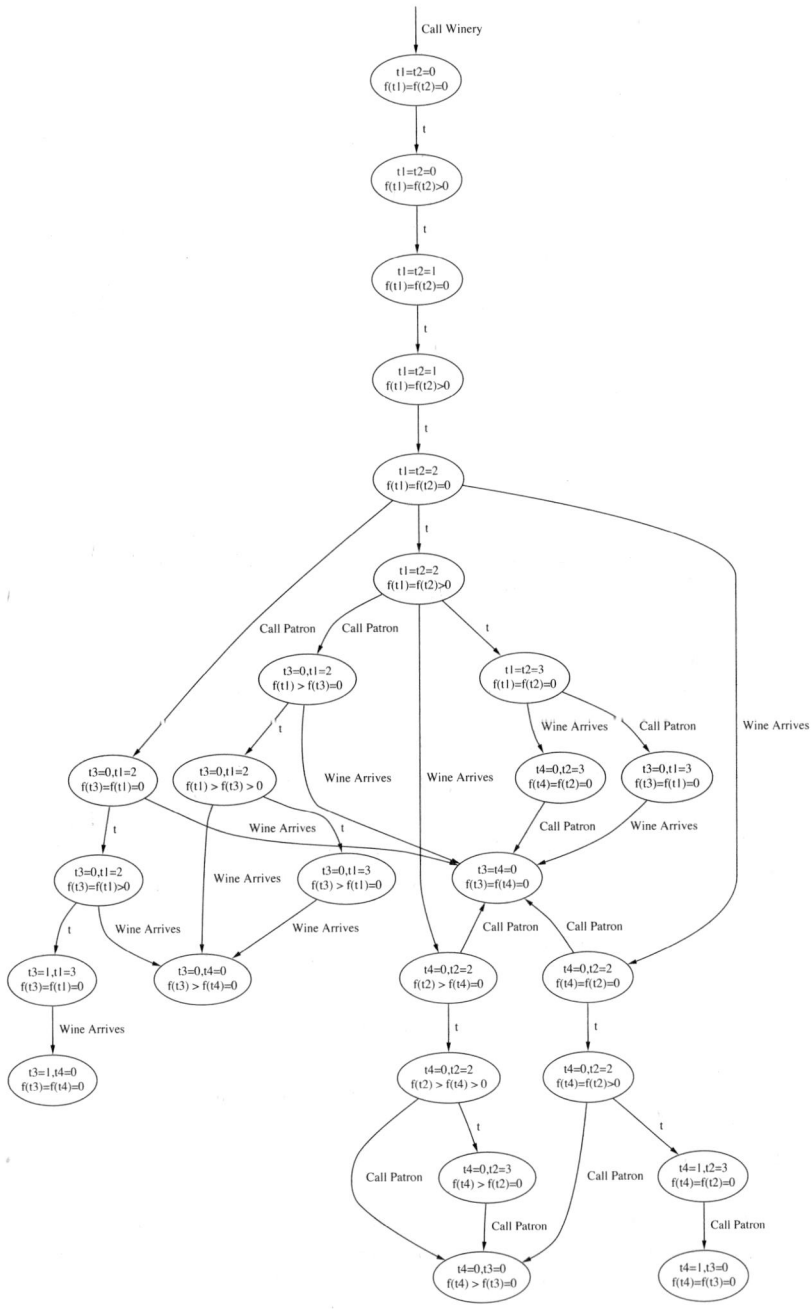

Fig. 7.4 Part of the timed state space using regions.

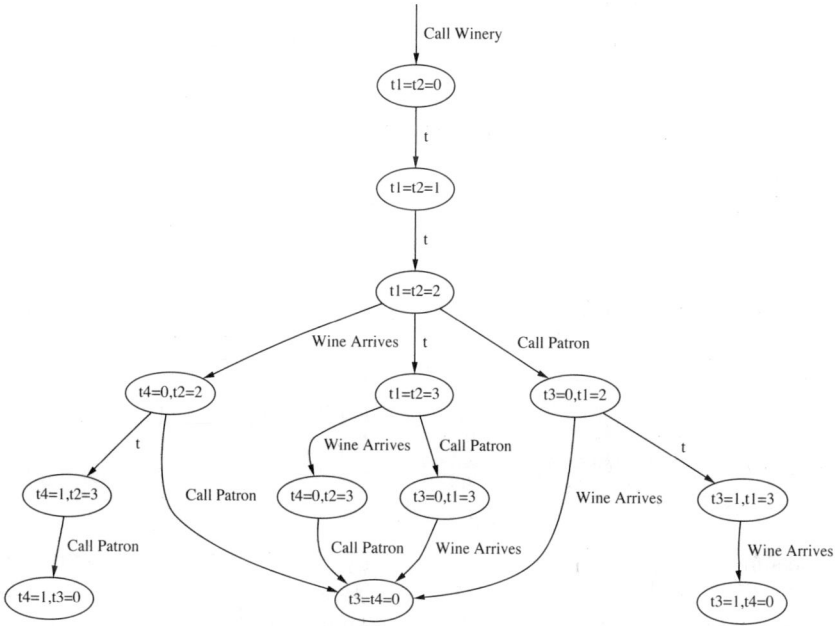

Fig. 7.5 Part of the timed state space using discrete time.

time the winery has been called until both the patron has been called and the wine has arrived. This is a reduction in half as compared with the 26 found using regions.

Unfortunately, the discrete-time technique is still exponential in the number of concurrent timers and size of the timing bounds. For example, if we change each timing bound of $[2,3]$ to $[19,31]$ and change $[5,inf]$ to $[53,inf]$, the total number of timed states grows from 69 to more than 3000. Changing the bounds to $[191,311]$ and $[531,inf]$ increases the number of discrete timed states to over 300,000!

7.4 ZONES

Another approach is to use convex polygons, called *zones*, to represent equivalence classes of timed states. For example, if there are two concurrent timers that can take on any value between 0 and 5, it can be represented using a single zone as shown in Figure 7.2(c). In other words, one zone is representing 171 regions or 36 discrete-time states.

Any convex polygon can be represented using a set of linear inequalities. To make them uniform, we introduce a *dummy timer* t_0 which always takes the value 0. For each pair of timers, we introduce an inequality of the form

$$t_j - t_i \leq m_{ij}$$

This set of inequalities is typically collected into a data structure called a *difference bound matrix* (DBM).

Example 7.4.1 For the example in Figure 7.2(c), the set of linear inequalities and the corresponding DBM are shown below.

$$
\begin{aligned}
t_0 - t_0 &\leq 0 \\
t_1 - t_0 &\leq 5 \\
t_2 - t_0 &\leq 5 \\
t_0 - t_1 &\leq 0 \\
t_1 - t_1 &\leq 0 \\
t_2 - t_1 &\leq 5 \\
t_0 - t_2 &\leq 0 \\
t_1 - t_2 &\leq 5 \\
t_2 - t_2 &\leq 0
\end{aligned}
$$

	t_0	t_1	t_2
t_0	0	5	5
t_1	0	0	5
t_2	0	5	0

The same zone can be represented by many different sets of linear inequalities and thus many different DBMs. During state space exploration, it is necessary to determine when you have encountered a timed state that you have seen before. In other words, when you find an untimed state that you have seen before, you must also check that the zone is the same as before. If there are multiple different representations, you may encounter the same timed state and not know it. Fortunately, there is a canonical DBM representation for each zone when all entries are maximally tight.

Example 7.4.2 Consider the following DBM:

	t_0	t_1	t_2
t_0	0	5	5
t_1	0	0	7
t_2	0	5	0

This DBM indicates the following relationship:

$$t_2 - t_1 \leq 7$$

However, the DBM also indicates these relationships:

$$t_2 - t_0 \leq 5$$

$$t_0 - t_1 \leq 0$$

If we add them together, we get the following relationship:

$$t_2 - t_1 \leq 5$$

Therefore, the first inequality is not maximally tight in that it is not possible for $t_2 - t_1$ actually to take any value in the range greater than 5 and less than 7. Therefore, the constraint can be tightened to produce our original DBM.

```
recanonicalization(M) {
    for k = 1 to n
        for i = 1 to n
            for j = 1 to n
                if (m_ij > m_ik + m_kj) then
                    m_ij = m_ik + m_kj;
}
```

Fig. 7.6 Floyd's all-pairs shortest-path algorithm.

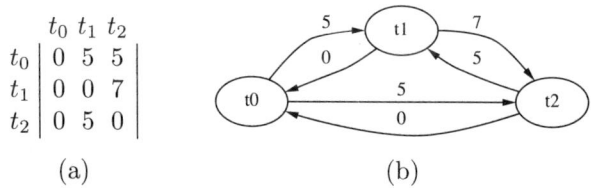

	t_0	t_1	t_2
t_0	0	5	5
t_1	0	0	7
t_2	0	5	0

(a) (b)

Fig. 7.7 (a) Example DBM. (b) Its digraph representation.

Finding the canonical DBM is equivalent to finding all pairs of shortest-paths in a graph. To cast it as the shortest path problem, we create a labeled digraph where there is a vertex for each timer t_i and an arc from t_i to t_j for each linear inequality of the form $t_j - t_i \leq m_{ij}$ when $i \neq j$. Each arc is labeled by m_{ij}. We use Floyd's all-pairs shortest-path algorithm to perform *recanonicalization*. Floyd's algorithm is an efficient algorithm for finding all pairs of shortest paths in a graph, and it is shown in Figure 7.6. This algorithm examines each pair of vertices t_i and t_j in turn and attempts to find an alternative route between them which passes through some t_k. If this alternative route is shorter than the direct route, we can reduce the distance of the direct route to this value without changing the shortest path. This is equivalent in the DBM case of tightening a loose bound.

Example 7.4.3 The digraph representation of the DBM shown in Figure 7.7(a) appears in Figure 7.7(b). For this graph, the algorithm would discover that the route from t_1 to t_2 through t_0 is only distance 5, while the route from t_1 directly to t_2 is distance 7. Therefore, the arc between t_1 and t_2 can be reduced to 5.

During timed state space exploration using zones, a state transition occurs as the result of a rule firing. A rule, r_j, can fire if its timer, t_j, can reach a value that makes it satisfied (i.e., $m_{0j} \geq l_j$). If this rule firing is the last one enabling an event, an event happens, leading to new rules being enabled. Whether or not an event happens, it is necessary to calculate a new zone from the preceding one. This is accomplished using the algorithm shown in Figure 7.8. This algorithm takes the DBM for the original zone, the rule that

```
update_zone(M,r_j,event_fired,R_en,R_new) {
    if m_{j0} > -l_j then m_{j0} = -l_j;        /* Restrict to lower bound.*/
    recanonicalize(M);                           /* Tighten loose bounds.*/
    project(M,r_j);         /* Project out timer for rule that fired.*/
    if (event_fired) then
        foreach r_i ∈ R_new {                    /* Extend DBM with new timers.*/
            m_{i0} = m_{0i} = 0;
            foreach r_k ∈ R_new
                m_{ik} = m_{ki} = 0;
            foreach r_k ∈ (R_en − R_new){
                m_{ik} = m_{0k};
                m_{ki} = m_{k0};
            }
        }
    foreach r_i ∈ R_en                           /* Advance time.*/
        m_{0i} = u_i;
    recanonicalize(M);                           /* Tighten loose bounds.*/
    normalize(M,R_en);                           /* Adjust infinite upper bounds.*/
}
```

Fig. 7.8 Algorithm to update the zone.

fired, a flag indicating whether an event fired, the set of currently enabled rules, and the set of those rules which are newly enabled by this rule firing.

As an example, consider the firing of a rule $r_j = \langle e_j, f_j, l_j, u_j \rangle$, where e_j is the enabling event, f_j is the enabled event, l_j is the lower bound of the corresponding timer t_j, and u_j is the upper bound on the timer. The first step is to *restrict* the DBM to indicate that for this rule to have fired, its timer must have reached its lower bound. In other words, if $t_0 - t_j > -l_j$, we must set m_{j0} to $-l_j$. This additional constraint may result in some of the other constraints no longer being maximally tight. Therefore, it is necessary to recanonicalize the DBM using Floyd's algorithm. The next step is to *project* the row and column corresponding to timer t_j since once this rule has fired, we no longer need to maintain information about its timer.

If the rule firing causes an event, this event may enable new rules. For each of these newly enabled rules, we must introduce a new timer which corresponds to a new row and column in the DBM. For each newly enabled rule (i.e., $r_i \in R_{new}$), we add a new timer t_i. We initialize the lower and upper bounds of this timer (i.e., m_{i0} and m_{0i}) to 0 to represent that the timer is initialized to 0. The time separation between the timer for each pair of rules that has been newly enabled is also set to 0, as these timers got initialized at the same time. Finally, the remaining new row entries, m_{ik}, are set equal to the upper bounds of their timers t_k (i.e., m_{0k}), and the remaining new column entries, m_{ki}, are set equal to the lower bounds of their timers t_k (i.e., m_{k0}).

ZONES   271

```
normalize(M, R_en) {
  foreach r_i ∈ R_en
    if (m_i0 < -premax(r_i)) then     /* Reduce timer to premax value.*/
      foreach r_j ∈ R_en {
        m_ij = m_ij - (m_i0 + premax(r_i));
        m_ji = m_ji + (m_i0 + premax(r_i));
      }
  foreach r_i ∈ R_en                  /* Adjust maximums.*/
    if (m_0i > premax(r_i)) then
      m_0i = max_j(min(m_0j, premax(r_j)) - m_ij);
  recanonicalize(M);                  /* Tighten loose bounds.*/
}
```

Fig. 7.9 Normalization algorithm.

The next step is to *advance time* by setting all timers to their upper bound (i.e., $m_{0i} = u_i$). The resulting DBM may now again contain entries that are not maximally tight. Therefore, we again recanonicalize the DBM.

The final step is to *normalize* the DBM to account for rules with infinite upper bounds. This is necessary to keep the state space finite. The algorithm for normalization is shown in Figure 7.9. This algorithm uses the function *premax*, which takes a rule r_i and returns u_i if it is finite and returns l_i if u_i is infinite. This algorithm works on the assumption that the value of a timer of a rule with an infinite upper bound becomes irrelevant once it has exceeded its lower bound. It simply must remember that it reached its lower bound. The algorithm adjusts timers accordingly, in three steps. First, if the lower bound of the timer on a rule r_i has already exceeded its *premax* value, the zone is adjusted to reduce it back to its *premax* value. This may reduce some timers below their *premax* or current maximum value, so it may be necessary to allow the timers to take a value exceeding *premax*. Therefore, for each timer that has exceeded its *premax* value, we find the minimum maximum value needed which does not constrain any other rules. Finally, the DBM is recanonicalized again.

Example 7.4.4 Let us illustrate timed state space exploration with zones using the example shown in Figure 7.1. The initial state has only one timer t_6, so our initial DBM is only two by two (i.e., t_0 and t_6). We begin by initializing the DBM to all zeros to represent that all timers are initially zero. We then advance time by setting the top row entry for t_6 to the maximum value allowed for t_6, which is 3. We then recanonicalize and normalize, which simply results in the same DBM. This sequence of steps is shown in Figure 7.10.

The only possible rule that can fire in this initial zone is r_6. Since the lower bound on this rule is 2, we need to restrict the timer t_6 in the DBM such that it cannot be less than 2 (see Figure 7.11). Again, recanonicalization has no effect. Next, we project out the row and column associated with timer t_6. The firing of this rule results in the

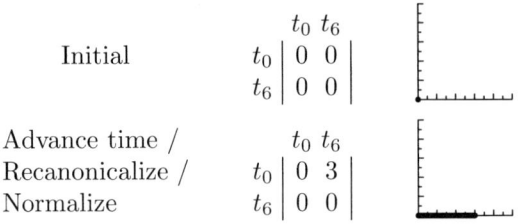

Initial		t_0	t_6
	t_0	0	0
	t_6	0	0

Advance time / Recanonicalize / Normalize		t_0	t_6
	t_0	0	3
	t_6	0	0

Fig. 7.10 Initial zone.

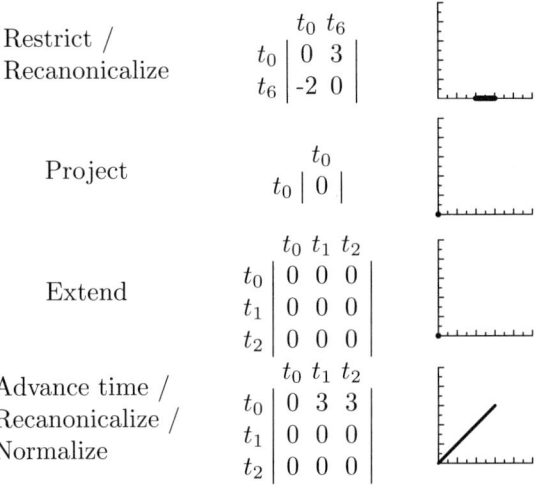

Restrict / Recanonicalize		t_0	t_6
	t_0	0	3
	t_6	-2	0

Project		t_0
	t_0	0

Extend		t_0	t_1	t_2
	t_0	0	0	0
	t_1	0	0	0
	t_2	0	0	0

Advance time / Recanonicalize / Normalize		t_0	t_1	t_2
	t_0	0	3	3
	t_1	0	0	0
	t_2	0	0	0

Fig. 7.11 Example of zone creation after the winery is called.

event *Call Winery*, which enables two new rules. So we need to extend the DBM to include the timers t_1 and t_2. We then advance time by setting the 0th row of the DBM to the upper bounds of these timers, three in each case (see Figure 7.11).

In this new zone, there are two possible rule firings. We can fire either the rule r_1 or r_2. Let us first fire r_1. Note that we must remember that we had a choice and come back and consider firing these rules in the other order. Since the lower bound on this rule is 2, we must restrict timer t_1 to be at least 2. This matrix is not maximally tight, so we must recanonicalize. The entry for timer t_2 in the 0th column becomes -2, since from t_2 to t_1 is 0 and t_1 to t_0 is -2. Next, we eliminate the columns corresponding to timer t_1. This rule firing results in the event *Wine Arrives*, so a new timer, t_4, is introduced. We extend the DBM to include t_4. We initialize its lower and upper bounds to 0 (i.e., $m_{40} = m_{04} = 0$). The entry m_{42} is filled in with 3 since timer t_2 could be as large as 3 (i.e., $m_{02} = 3$). The entry m_{24} is filled in with -2 since timer t_2 is at least 2 (i.e., $m_{20} = -2$). Next, we advance

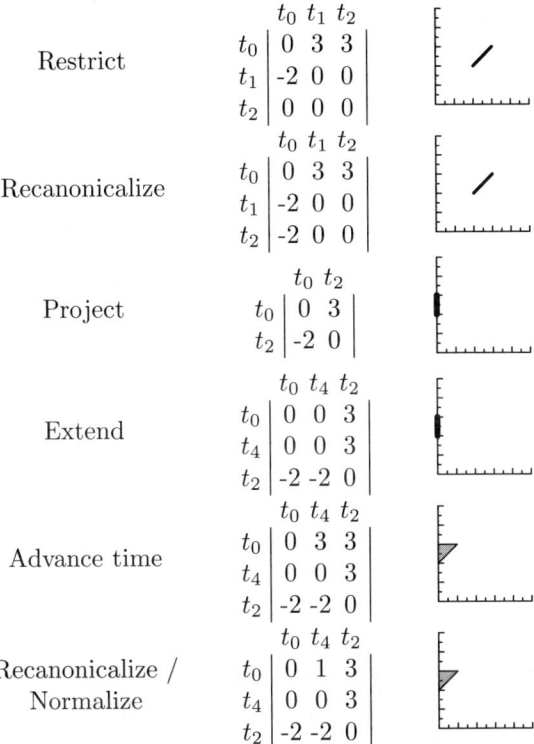

Fig. 7.12 Example of zone creation after the wine arrives.

time by setting the 0th row to the maximum value of all the timers. After recanonicalization, though, the entry m_{04} is reduced to 1, since timer t_4 cannot advance beyond 1 without forcing timer t_2 to expire. In other words, the constraint between t_0 and t_2 is 3 and t_2 and t_4 is -2, which sum to 1. There are no enabled rules with infinite maximums, so normalization has no effect (see Figure 7.12).

In this zone, there are two enabled rules, but only one of the rules is satisfied and can fire. The timer t_4 has a maximum value of 1 (see entry m_{04} in Figure 7.13), so it is not satisfied. Therefore, the only rule that can fire is r_2. Timer t_2 is already at its lower bound, so the restrict and first recanonicalization have no effect. Next, we eliminate the row and column associated with timer t_2. The firing of this rule results in the event *Call Patron*, which enables a new rule, so we must extend the DBM with the new timer t_3. The new entries are all 0 except that m_{34} becomes 1 since timer t_4 may be as high as 1 at this point (see m_{04}). We again advance time by introducing the upper bounds into the 0th row. In this case, one of these entries is infinity. However, after recanonicalization, this infinity is reduced to 3 since the timer t_3 cannot exceed 3 without forcing timer t_4 to expire. There is now a rule with an

274 TIMED CIRCUITS

Restrict / Recononicalize

$$\begin{array}{c|ccc} & t_0 & t_4 & t_2 \\ \hline t_0 & 0 & 1 & 3 \\ t_4 & 0 & 0 & 3 \\ t_2 & -2 & -2 & 0 \end{array}$$

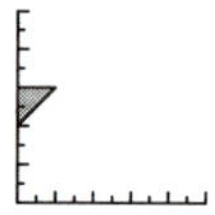

Project

$$\begin{array}{c|cc} & t_0 & t_4 \\ \hline t_0 & 0 & 1 \\ t_4 & 0 & 0 \end{array}$$

Extend

$$\begin{array}{c|ccc} & t_0 & t_4 & t_3 \\ \hline t_0 & 0 & 1 & 0 \\ t_4 & 0 & 0 & 0 \\ t_3 & 0 & 1 & 0 \end{array}$$

Advance time

$$\begin{array}{c|ccc} & t_0 & t_4 & t_3 \\ \hline t_0 & 0 & 3 & \infty \\ t_4 & 0 & 0 & 0 \\ t_3 & 0 & 1 & 0 \end{array}$$

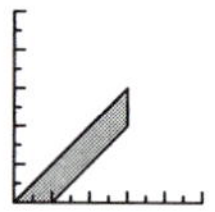

Recononicalize / Normalize

$$\begin{array}{c|ccc} & t_0 & t_4 & t_3 \\ \hline t_0 & 0 & 3 & 3 \\ t_4 & 0 & 0 & 0 \\ t_3 & 0 & 1 & 0 \end{array}$$

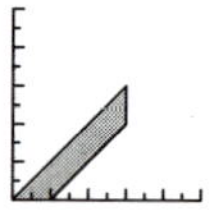

Fig. 7.13 Example of zone creation after wine arrives and patron has been called.

infinite maximum, r_3, but it has not yet reached its lower bound, and so normalization has no effect. These steps are shown in Figure 7.13.

In this zone, again only one of the two enabled rules can fire. The rule r_3 cannot fire since the upper bound of its timer is only 3. Therefore, the only rule that can fire is r_4. To fire this rule, we first restrict timer t_4 to be at least 2, the lower bound of this rule. After recanonicalization, the m_{30} entry becomes -1, indicating that timer t_3 must be at least 1 at this point. Next, we project out the rows and columns associated with timer t_4. Note that the firing of this rule does not result in the occurrence of any event since the rule r_5 has not fired yet. Therefore, no new timers need to be introduced. We advance time by setting the entry m_{03} to ∞, and recanonicalize. The timer t_3 has now exceeded its lower bound, and it has an infinite upper bound. Therefore, normalization changes this timer back to its lower bound. The result is shown in Figure 7.14.

In this zone, the only possible rule that can fire is r_3. We restrict the timer t_5 to its lower bound, 5, and recanonicalize. We then eliminate the row and column associated with timer t_5. The firing of this rule results in the event *Patron Arrives*, which enables a new rule, so we need to extend the DBM to include the new timer t_5. We then advance time on this timer, recanonicalize, and normalize (see Figure 7.15).

Restrict	$\begin{array}{c|ccc} & t_0 & t_4 & t_3 \\ \hline t_0 & 0 & 3 & 3 \\ t_4 & -2 & 0 & 0 \\ t_3 & 0 & 1 & 0 \end{array}$	
Recononicalize	$\begin{array}{c|ccc} & t_0 & t_4 & t_3 \\ \hline t_0 & 0 & 3 & 3 \\ t_4 & -2 & 0 & 0 \\ t_3 & -1 & 1 & 0 \end{array}$	
Project	$\begin{array}{c|cc} & t_0 & t_3 \\ \hline t_0 & 0 & 3 \\ t_3 & -1 & 0 \end{array}$	
Advance time / Recononicalize	$\begin{array}{c|cc} & t_0 & t_3 \\ \hline t_0 & 0 & \infty \\ t_3 & -1 & 0 \end{array}$	
Normalize	$\begin{array}{c|cc} & t_0 & t_3 \\ \hline t_0 & 0 & 5 \\ t_3 & -1 & 0 \end{array}$	

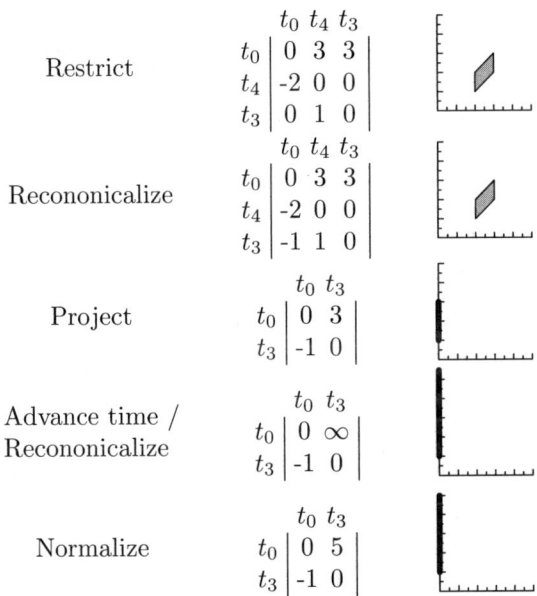

Fig. 7.14 Example of zone creation after a rule expires.

Restrict / Recononicalize	$\begin{array}{c|cc} & t_0 & t_3 \\ \hline t_0 & 0 & 5 \\ t_3 & -5 & 0 \end{array}$	
Project	$\begin{array}{c|c} & t_0 \\ \hline t_0 & 0 \end{array}$	
Extend	$\begin{array}{c|cc} & t_0 & t_5 \\ \hline t_0 & 0 & 0 \\ t_5 & 0 & 0 \end{array}$	
Advance time / Recononicalize / Normalize	$\begin{array}{c|cc} & t_0 & t_5 \\ \hline t_0 & 0 & 3 \\ t_5 & 0 & 0 \end{array}$	

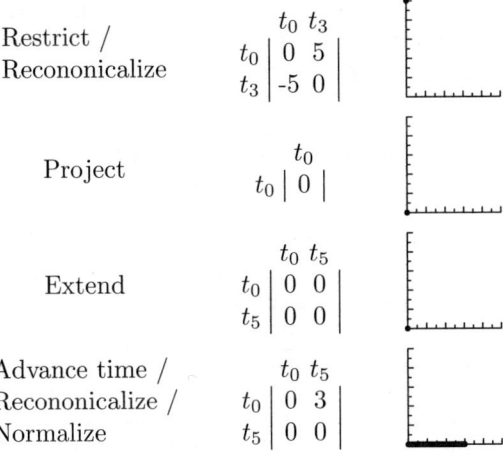

Fig. 7.15 Example of zone creation after the patron arrives.

276 TIMED CIRCUITS

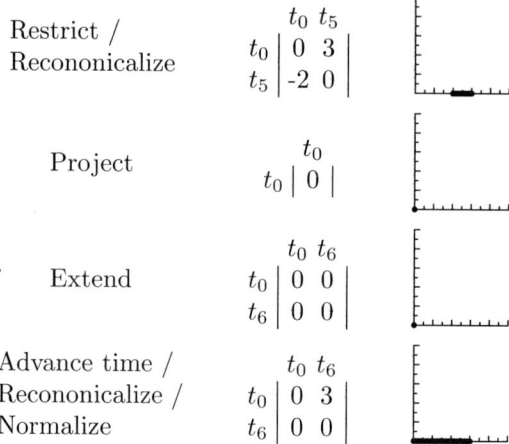

Fig. 7.16 Example of zone creation after the wine is purchased.

In this zone, the rule r_5 fires. We restrict the zone, recanonicalize, and project out timer t_5 (see Figure 7.16). The firing of this rule results in the event *Wine Is Purchased*, since the wine has already arrived and the corresponding rule has fired. Therefore, we extend the DBM to include the newly started timer t_6, we advance time, recanonicalize, and normalize (see Figure 7.16). Note that this timed state matches our initial time state, so we are done exploring down this path.

Remember that earlier we had two satisfied rules, and we chose to let the wine arrive before we called the patron. We must now backtrack to this point and consider the alternative order. This time we call the patron before the wine arrives. In other words, we fire the rule r_2 first (see Figure 7.17). This time, we restrict timer t_2 to its lower bound, 2. We then recanonicalize and project out timer t_2. The event *Call Patron* occurs as a result of this rule firing, which enables a new rule. We must extend the DBM to include the new timer, t_3. The entry m_{13} is set to -2 since timer t_1 is at least 2 at this point, and entry m_{31} is set to 3 since timer t_2 is no more than 3. We then advance time, recanonicalize, and normalize (see Figure 7.17).

In this zone, we can only fire the rule r_1. Its timer is already at its lower bound, so the restrict and first recanonicalization have no effect. We then project out the row and column associated with timer t_1. The event *Wine Arrives* fires, enabling a new rule and introducing the timer t_4. We extend the DBM to include this new timer. All new entries are set to 0 except that m_{43} is 1 since timer t_3 may be as high as 1. We then advance time and recanonicalize (see Figure 7.18). Normalization has no effect since t_3 has not yet reached its lower bound.

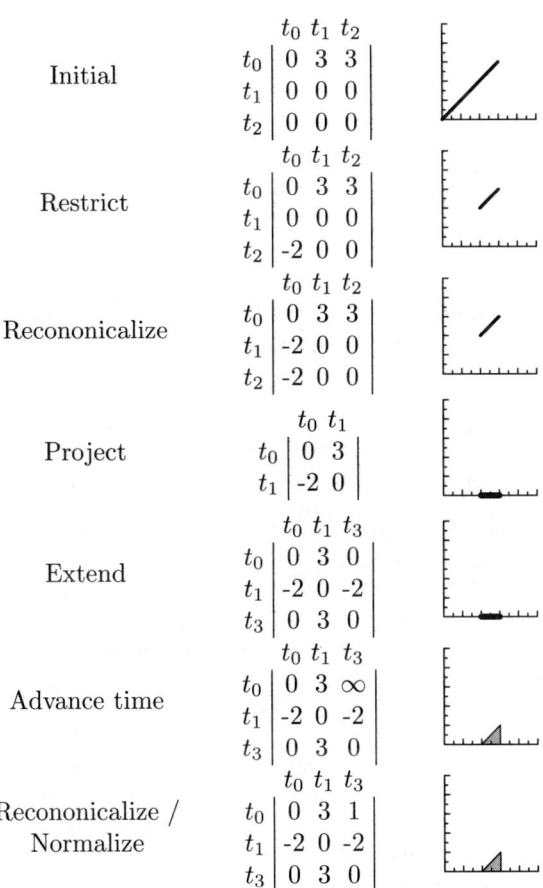

Fig. 7.17 Example of zone creation after the patron is called.

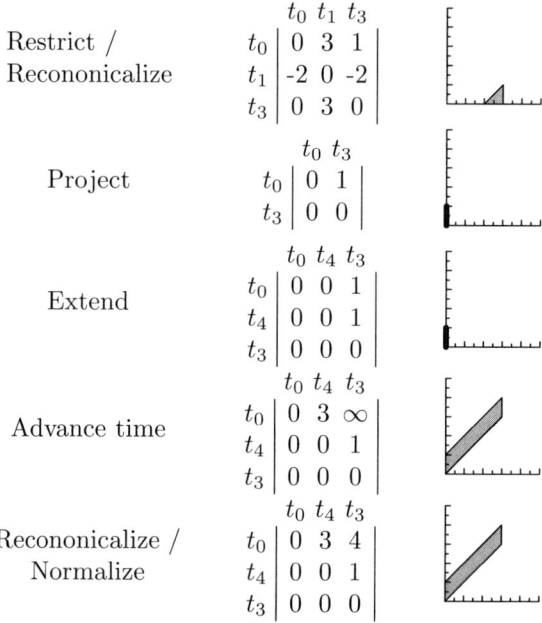

Fig. 7.18 Example of zone creation after the patron has been called and wine arrives.

In this zone, we can only fire the rule r_4. We restrict the timer t_4 to be at least 2, recanonicalize, and project out the timer t_4 (see Figure 7.19). We then advance time and again recanonicalize. Since t_3 has now exceeded its lower bound, normalization brings it back down to its lower bound. The resulting DBM is shown at the bottom of Figure 7.19. Compare this with the one shown at the bottom of Figure 7.14. This new DBM represents a zone that is a subset of the one we found before. This can be seen either in the picture or by the fact that all the entries in the DBM are less than or equal to those found before. Therefore, we do not need to continue state space exploration beyond this point because any possible future would have been found by our exploration beginning with the zone found in Figure 7.14.

Since there are no more unexplored paths, the entire timed state has been found, and it is shown in Figure 7.20. It takes only eight zones to represent the entire timed state space, whereas it took 26 regions or 13 discrete-time states to represent just a part of it. Another important consideration is that if we change the timers such that bounds of $[2, 3]$ are set to $[19, 31]$ and the bounds of $[5, \infty]$ are set to $[53, \infty]$, the number of timed states does not change when represented as zones.

ZONES   279

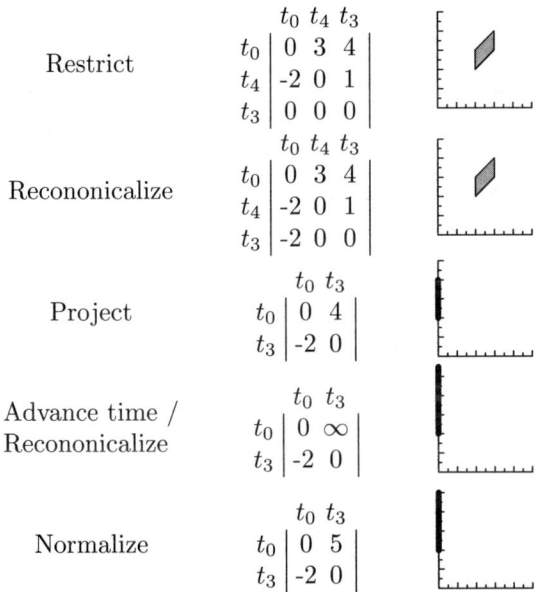

Restrict

$$\begin{array}{c|ccc} & t_0 & t_4 & t_3 \\ \hline t_0 & 0 & 3 & 4 \\ t_4 & -2 & 0 & 1 \\ t_3 & 0 & 0 & 0 \end{array}$$

Recononicalize

$$\begin{array}{c|ccc} & t_0 & t_4 & t_3 \\ \hline t_0 & 0 & 3 & 4 \\ t_4 & -2 & 0 & 1 \\ t_3 & -2 & 0 & 0 \end{array}$$

Project

$$\begin{array}{c|cc} & t_0 & t_3 \\ \hline t_0 & 0 & 4 \\ t_3 & -2 & 0 \end{array}$$

Advance time / Recononicalize

$$\begin{array}{c|cc} & t_0 & t_3 \\ \hline t_0 & 0 & \infty \\ t_3 & -2 & 0 \end{array}$$

Normalize

$$\begin{array}{c|cc} & t_0 & t_3 \\ \hline t_0 & 0 & 5 \\ t_3 & -2 & 0 \end{array}$$

Fig. 7.19 Example of zone creation after a rule expires.

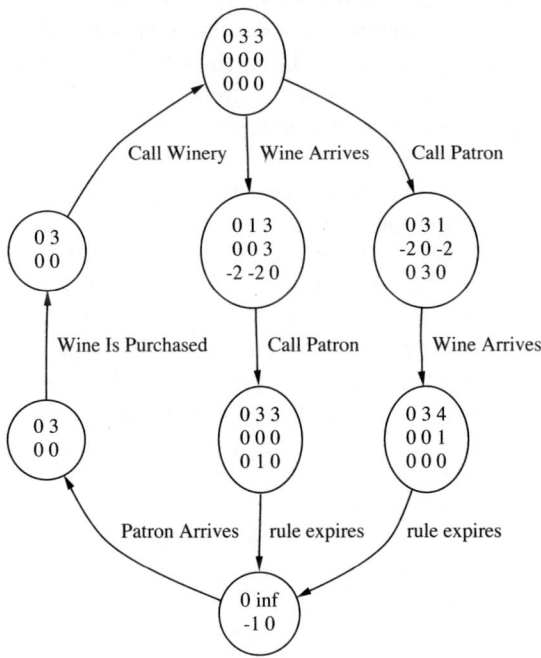

Fig. 7.20 Timed state space using zones.

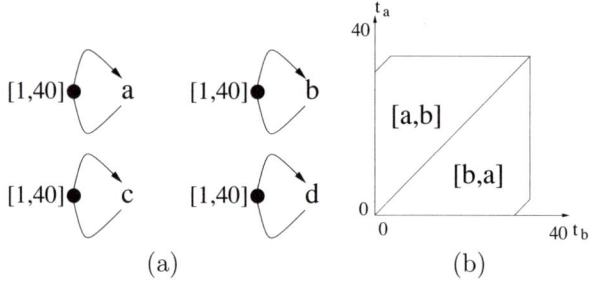

Fig. 7.21 (a) Adverse example. (b) Zones for two firing sequences, $[a, b]$ and $[b, a]$.

7.5 POSET TIMING

The zone approach works well for a lot of examples, but when there is a high degree of concurrency, there can be more zones than discrete-time states. Consider the simple example shown in Figure 7.21(a). This example obviously has only one untimed state. There are an astronomical 2,825,761 discrete-time states. Even worse, there are 219,977,777 zones!

The reason for the explosion in the number of zones can be illustrated by considering zones found for two possible sequences, $[a, b]$ and $[b, a]$, shown in Figure 7.21(b). The upper zone is found for the sequence $[a, b]$, while the lower zone is found for the sequence $[b, a]$. Even though these two sequences result in the same untimed state, they result in different zones. The zone reflects the order in which the two concurrent events occurred in the sequence. In fact, as the length of the sequence, n, increases, the number of zones grows like $n!$. When linear sequences of events are used to find the timed state space, it is not possible to distinguish concurrency from causality.

In order to separate concurrency from causality, *POSET timing* finds the timed state space by considering *partially ordered sets* (POSETs) of events rather than linear sequences. A graphical representation of a POSET is shown in Figure 7.22(a). This POSET represents both the sequence $[a, b]$ and the sequence $[b, a]$. For each POSET we derive a *POSET matrix* which includes the time separation between each pair of events in the POSET. The POSET matrix shown in Figure 7.22(b) indicates that a and b can occur between 1 and 40 time units after the reset event, r, and that either a or b could occur as much as 39 times units after the other.

After a rule is fired during timed state space exploration using zones and POSETs, the algorithm shown in Figure 7.23 is used to update the POSET matrix and find the corresponding zone. This algorithm takes the old POSET matrix and zone, the rule that fired, a flag that indicates if an event has fired, and the set of enabled rules. If the rule firing resulted in an event firing, we must update the POSET matrix and derive a new zone. First, the newly fired event, f_j, is added to the POSET matrix. Timing relationships are added

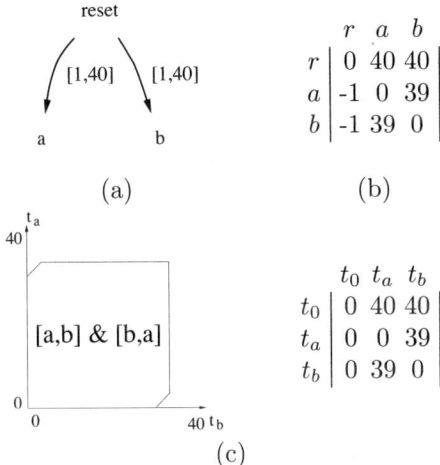

Fig. 7.22 (a) POSET graph. (b) POSET matrix for sequences [a, b] and [b, a]. (c) Zone found for either the sequence [a, b] or [b, a] using the POSET matrix.

between this event and all other events in the POSET matrix. If an event, e_i, in the POSET matrix is *causal* to event f_j, the separation is set to the bounds (i.e., $p_{ji} = -l$ and $p_{ij} = u$). We say that an event e_i is causal to another event f_j if this event is the enabling event for the last rule that fires before f_j fires. If event e_i must directly precede f_j but it is not causal to e_j, the minimum separation is set to the lower bound and the maximum left unbounded (i.e., $p_{ji} = -l$ and $p_{ij} = \infty$). If e_i does not precede f_j directly, no timing relationship is established between these events. The algorithm then recanonicalizes the new POSET matrix and projects out any events that are no longer needed. Events can be projected once there no longer exists any enabled rules that have that event as an enabling event. Next, the algorithm uses the POSET matrix, P, to derive the zone. The algorithm begins by setting the minimums to 0 (i.e., $m_{i0} = 0$) and setting the maximums to the upper bound (i.e., $m_{0j} = u_j$). It then copies the relevant time separations from the POSET matrix to the zone. Consider two timers t_i and t_j, which correspond to two rules with enabling events e_i and e_j, respectively. The m_{ij} entry is found by copying the p_{ij} entry from the POSET matrix. This new zone is then recanonicalized and normalized. If the rule firing did not result in an event firing, the algorithm simply updates the previous zone by projecting out the rule that fired, advancing time, recanonicalizing, and normalizing.

The zone for the POSET in Figure 7.22(b) is shown in Figure 7.22(c). Note that the zone now includes both zones that had been found previously. In fact, if we use this approach to analyze the example in Figure 7.21, we find exactly one zone for the one untimed state.

```
update_poset(P,M,r_j,event_fired,R_en) {
  if (event_fired) then {
    foreach e_i ∈ P                    /* Update POSET matrix.*/
      if e_i is causal to f_j then {
        p_ji = -l;
        p_ij = u;
      } else if e_i directly precedes e_j then {
        p_ji = -l;
        p_ij = ∞;
      } else {
        p_ji = ∞;
        p_ij = ∞;
      }
    recanonicalize(P);                 /* Tighten loose bounds.*/
    project(P);                        /* Project events no longer needed.*/
    foreach r_i ∈ R_en {               /* Create a zone.*/
      m_i0 = 0;                        /* Set minimums to 0.*/
      m_0i = u_i;                      /* Set maximums to upper bound.*/
      foreach r_j ∈ R_en
        m_ij = p_ij;                   /* Copy relevant timing from POSET matrix.*/
  } else {
    project(M,r_j);                    /* Project out timer for rule that fired.*/
    foreach r_i ∈ R_en                 /* Advance time.*/
      m_0i = u_i;
  }
  recanonicalize(M);                   /* Tighten loose bounds.*/
  normalize(M,R_en);                   /* Adjust infinite upper bounds.*/
}
```

Fig. 7.23 Algorithm to update the POSET and zone.

Initial POSET

$r \begin{array}{|c|} r \\ \hline 0 \end{array}$

reset

Initial zone /
Recononicalize /
Normalize

$\begin{array}{c|cc} & t_0 & t_6 \\ \hline t_0 & 0 & 3 \\ t_6 & 0 & 0 \end{array}$

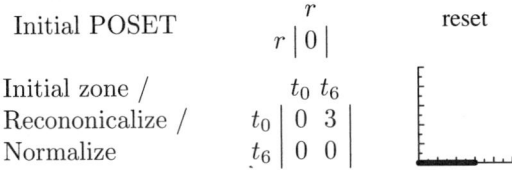

Fig. 7.24 Initial zone using POSETs.

Extend POSET

$\begin{array}{c|cc} & r & cw \\ \hline r & 0 & 3 \\ cw & -2 & 0 \end{array}$

reset
↓
[2,3]
↓
Call Winery

Project POSET

$\begin{array}{c|c} & cw \\ \hline cw & 0 \end{array}$

Initial zone /
Recononicalize /
Normalize

$\begin{array}{c|ccc} & t_0 & t_1 & t_2 \\ \hline t_0 & 0 & 3 & 3 \\ t_1 & 0 & 0 & 0 \\ t_2 & 0 & 0 & 0 \end{array}$

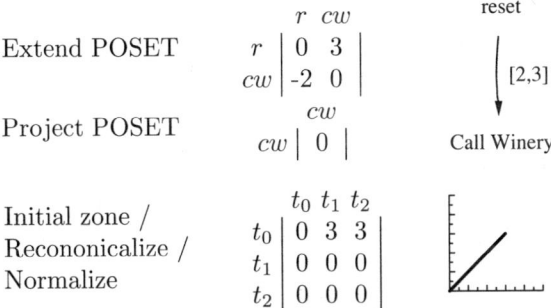

Fig. 7.25 Example of zone creation after the winery is called using POSETs.

Example 7.5.1 Let us again consider the example shown in Figure 7.1. The initial POSET includes only the special *reset* event, so the initial POSET matrix is only one-dimensional. The initial state has only one timer t_6, so our initial DBM is only two by two (i.e., t_0 and t_6). We set the top row entry for t_6 to the maximum value allowed for t_6, which is 3, and the other entries are set to 0. Recanonicalization and normalization result in the same DBM. This sequence of steps is shown in Figure 7.24.

The only possible rule that can fire in this initial zone is r_6. The firing of this rule results in the event *Call Winery*, so we update the POSET to include this event (i.e., cw). We no longer need to keep the reset event in the POSET matrix, since all enabled rules have *Call Winery* as their enabling event. Therefore, the POSET matrix is again the trivial one-dimensional matrix. The new zone includes timers t_1 and t_2. We set the 0th row of the DBM to the upper bounds of these timers, three in each case. This sequence of steps is shown in Figure 7.25.

In this new zone, there are two possible rule firings: r_1 or r_2. Let us first fire r_1. We extend the POSET matrix to include *Wine Arrives* (i.e., wa), which can happen between 2 and 3 time units after *Call Winery*. We create a zone including the two timers t_4 and t_2. We fill in the upper bounds in the top row, and copy the information from the POSET matrix for the core. We then recanonicalize and normalize. The result is shown in Figure 7.26.

284 TIMED CIRCUITS

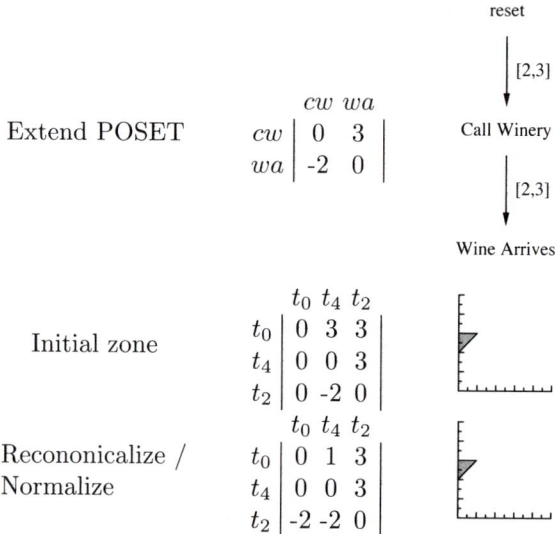

Fig. 7.26 Example of zone creation after the wine arrives using POSETs.

In this zone, the only satisfied rule is r_2. The firing of this rule results in the event *Call Patron*. We add *Call Patron* (i.e., cp) to the POSET matrix. After recanonicalization, we determine that either *Call Patron* or *Wine Arrives* can occur up to 1 time unit after the other. We can now remove *Call Winery* from the POSET matrix since it is no longer an enabling event for any active rules. We create a zone including the active timers t_4 and t_3. We copy the upper bounds into the top row, determine the core from the POSET matrix, recanonicalize, and normalize. The result is shown in Figure 7.27. This zone is the union of the two shown in Figures 7.13 and 7.18. In other words, even though we found this by considering that *Wine Arrives* first, it produces a zone that includes the case where the *Call Patron* event happens first.

In this zone, the only rule that can fire is r_4. The firing of this rule does not result in an event firing. Therefore, we simply update the zone. The zone is updated by projecting t_4 from the zone shown at the bottom of Figure 7.27, advancing time, recanonicalizing, and normalizing. The result is shown in Figure 7.28.

In this zone, the only possible rule that can fire is r_3. This results in the event *Patron Arrives*, which must be added to the POSET matrix, and we recanonicalize the POSET matrix. The events *Wine Arrives* and *Call Patron* can be removed from the matrix since they are not needed for any active rules. We then create a zone for the one active timer t_5 as shown in Figure 7.29.

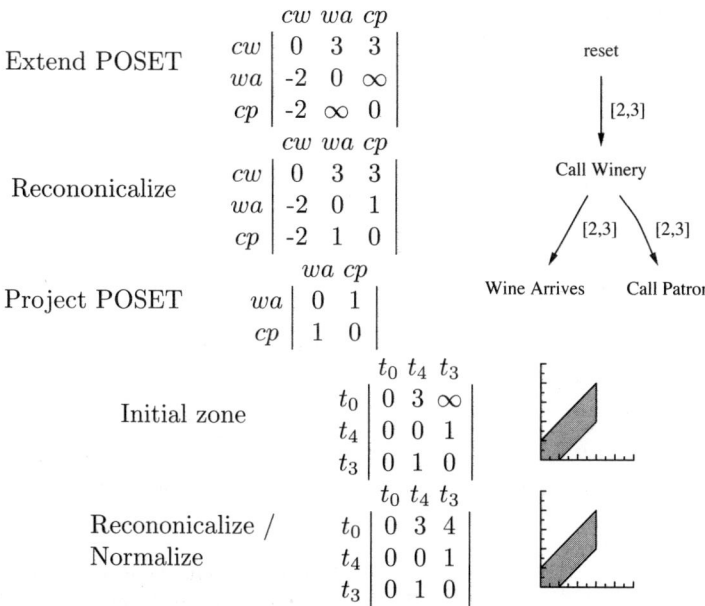

Fig. 7.27 Example of zone creation after wine arrives and patron has been called using POSETs.

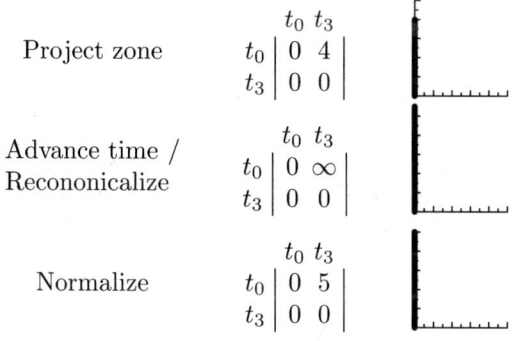

Fig. 7.28 Example of zone creation after a rule expires using POSETs.

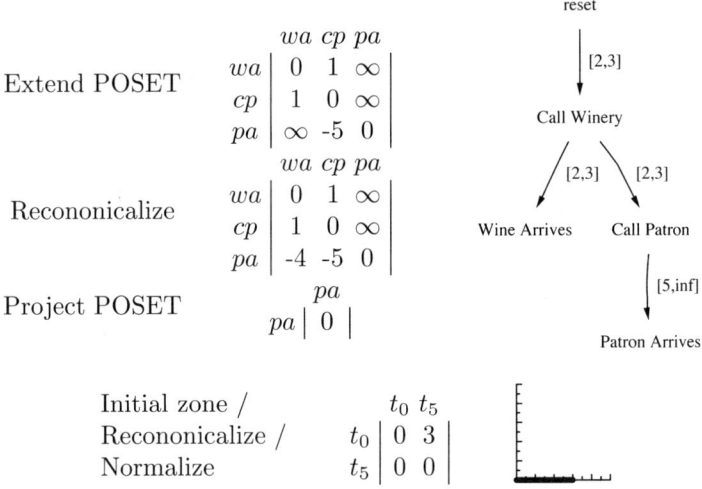

Fig. 7.29 Example of zone creation after the patron arrives using POSETs.

In this zone, the rule r_5 fires. The result is that we add the *Wine Is Purchased* event to the POSET matrix and remove the event *Patron Arrives*. We then derive a new zone for the active timer t_6, as shown in Figure 7.30. The resulting zone is the same as the initial zone, so we are done exploring down this path.

We now backtrack and consider calling the patron before the wine arrives. The POSET for this case is shown in Figure 7.31. Using the POSET matrix, we derive the new zone, recanonicalize, and normalize as shown in Figure 7.31.

In this zone, we fire the rule r_1. This result, shown in Figure 7.32, is the same POSET in Figure 7.27. Therefore, as expected, it results in the same zone. We can now backtrack at this point. Note that we get to backtrack one state sooner than when we used zones alone.

Since there are no more alternative paths that we have not explored, we have completed the timed state space exploration. The entire timed state space is shown in Figure 7.33. Using POSETs, we found seven zones to represent the entire timed state space, compared with the eight we found before. Although this is a very modest improvement, for highly concurrent examples the amount of improvement can be orders of magnitude.

POSET TIMING 287

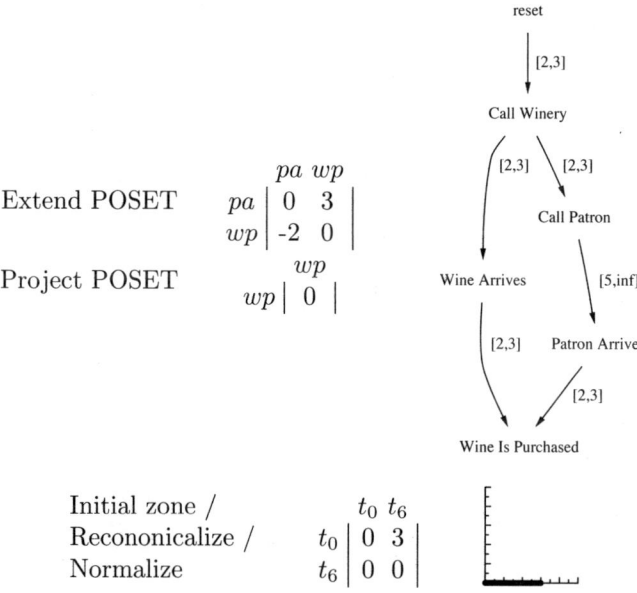

Extend POSET $\quad \begin{array}{c|cc} & pa & wp \\ \hline pa & 0 & 3 \\ wp & -2 & 0 \end{array}$

Project POSET $\quad \begin{array}{c|c} & wp \\ \hline wp & 0 \end{array}$

Initial zone /
Recononicalize / $\quad \begin{array}{c|cc} & t_0 & t_6 \\ \hline t_0 & 0 & 3 \\ t_6 & 0 & 0 \end{array}$
Normalize

Fig. 7.30 Example of zone creation after the wine is purchased using POSETs.

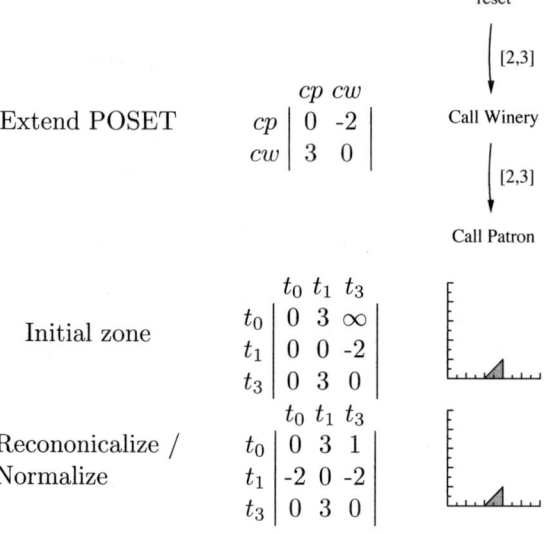

Extend POSET $\quad \begin{array}{c|cc} & cp & cw \\ \hline cp & 0 & -2 \\ cw & 3 & 0 \end{array}$

Initial zone $\quad \begin{array}{c|ccc} & t_0 & t_1 & t_3 \\ \hline t_0 & 0 & 3 & \infty \\ t_1 & 0 & 0 & -2 \\ t_3 & 0 & 3 & 0 \end{array}$

Recononicalize /
Normalize $\quad \begin{array}{c|ccc} & t_0 & t_1 & t_3 \\ \hline t_0 & 0 & 3 & 1 \\ t_1 & -2 & 0 & -2 \\ t_3 & 0 & 3 & 0 \end{array}$

Fig. 7.31 Example of zone creation after the patron is called using POSETs.

288   TIMED CIRCUITS

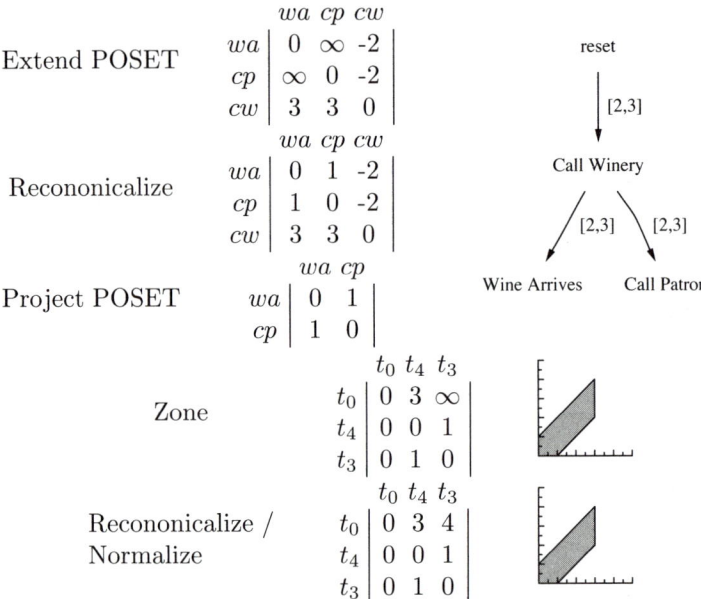

Fig. 7.32 Example of zone creation after the patron has been called and wine arrives using POSETs.

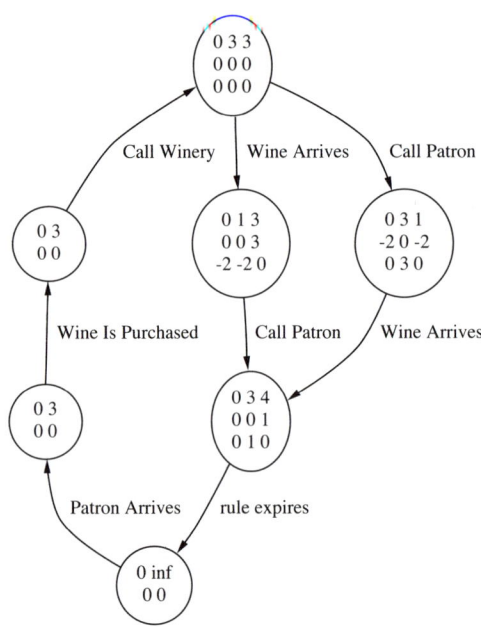

Fig. 7.33 Timed state space using POSETs.

Table 7.1 Timing assumptions for passive/active wine shop.

Assumption	Delay
Winery delays	2 to 3 minutes
Patron responds	5 to ∞ minutes
Patron resets	2 to 3 minutes
Inverter delay	0 to 6 seconds
AND gate delay	6 to 12 seconds
OR gate delay	6 to 12 seconds
C-element delay	12 to 18 seconds

7.6 TIMED CIRCUITS

We conclude this chapter by showing the effect that timing information can have on the timed circuit that is synthesized.

Example 7.6.1 We return to the passive/active wine shop example discussed in Chapter 6. It was found at the end of Chapter 6 that if the bubbles at the inputs to the AND gates in the circuit shown in Figure 6.23 are changed to explicit inverters, the circuit is no longer hazard-free under the speed-independent delay model. However, if we make the timing assumptions shown in Table 7.1 (granted a pretty slow logic family), we can determine using timed state space exploration that this circuit is hazard-free.

Not only can we verify that our gate-level circuit is hazard-free, we can use the timing assumptions to further improve the quality of the design. We begin using a TEL structure shown in Figure 7.34. We assume the delays given in Table 7.1 for the winery and patron and that the total circuit delays for any output are never more than 1 minute.

If we ignore the timing information provided, we would find the state graph shown in Figure 7.35, which has 18 states. Using the timing information, we derive a state graph that has only 12 states, as shown in Figure 7.36. Using the speed-independent state graph, the resulting circuit implementation is shown in Figure 7.37. The timed circuit is shown in Figure 7.38. The timed circuit implementation for *CSC0* is the same and for *req_patron* is reduced by one literal. The circuit for *ack_wine* is substantially reduced to only a three-input AND gate. Furthermore, we do not need to assume that we have a library of AND gates with inverted inputs, as the design using explicit inverters is also hazard-free.

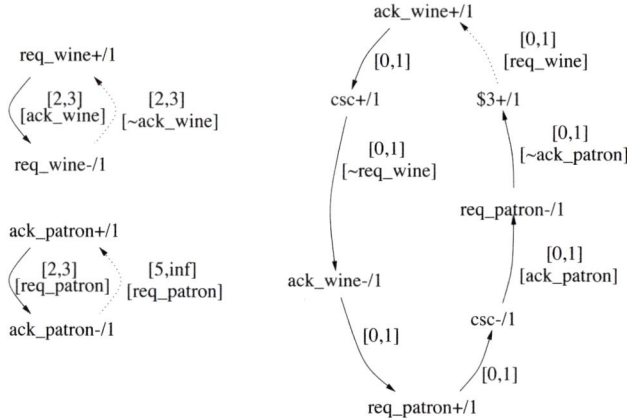

Fig. 7.34 TEL structure for the wine shop example.

Fig. 7.35 Speed-independent state graph for the wine shop example (state vector is ⟨*req_wine, ack_patron, ack_wine, req_patron, CSC0*⟩).

TIMED CIRCUITS 291

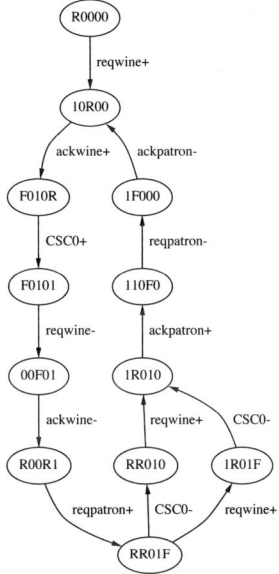

Fig. 7.36 Reduced state graph for the wine shop example (state vector is ⟨*req_wine, ack_patron, ack_wine, req_patron, CSC0*⟩).

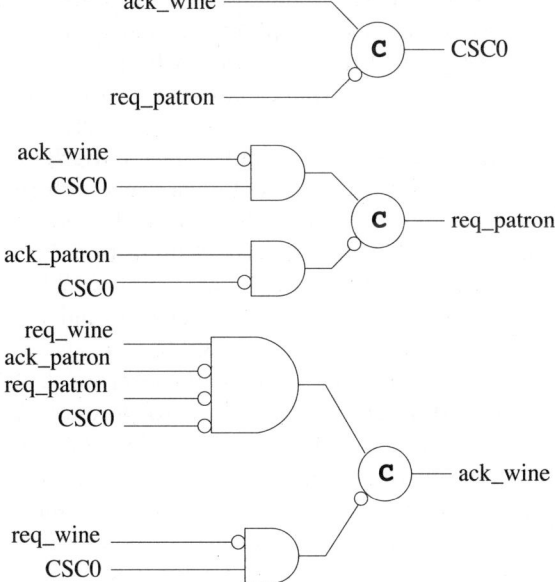

Fig. 7.37 Speed-independent circuit for the wine shop example.

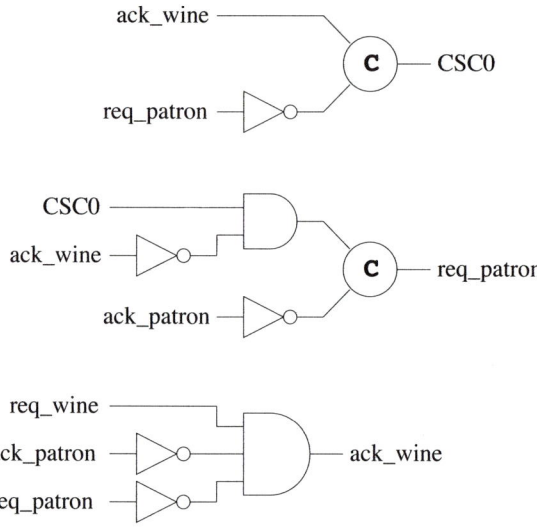

Fig. 7.38 Timed circuit for the wine shop example.

7.7 SOURCES

The region-based timing analysis method described in Section 7.2 was developed by Alur et al. [7, 8, 9]. The discrete-time analysis technique described in Section 7.3 was first applied to timed circuits by Burch [62, 63]. Recently, Bozga et al. proposed a discrete-time analysis approach which utilized binary decision diagrams [45]. Dill devised the zone method of representing timed states described in Section 7.4 [114]. The KRONOS tool [417] has been applied to timed circuit analysis, and it includes both the discrete-time method from [45] and a zone-based method. Rokicki and Myers developed the POSET timing analysis method described in Section 7.5 [325, 326]. Belluomini et al. generalized the POSET method to apply to TEL structures and applied it to the verification of numerous timed circuits and systems from both synchronous and asynchronous designs [31, 32, 33, 34]. An implicit method using multiterminal BDDs was proposed by Thacker et al. [377]. Myers et al. first applied timed state space exploration to produce optimized timed asynchronous circuits as described in Section 7.6 [283, 285, 286, 287].

A different approach to timing analysis and optimization is based on the idea of *time separation of events* (TSE). A TSE analysis algorithm finds the minimum and maximum separation between any two specified events. This information can later be used during state space exploration to determine whether or not two concurrently enabled events can occur in either order or must be ordered. McMillan and Dill developed a TSE algorithm for acyclic graphs [265]. Myers and Meng utilized a similar algorithm to determine an

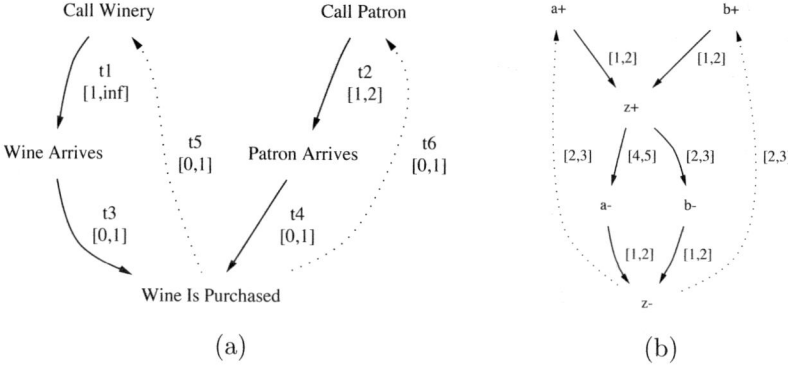

Fig. 7.39 TEL structures for homework problems.

estimate of the TSE in cyclic, choice-free graphs in polynomial time. They also applied this algorithm to the synthesis of timed circuits [285]. Jung and Myers also used this algorithm in a direct synthesis method from STGs to timed circuits [186]. Hulgaard et al. developed an exact TSE algorithm for cyclic graphs, including some types of choice behavior [173, 174, 175].

Problems

7.1 Modeling Timing
Give two timed firing sequences for the TEL structure shown in Figure 7.39(a) that end with the *Wine Is Purchased* event.

7.2 Modeling Timing
Give two timed firing sequences for the TEL structure shown in Figure 7.39(b) that end with the $z+$ event.

7.3 Regions
Using regions to represent timing, find the timed state space for one timed firing sequence for the TEL structure shown in Figure 7.39(a) that ends with the *Wine Is Purchased* event.

7.4 Regions
Using regions to represent timing, find the timed state space for one timed firing sequence for the TEL structure shown in Figure 7.39(b) that ends with the $z+$ event.

7.5 Discrete Time
Using discrete-time states, find the timed state space for one timed firing sequence for the TEL structure shown in Figure 7.39(a) that ends with the *Wine Is Purchased* event.

7.6 Discrete Time
Using discrete-time states, find the timed state space for one timed firing sequence for the TEL structure shown in Figure 7.39(b) that ends with the $z+$ event.

7.7 Zones
Using zones to represent timing, find the entire timed state space for the TEL structure shown in Figure 7.39(a).

7.8 Zones
Using zones to represent timing, find the entire timed state space for the TEL structure shown in Figure 7.39(b).

7.9 POSET Timing
Using POSETs and zones, find the entire timed state space for the TEL structure shown in Figure 7.39(a).

7.10 POSET Timing
Using POSETs and zones, find the entire timed state space for the TEL structure shown in Figure 7.39(b).

7.11 Timed Circuits
Consider the TEL structure shown in Figure 7.39(b).

7.11.1. Find a SG ignoring timing and use it to find a Muller circuit for output signal z.

7.11.2. Find a SG considering timing and use it to find a timed circuit for output signal z.

8

Verification

Prove all things; hold fast that which is good.
—New Testament, I Thessalonians

All technology should be assumed guilty until proven innocent.
—David Brower

No amount of experimentation can ever prove me right; a single experiment can prove me wrong.
—Albert Einstein

A common mistake that people make when trying to design something completely foolproof is to underestimate the ingenuity of complete fools.
—Douglas Adams, *Mostly Harmless*

When writing a specification of a circuit, we are usually trying to accomplish certain goals. For example, we may want to be sure that the protocol never leads to a deadlock or that whenever there is a request, it is followed by an acknowledgment possibly in a bounded amount of time. In order to validate that a specification will lead to a circuit that achieves these goals, simulation can be used, but this cannot guarantee complete coverage. This chapter, therefore, describes methods to verify that specifications meet their goals under all permissible delay behaviors.

After designing a circuit using one of the methods described in the previous chapters, we check the circuit by simulating a number of important cases until we are confident that it is correct. Unfortunately, anything short of exhaustive simulation will not guarantee the correctness of our design. This is especially problematic in asynchronous design, where a hazard may manifest as a failure only under a very particular set of delays. Therefore, it is necessary to use *verification* to check if a circuit operates correctly under all the allowed combinations of delay. In this chapter we describe methods for verifying a circuit's correctness under all permissible delay behavior.

8.1 PROTOCOL VERIFICATION

In this section we describe a method for specifying and verifying whether a protocol or circuit has certain desired properties. This type of verification is often called *model checking*. To specify the desired behavior of a combinational circuit, one typically uses *propositional logic*. For sequential circuits, it is necessary to describe behavior of a circuit over time, so one must use a *temporal logic*. In this section we introduce *linear-time temporal logic* (LTL)..

8.1.1 Linear-Time Temporal Logic

A temporal logic is a propositional logic which has been extended with operators to reason about future states of a system. LTL is a *linear-time* temporal logic in which truth values are determined along paths of states. The set of LTL formulas can be described recursively as follows:

1. Any signal u is a LTL formula.

2. If f and g are LTL formulas, so are:

    (a) $\neg f$ (*not*)

    (b) $f \wedge g$ (*and*)

    (c) $\bigcirc f$ (*next state operator*)

    (d) $f \mathbf{U} g$ (*strong until operator*)

We can describe the set of all allowed sequences using a SG, $\langle S, \delta, \lambda_S \rangle$, as described in Chapter 4. We require that the state transition relation is *total*. In other words, if any state has no successor, that state is assumed to have itself as a successor. Recall that $\delta \subseteq S \times T \times S$ is the set of possible state transitions. If there exists a state s_i for which there does not exist any transition t_j and state s_k such that $(s_i, t_j, s_k) \in \delta$, add $(s_i, \$, s_i)$ into δ, where $\$$ is a sequencing transition.

The truth of a LTL formula f can be defined with respect to a state s_i (denoted $s_i \models f$). A signal u is true in a state if the signal u is labeled with a 1 in that state [i.e., $\lambda_S(s_i)(u) = 1$]. The formula $\neg f$ is true in a state s_i when f is false in that state. The formula $f \wedge g$ is true when both f and g are true in s_i. The formula $\bigcirc f$ is true in state s_i when f is true in all next states s_j

reachable in one transition. The formula $f \mathbf{U} g$ is true in a state s_i when in all allowed sequences starting with s_i, f is true until g becomes true. A state in which g becomes true must always be reached in every allowed sequence. For this reason, this is often called the *strong until operator*. This relation can be defined more formally as follows:

$$\begin{aligned}
s_i \models u &\quad \text{iff} \quad \lambda_S(s_i)(u) = 1 \\
s_i \models \neg f &\quad \text{iff} \quad s_i \not\models f \\
s_i \models f \wedge g &\quad \text{iff} \quad s_i \models f \text{ and } s_i \models g \\
s_i \models \bigcirc f &\quad \text{iff} \quad \text{for all states } s_j \text{ such that } (s_i, t, s_j) \in \delta \ . \ s_j \models f \\
s_i \models f \mathbf{U} g &\quad \text{iff} \quad \text{for all allowed sequences } (s_i, s_{i+1}, \ldots), \\
& \qquad \exists j \ . \ j \geq i \wedge s_j \models g \wedge (\forall k \ . \ i \leq k < j \Rightarrow s_k \models f)
\end{aligned}$$

It is often convenient to use some abbreviations to express some common LTL formulas. For example, in propositional logic, the formula $f \vee g$ can be used to express $\neg((\neg f) \wedge (\neg g))$. We also often use $f \Rightarrow g$ to express $(\neg f) \vee g$. New temporal operators include $\Diamond f$, which is used to express that f will eventually become true in all allowed sequences starting in the current state. The formula $\Box f$ is used to say that f is always true in all allowed sequences starting in the current state. Finally, a *weak until operator* (denoted $f \mathbf{W} g$) is used to state that f remains true until a state is reached in which g is true, but g never needs to hold as long as f holds. These formulas can be defined in terms of the previously defined ones as follows:

$$\begin{aligned}
\Diamond f &\equiv \text{true } \mathbf{U} \ f \ (\textit{eventually operator}) \\
\Box f &\equiv \neg \Diamond (\neg f) \ (\textit{always operator}) \\
f \mathbf{W} g &\equiv (f \mathbf{U} g \vee \Box f \ (\textit{weak until operator})
\end{aligned}$$

Example 8.1.1 Let's return to our favorite wine shop example and write some formulas for the desired properties of a passive/active wine shop. First, we should not raise ack_wine until req_wine goes high:

$$\Box(\neg ack\_wine \Rightarrow (\neg ack\_wine \ \mathbf{U} \ req\_wine)) \qquad (8.1)$$

This property states that in all states in which ack_wine is low, it should stay low until req_wine goes high. We also require that once we set ack_wine high, it must stay high until req_wine goes low again:

$$\Box(ack\_wine \Rightarrow (ack\_wine \ \mathbf{U} \ \neg req\_wine)) \qquad (8.2)$$

On the patron side, once the shop has set req_patron high, it must hold it high until ack_patron goes high:

$$\Box(req\_patron \Rightarrow (req\_patron \ \mathbf{U} \ ack\_patron)) \qquad (8.3)$$

Once req_patron is set low, it must stay low until ack_patron goes low:

$$\Box(\neg req\_patron \Rightarrow (\neg req\_patron \ \mathbf{U} \ \neg ack\_patron)) \qquad (8.4)$$

Another important property is once the request and acknowledge wires on either side go high, they must be reset again:

$$\Box((\mathit{req\_wine} \land \mathit{ack\_wine}) \Rightarrow \Diamond(\neg \mathit{req\_wine} \land \neg \mathit{ack\_wine})) \quad (8.5)$$

$$\Box((\mathit{req\_patron} \land \mathit{ack\_patron}) \Rightarrow \Diamond(\neg \mathit{req\_patron} \land \neg \mathit{ack\_patron})) (8.6)$$

We also don't want the wine to stay on the shelf forever, so after each bottle arrives, the patron should be called.

$$\Box(\mathit{ack\_wine} \Rightarrow \Diamond \mathit{req\_patron}) \quad (8.7)$$

Finally, the patron should not arrive expecting wine in the shop before the wine has actually arrived.

$$\Box(\neg \mathit{ack\_patron} \Rightarrow (\neg \mathit{ack\_patron} \; \mathbf{U} \; \mathit{ack\_wine})) \quad (8.8)$$

We can check whether or not a LTL formula is valid for a SG through a simple analysis. We can consider an LTL formula as being composed of propositional and temporal terms. Our first step is to mark each state that satisfies each of the propositional terms of the formula. Next, for each term of the form $\bigcirc f$ where f is a propositional formula, we mark each state in which f is true in all successors as satisfying $\bigcirc f$. To address terms of the form $f \; \mathbf{U} \; g$ requires a two-step procedure. First, we mark each state that satisfies g as also satisfying $f \; \mathbf{U} \; g$. Second, we mark each predecessor of a state marked as satisfying $f \; \mathbf{U} \; g$ that also satisfies f as satisfying $f \; \mathbf{U} \; g$. We then recursively mark the predecessors of these states until there is no change. Other temporal operators can be converted to their basic form for treatment. Now, to check the entire formula, we apply these procedures recursively to build up the formula.

Example 8.1.2 Let's check the validity of LTL formula 8.8 on a couple of potential protocols. The SG for the first is shown in Figure 8.1. This formula is trivially satisfied in the states in which $\mathit{ack\_patron}$ is high: $\{\langle F111 \rangle, \langle 01F1 \rangle, \langle R10F \rangle, \langle 110F \rangle, \langle RF00 \rangle, \langle 1F00 \rangle\}$. In the remaining states, we must verify that the formula $(\neg \mathit{ack\_patron} \; \mathbf{U} \; \mathit{ack\_wine})$ holds. To check this, we first mark the states where $\mathit{ack\_wine}$ is true, since this formula holds in these states (i.e., $\langle F01R \rangle$, $\langle 001R \rangle$, $\langle FR11 \rangle$, $\langle 0R11 \rangle$, $\langle F111 \rangle$, and $\langle 01F1 \rangle$). We then search for states which can reach the states we just marked in one step, and we check that in that state $\mathit{ack\_patron}$ is low. There is one such state, $\langle 10R0 \rangle$, where indeed $\mathit{ack\_patron}$ is low, so we can mark that it satisfies the formula. We then continue our search, and we find that state $\langle R000 \rangle$ reaches $\langle 10R0 \rangle$ and has $\mathit{ack\_patron}$ low, so it also satisfies the formula. We have now marked all states as satisfying the formula, so the LTL formula is valid for this protocol. It can quickly be verified that this protocol satisfies all the other desired properties as well.

An SG for an alternative protocol is shown in Figure 8.2. We again begin by marking states with $\mathit{ack\_patron}$ high as trivially satisfying this formula (i.e., $\langle 11R1 \rangle$, $\langle F111 \rangle$, $\langle 011F \rangle$, $\langle 0FF0 \rangle$, $\langle RF00 \rangle$, and $\langle 1F00 \rangle$).

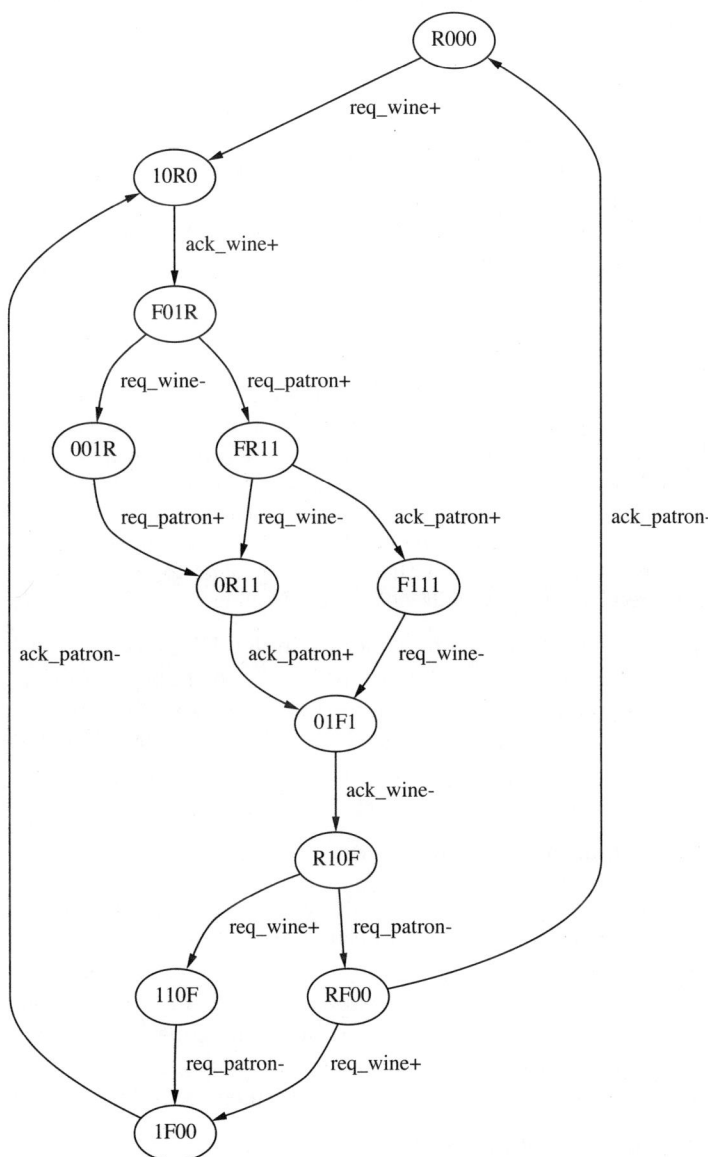

Fig. 8.1 SG for legal protocol for the passive/active shop (state vector is ⟨*req_wine*, *ack_patron*, *ack_wine*, *req_patron*⟩).

Again, we are left to check that in the remaining states the formula ($\neg ack\_patron$ **U** ack_wine) holds. We start by marking the states where ack_wine is true (i.e., $\langle FR11\rangle$, $\langle 0R11\rangle$, $\langle F111\rangle$, $\langle 011F\rangle$, $\langle 0FF0\rangle$, and $\langle 00F0\rangle$). Next, we look for states which can reach this set of states in one step, and we find $\langle 11R1\rangle$. In this state, ack_patron is high, so ($\neg ack\_patron$ **U** ack_wine) does not hold in this state. This in itself is not bad, since the entire formula does hold in this state. Consider, however, state $\langle 1RR1\rangle$, in which ack_patron is low. This state can reach $\langle 11R1\rangle$, so it also does not satisfy ($\neg ack\_patron$ **U** ack_wine). In this case, we have found a state in which the entire formula does not hold, so the LTL formula is not valid for this protocol. A sequence of events which leads to this failure would be

$$req\_wine+, req\_patron+, ack\_patron+$$

Another alternative protocol is shown in Figure 8.3. A careful analysis determines that the patron will never arrive at a shop that has no wine. Unfortunately, the LTL formula that we used to express this property does not hold for this protocol. Consider state $\langle 1R01\rangle$, in which ack_patron is low. According to the formula, ack_patron must remain low until ack_wine goes high. However, the state $\langle 110F\rangle$ can be reached next, in which ack_patron is high. This violates the formula. This is okay, though, since the last bottle of wine is still sitting on the shelf. In fact, any protocol that allows ack_wine to reset before the patron receives the wine will violate this formula.

It is quite difficult (if not impossible) to find an LTL formula which admits only valid protocols and all valid protocols. Even if such a formula exists, how does one know when they found it? This is the key difficulty with model checking in that when verification succeeds, it does not mean that the protocol is correct but only that it has the property being checked.

8.1.2 Time-Quantified Requirements

The LTL formula $\Diamond f$ states that eventually f becomes true, but it puts no guarantee on how long before f will become true. If we wish to express properties such as *bounded response time*, it is necessary to extend the temporal logic that we use to specify timing bounds. One possible way of doing this is to annotate each temporal operator with a timing constraint. For example, $\Diamond_{<5} f$ states that f becomes true in less than 5 time units.

The set of *timed LTL* formulas can be described recursively as follows:

1. Any signal u is a timed LTL formula.

2. If f and g are timed LTL formulas, so are:

    (a) $\neg f$ (not)
    (b) $f \wedge g$ (and)
    (c) $f\ \mathbf{U}_{\sim c}\ g$ (*timed until operator*)

PROTOCOL VERIFICATION 301

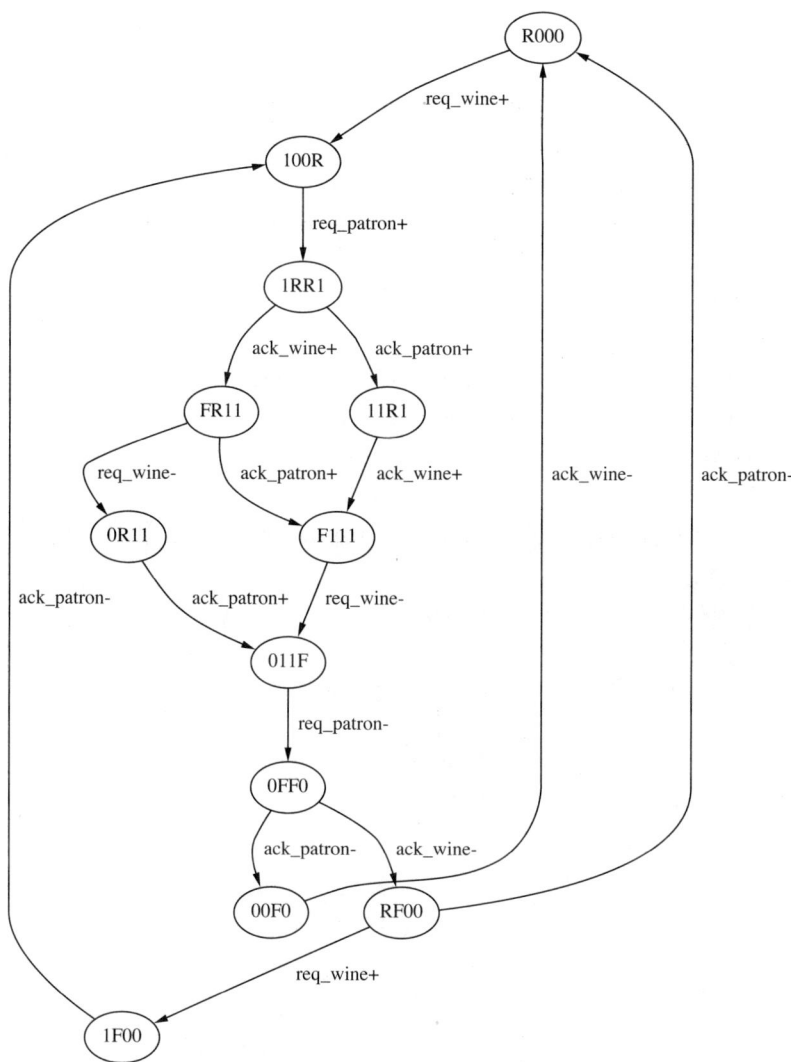

Fig. 8.2 SG for illegal protocol for the passive/active shop (state vector is ⟨*req_wine, ack_patron, ack_wine, req_patron*⟩).

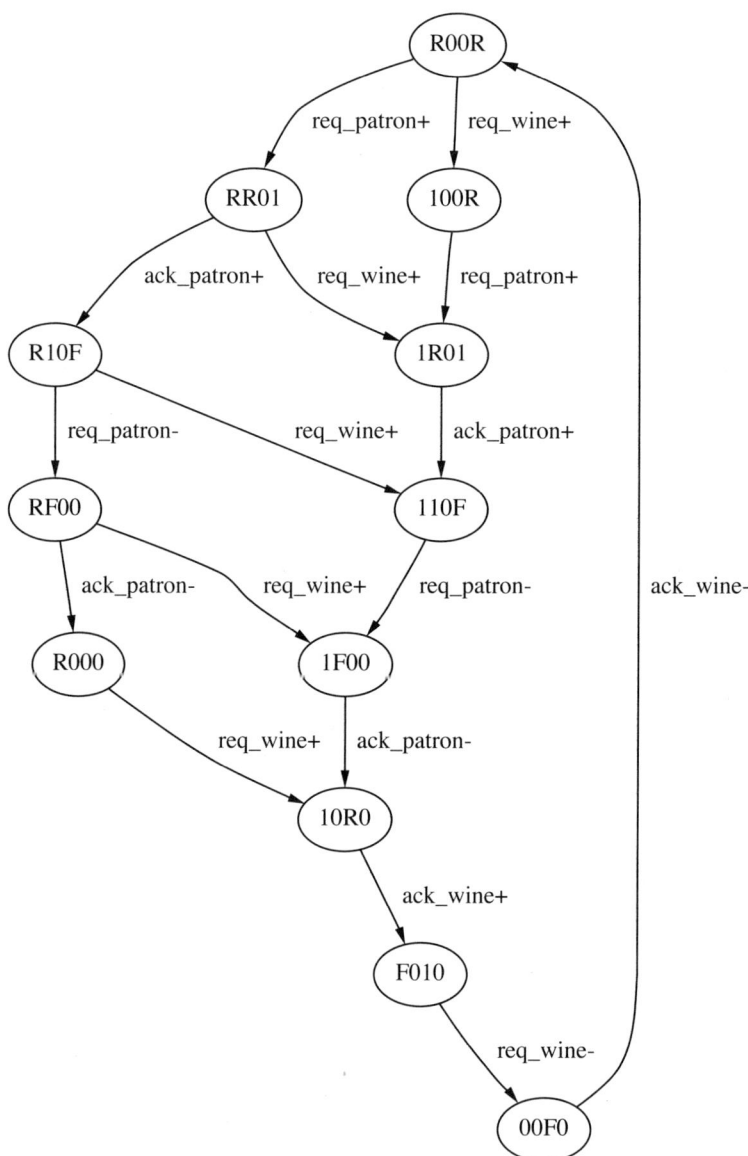

Fig. 8.3 SG for legal protocol for the passive/active shop that violates the property (state vector is ⟨*req_wine, ack_patron, ack_wine, req_patron*⟩).

where $\sim$ is $<, \leq, =, \geq, >$. Note that there is no next time operator, since when time is dense, there can be no unique next time.

Again, in timed LTL, we have the following set of abbreviations:

$$\Diamond_{\sim c} f \equiv true\ \mathbf{U}_{\sim c}\ f\ (timed\ eventually\ operator)$$
$$\Box_{\sim c} f \equiv \neg \Diamond_{\sim c}(\neg f)\ (timed\ always\ operator)$$

Using the basic timed LTL primitives, we can also define temporal operators subscripted with time intervals.

$$\Diamond_{(a,b)} f \equiv \Diamond_{=a} \Diamond_{<(b-a)} f$$

Example 8.1.3 Let's again consider the passive/active wine shop and specify a few bounded response time properties. First, once the request and acknowledge wires on either side go high, they must be reset again within 10 minutes:

$$\Box((req\_wine \wedge ack\_wine) \Rightarrow \Diamond_{\leq 10} (\neg req\_wine \wedge \neg ack\_wine))$$

$$\Box((req\_patron \wedge ack\_patron) \Rightarrow \Diamond_{\leq 10} (\neg req\_patron \wedge \neg ack\_patron))$$

We also don't want the wine to age too long on the shelf, so after each bottle arrives, the patron should be called within 5 minutes:

$$\Box(ack\_wine \Rightarrow \Diamond_{\leq 5}\ req\_patron)$$

In order to check a timed LTL property, the timed state space exploration algorithms described in Chapter 7 must be used. For example, to check the formula given above using the discrete-time analysis method, one can simply analyze the timed state space after it has been found. In particular, one could search for any state in which *ack_wine* is high and *req_patron* is low, and there must not exist any path from that state which takes 5 time steps before *req_patron* goes high. Using region- or zone-based analysis methods to check this property is a bit more involved and beyond the scope of this chapter.

8.2 CIRCUIT VERIFICATION

While model checking can be used for circuit verification, it has been found to be a bit unwieldy. In this section we present an alternative verification methodology based on *trace theory* for verifying whether a circuit implements a given specification.

8.2.1 Trace Structures

In order to verify that a circuit implements, or *conforms* to, a specification, it is necessary to check that all the possible behaviors of the circuit are allowed behaviors in the specification. To define these behaviors, we will use *traces*

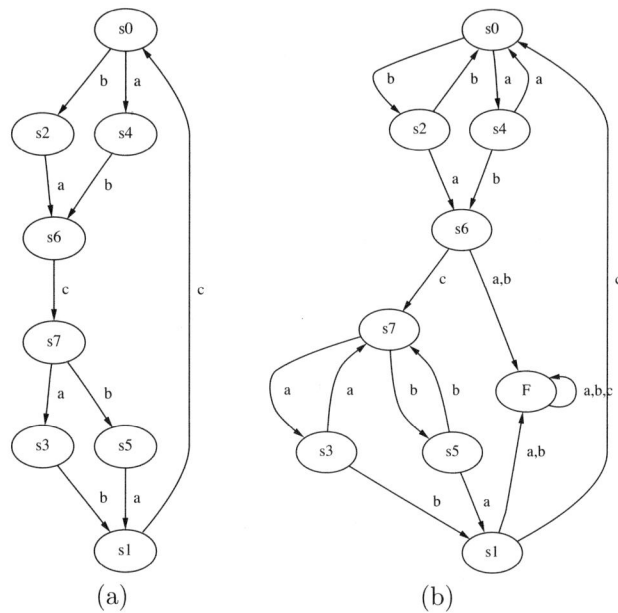

Fig. 8.4 (a) SG for a C-element. (b) Receptive SG for a C-element.

of events on signals. A trace is similar to an allowed sequence described earlier, but rather than keeping track of the states that the system has passed through, we track the signals which have had events on them.

> **Example 8.2.1** Consider the SG for a C-element in Figure 8.4(a). In this SG, one possible allowed sequence is $(s_0, s_4, s_6, s_7, \ldots)$. The corresponding trace would be $(a, b, c, \ldots)$.

The set of all traces is represented using a *trace structure*. In order to verify that a circuit is hazard-free, we will use a class of trace structures called *prefix-closed trace structures*. A prefix-closed trace structure can be described using a four-tuple: $\langle I, O, S, F \rangle$. The set I is the set of input signals, those signals controlled by the environment. The set O is the set of output signals, those controlled by the circuit. The set S is all traces which are considered successful. The set F is all traces which are considered a failure. At times, we may use A to indicate the set of all possible events and P to indicate the set of all possible traces. A trace structure must be *receptive*. In other words, the state of a circuit cannot prevent an input from happening (i.e., $PI \subseteq P$).

> **Example 8.2.2** If the signals a and b are inputs in the SG shown in Figure 8.4(a), the SG is not receptive since, for example, a is not allowed to occur in states s_4, s_6, s_3, and s_1. A receptive SG is shown in Figure 8.4(b). Note that now we must add a failure state, F, since if c is enabled to change and one of the inputs change instead, the C-element is hazardous and could glitch.

8.2.2 Composition

Due to the receptiveness requirement, if we wish to compose two circuits together and determine the corresponding trace structure, we must first make their signal sets match. Consider composing two trace structures: $T_1 = \langle I_1, O_1, S_1, F_1 \rangle$ and $T_2 = \langle I_2, O_2, S_2, F_2 \rangle$. If N is the set of signals in A_2 which are not in A_1, we must add N to I_1 and extend S_1 and F_1 to allow events on signals in N to occur at any time. Similarly, we must extend T_2 with those signals in A_1 but not in A_2. The function to perform this signal extension is called *inverse delete*, denoted $\text{del}(N)^{-1}(X)$, where N is a set of signals and X is a set of traces. This function inserts elements of N^* between any consecutive signals in a trace in X (where N^* indicates zero or more events on signals in N). This function can be extended to a trace structure as follows:

$$\text{del}(N)^{-1}(T) = \langle I \cup N, O, \text{del}(N)^{-1}(S), \text{del}(N)^{-1}(F) \rangle$$

Given two trace structures with *consistent signal sets* (i.e., $A_1 = A_2$ and $O_1 \cap O_2 = \emptyset$), they can be intersected as follows:

$$T_1 \cap T_2 = \langle I_1 \cap I_2, O_1 \cup O_2, S_1 \cap S_2, (F_1 \cap P_2) \cup (F_2 \cap P_1) \rangle$$

The implication of this definition is that a trace is considered a success in the composite only when it is a success in both circuits. It is considered a failure when it is a failure in either of the original circuits. One last thing to note is that the set of possible traces may be reduced through composition, since the set of possible traces in the composite is $P_1 \cap P_2$.

Using this definition and that of inverse deletion, composition can be defined as follows:

$$T_1 \| T_2 = \text{del}(A_2 - A_1)^{-1}(T_1) \cap \text{del}(A_1 - A_2)^{-1}(T_2)$$

Example 8.2.3 Let us compose the trace structure for the C-element shown in Figure 8.4(b) with the trace structures for two inverters to form the circuit shown in Figure 8.5(a). The trace structure for an inverter is shown in Figure 8.6(a). The first thing we must do is introduce the notion of *renaming*. The trace structure for the inverter shown in Figure 8.6(a) has x as an input and y as an output. To make the signals match those in the diagram for the first inverter, we must rename x to c and y to a. We must also extend the trace structure to include the signal b as an additional input using the inverse delete function before it can be composed with the C-element. The result is shown in Figure 8.6(b).

The trace structure for the inverter shown in Figure 8.6(b) can now be composed with the trace structure for the C-element, and the result is shown in Figure 8.7. Note that the set of input signals, I, is now only b, while the set of output signals, O, is $\{a, c\}$. The failure state from the inverter can no longer be reached after composition, since a second c change cannot come before an a change. Also, sequences that involve two a changes in a row have also been removed.

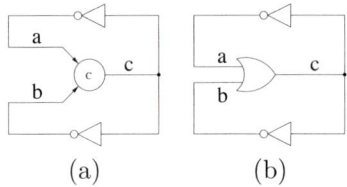

Fig. 8.5 (a) Simple C-element circuit. (b) Simple OR gate circuit.

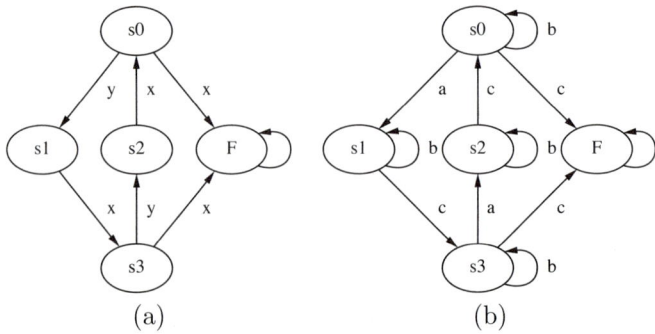

Fig. 8.6 (a) Receptive SG for an inverter. (b) SG for an inverter with input c, output a, and unconnected input b.

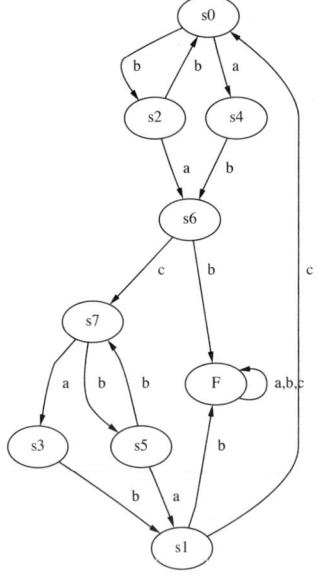

Fig. 8.7 (b) SG after composing one inverter with the C-element.

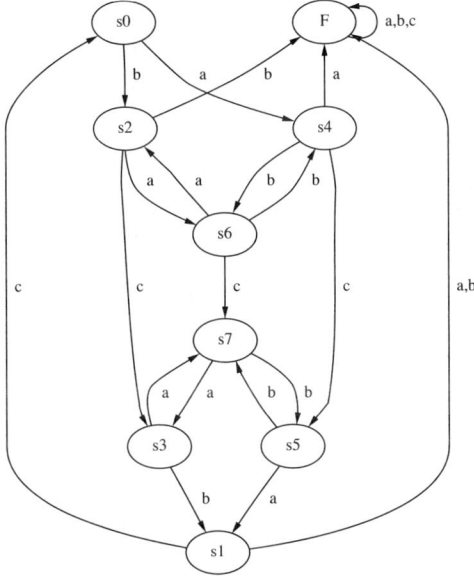

Fig. 8.8 Receptive SG for an OR gate.

If we rename and extend the trace structure in a similar fashion to create the second inverter and compose it with the trace structure shown in Figure 8.7, we get the trace structure shown in Figure 8.4(a). Notice that the failure state is no longer reachable. In other words, in this trace structure the failure set, F, is empty. Therefore, the circuit shown in Figure 8.5(a) is *failure-free*.

Example 8.2.4 Consider the composition of the trace structure for an OR gate shown in Figure 8.8 with the trace structure for two inverters [see Figure 8.6(a)] to form the circuit shown in Figure 8.5(b). First, we again do renaming to obtain the trace structure for the inverter shown in Figure 8.6(b). The trace structure for the inverter shown in Figure 8.6(b) can now be composed with the trace structure for the OR gate, and the result is shown in Figure 8.9(a). Note that the failure state can now be reached starting in state $s0$ after a b and c change, since the inverter driving signal a can become disabled.

If we rename and extend the trace structure in a similar fashion to create the second inverter and compose it with the trace structure shown in Figure 8.9(a), we get the trace structure shown in Figure 8.9(b). Notice that the failure state is still reachable. Starting in state $s0$, the failure state can be reached either by an a change followed by a c change or a b change followed by a c change. In each case, one of the inverters becomes disabled, causing a hazard. Note that this circuit is actually speed-independent as defined in Chapter 6, but it is not semi-modular.

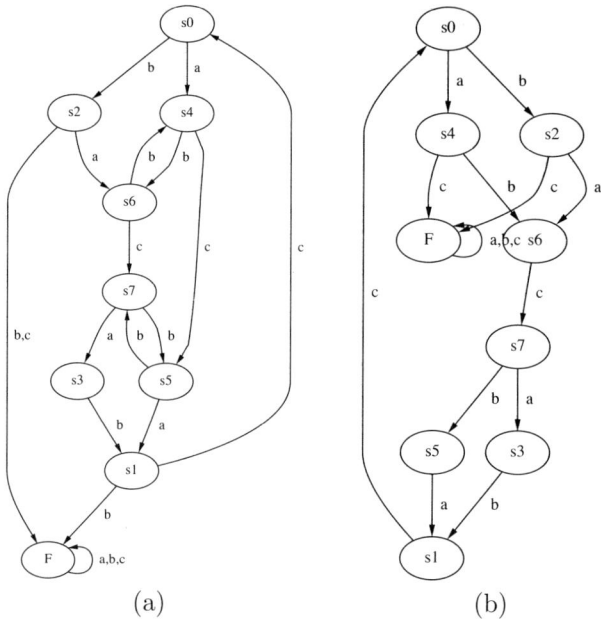

Fig. 8.9 (a) SG after composing one inverter with the OR gate. (b) SG after composing both inverters with the OR gate.

8.2.3 Canonical Trace Structures

To verify that a circuit correctly implements a specification, we can determine a trace structure for the circuit, T_I, and another for the specification, T_S, and show that T_I *conforms to* T_S (denoted $T_I \preceq T_S$). In other words, we wish to show that in any *environment* where the specification is failure-free, the circuit is also failure-free. An environment can be modeled by a trace structure, T_E, that has complementary inputs and outputs (i.e., $I_E = O_I = O_S$ and $O_E = I_I = I_S$). To check conformance, it is necessary to show that for every possible T_E, if $T_E \cap T_S$ is failure-free, so is $T_E \cap T_I$.

One important advantage of checking conformance is that it allows for *hierarchical verification*. In other words, once we have shown that an implementation conforms to a specification, we can use the specification in place of the implementation in verifying a larger system that has this circuit as a component. If there are many internal signals, this can be a huge benefit.

We say that two trace structures T_1 and T_2 are *conformation equivalent* (denoted $T_1 \sim_C T_2$) when $T_1 \preceq T_2$ and $T_2 \preceq T_1$. Unfortunately, if $T_1 \sim_C T_2$, it does not imply that $T_1 = T_2$. To make this true, we reduce prefix-closed trace structures to a canonical form using two transformations.

The first transformation is *autofailure manifestation*.

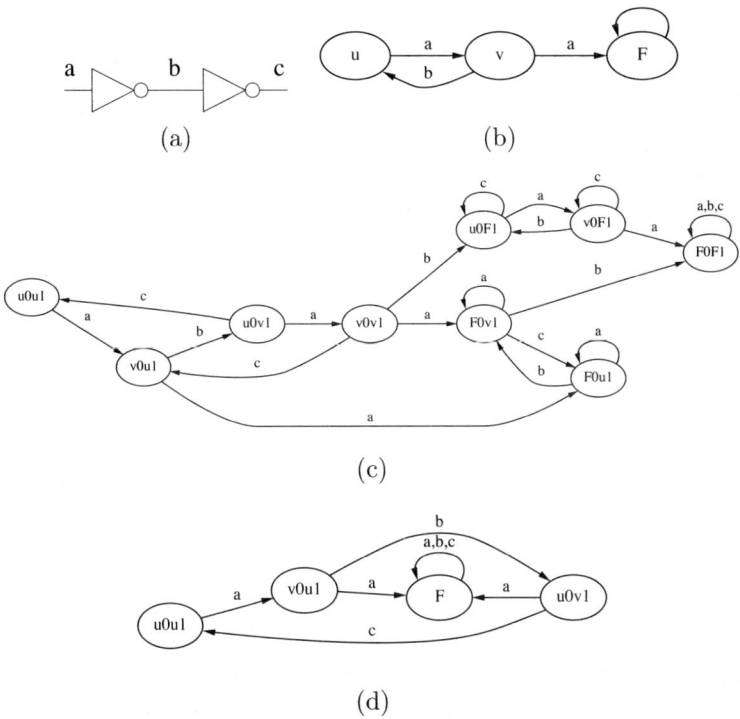

Fig. 8.10 (a) SG for an inverter. (b) Two inverters in series. (c) SG for two inverters in series. (d) SG for two inverters in series after simplification.

Example 8.2.5 Consider composing two inverters in series as in Figure 8.10(a). The state graph for a single inverter is shown in Figure 8.10(b). To simplify the explanation, we have reduced it from the five states in Figure 8.6(a) to 3 by folding states s_1 and s_2 to form state u and states s_0 and s_3 to form state v. After composing the two inverters, we find the SG shown in Figure 8.10(c). Any trace that ends in a state that includes F_0 or F_1 in its label is a failure trace. All other traces are considered successes.

Consider the partial trace (a, b, a), which ends in state $v0v1$. While this trace is successful, a possible next event is a transition on the output signal b which would lead to a failure. In this state, there is no way the environment can prevent this failure from happening. While it is possible that the output c could happen first, there is no guarantee that this will happen. In verification, any circuit which has a potential for a failure should be considered as failing verification. This type of failure is called an *autofailure*, since the circuit itself causes the failure. Autofailure manifestation adds the trace (a, b, a) to the failure set.

More formally, an autofailure is a trace x which can be extended by a signal $y \in O$ such that $xy \in F$. Another way of denoting this is $F/O \subseteq F$, where F/O is defined to be $\{x \mid \exists y \in O . xy \in F\}$. We also add to the failure set any trace that has a failure as a prefix (i.e., $FA \subseteq F$). The result of these two changes (assuming that $S \neq \emptyset$) is that any failure trace has a prefix that is a success, and the signal transition that causes it to become a failure is on an input signal. In other words, the circuit fails only if the environment sends a signal change that the circuit is not prepared for, and in this case, the circuit is said to *choke*.

The second transformation is called *failure exclusion*. In this transformation, we make the success and failure sets disjoint. When a trace occurs in both, it means that the circuit may or may not fail, but this again indicates a dangerous circuit. Therefore, we remove from the success set any trace which is also a failure (i.e., $S = S - F$).

> **Example 8.2.6** After applying both transformations, our simplified SG for two inverters in series is shown in Figure 8.10(d).

A *canonical prefix-closed trace structure* is one which satisfies the following three requirements:

1. Autofailures are failures (i.e., $F/O \subseteq F$).

2. Once a trace fails, it remains a failure (i.e., $FA \subseteq F$).

3. No trace is both a success and a failure (i.e., $S \cap F = \emptyset$).

In a canonical prefix-closed trace structure, the failure set is not necessary, so it can be represented with the triple $T = \langle I, O, S \rangle$. We can determine the failure set as follows:

$$F = [(SI \cup \{\epsilon\}) - S]A^*$$

In other words, any successful trace when extended with an input signal transition and is no longer found in the success set is a failure. Furthermore, any such failure trace can be extended indefinitely with other input or output signal transitions, and it will always be a failure.

8.2.4 Mirrors and Verification

To check conformance of a trace structure T_I to another T_S, we said that it is necessary to check that in all environments in which T_S is failure-free, T_I is also failure-free. It is difficult to imagine performing such a check. Fortunately, we can construct a unique worst-case environment that we can use to perform this feat in a single test. This environment is called a *mirror* of T (denoted T^M). If we have a canonical prefix-closed trace structure, the mirror can be constructed simply by swapping the inputs and outputs (i.e., $I^M = O$, $O^M = I$, and $S^M = S$).

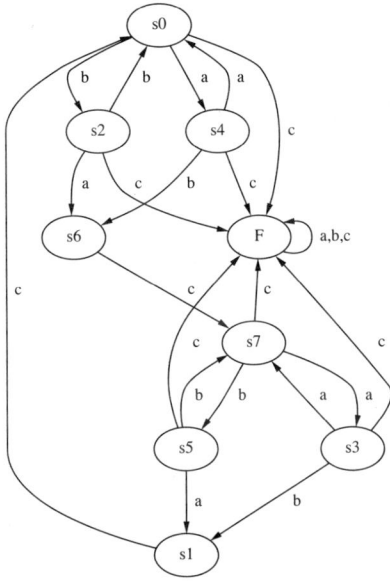

Fig. 8.11 Mirror for a C-element.

Example 8.2.7 The mirror of the C-element shown in Figure 8.4(b) is shown in Figure 8.11. Note that changing the inputs and outputs has the effect of changing the failure set. Recall that in canonical trace structures a failure occurs when an input happens at the wrong time. For the mirror of the C-element, the signal c is now the only input. If c changes in a state in which it was not expected to change, the trace is a failure. Also, note that changes on signals a and b are no longer allowed if they cause a failure.

Using the following theorem, we can use the mirror to check conformance of an implementation, T_I, to a specification, T_S.

Theorem 8.1 (Dill, 1989) *If $T_I \| T_S^M$ is failure-free, $T_I \preceq T_S$.*

Example 8.2.8 Consider the merge element shown in Figure 8.12(a). A general merge accepts an input on either a or b and produces an output on c [see Figure 8.12(b)]. An alternating merge accepts an input on a and produces a c, then accepts an input on b and produces a c [see Figure 8.12(c)]. There are two questions we can ask. First, if we require an alternating merge, can we replace it with a general merge? Second if we require a general merge, can we replace it with an alternating merge?

To answer the first question, the SG shown in Figure 8.12(b) is our implementation, T_I, the one in Figure 8.12(c) is our specification, T_S, and we want to check if $T_I \preceq T_S$. To do this, we construct the mirror for

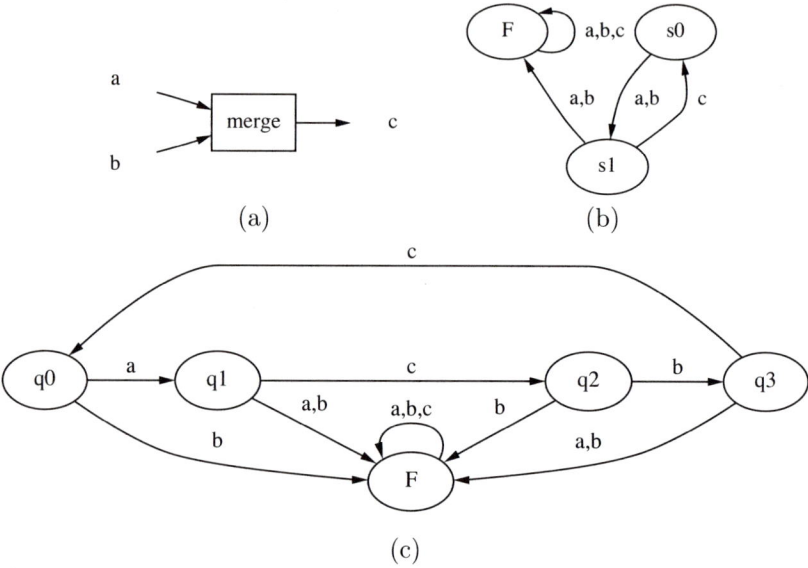

Fig. 8.12 (a) Merge element. (b) SG for a general merge element. (c) SG for an alternating merge.

T_g (i.e., the alternating merge), which is shown in Figure 8.13(a). We then compose this trace structure with the one for the general merge shown in Figure 8.12(b). The result is shown in Figure 8.13(b), and it is failure-free. This means that when an alternating merge is required, a general merge is a *safe substitute*.

To answer the second question, we must construct the mirror for the general merge, which is shown in Figure 8.13(c). We compose this trace structure with the one for the alternating merge shown in Figure 8.12(c), and the result is shown in Figure 8.13(d). This trace structure is not failure-free, so it is not safe to substitute an alternating merge when a general merge is required. Consider the initial state $q0s0$. In a general merge, it must be able to accept either an a or a b, while an alternating merge can only accept an a. If it receives a b, it would fail.

8.2.5 Strong Conformance

One limitation with this approach to verification is that it checks only safety properties. In other words, if a circuit verifies, it means that it does nothing bad. It does not mean, however, that it does anything good. For example, consider the trace structure for a "block of wood." A block of wood would accept any input, but it would never produce any output (i.e., $T - \langle I, O, I^* \rangle$). Assuming that the inputs and outputs are made to match, a block of wood would conform to any specification.

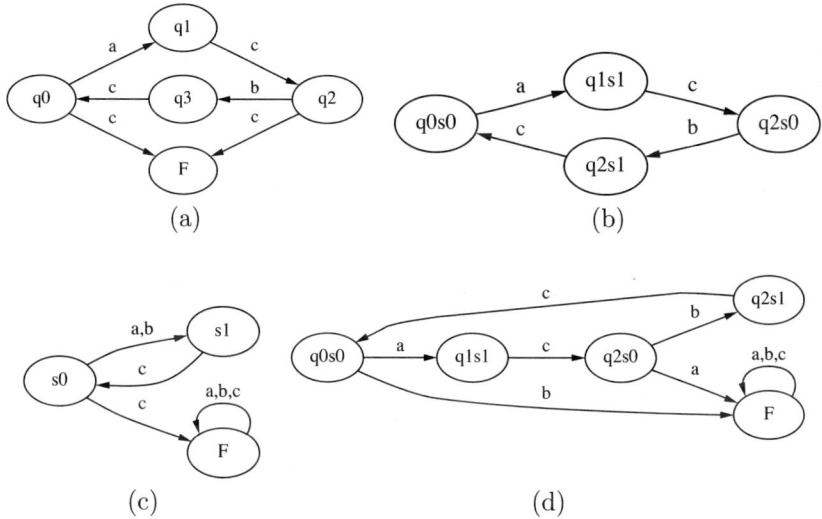

Fig. 8.13 (a) SG for mirror of alternating merge. (b) SG when checking if general merge conforms to alternating merge. (c) SG for mirror of general merge. (d) SG when checking if alternating merge conforms to general merge.

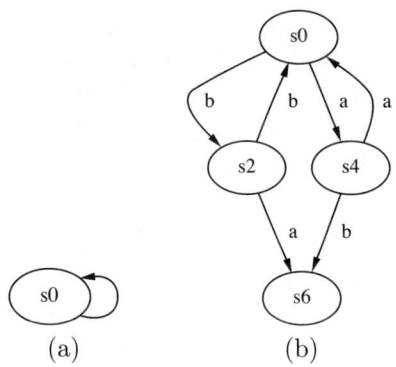

Fig. 8.14 (a) SG for a block of wood. (b) SG for a block of wood composed with the mirror of a C-element specification.

Example 8.2.9 Consider composing a block of wood whose behavior is shown in Figure 8.14(a) with the mirror of a C-element shown in Figure 8.11. The result is shown in Figure 8.14(b), and it is failure-free. Therefore, the block of wood conforms to the specification of a C-element. It is clear, however, that the block of wood is not a very good implementation of a C-element.

The notion of *strong conformance* removes this problem. T_1 *conforms strongly to* T_2 (denoted $T_1 \sqsubseteq T_2$) if $T_1 \preceq T_2$ and $S_1 \supseteq S_2$. In other words, all successful traces of T_2 must be included in the successful traces of T_1.

Example 8.2.10 Consider again the block-of-wood implementation of a C-element. The C-element has a successful trace (a, b, c), which is not in the set of successful traces of the block of wood. Therefore, the block of wood does not strongly conform to the specification of a C-element.

8.2.6 Timed Trace Theory

A *timed trace* is a sequence of $x = (x_1, x_2, \ldots)$ where each x_i is an event/time pair of the form (e_i, τ_i) such that:

- $e_i \in A$, where A is the set of signals.
- $\tau_i \in \mathbf{Q}$, where $\mathbf{Q}$ is the set of nonnegative rational numbers.

A timed trace must satisfy the following two properties:

- *Monotonicity*: for all i, $\tau_i \leq \tau_{i+1}$.
- *Progress*: if x is infinite, then for every $\tau \in \mathbf{Q}$ there exists an index i such that $\tau_i > \tau$.

Given a module M defined by a trace structure $\langle I, O, S \rangle$ and a trace $x \in S$, we say that the module M allows time to advance to time τ if for each $w' \in I \cup O$ and $\tau' < \tau$ such that $x(w', \tau') \in S$ implies that $x(w', \tau'') \in S$ for some $\tau'' \geq \tau$. Intuitively, this means that after trace x has happened, module M can allow time to advance to τ without needing an input or producing an output. We denote this by the predicate *advance_time*(M, x, τ).

Recall that we defined a failure to mean that some module produces an output when some other module is not ready to receive this as an input. In the timed case, this is complicated further by the fact that it must also be checked that the output is produced at an acceptable time. Consider a module $M = \langle I, O, S \rangle$ composed of several modules $\{M_1, \ldots, M_n\}$, where $M_k = \langle I_k, O_k, S_k \rangle$. Consider also a timed trace $x = (x_1, \ldots, x_m)$, where $x_m = (w, \tau)$ and $w \in O_k$ for some $k \leq n$. This trace causes a failure if *advance_time*$(M, (x_1, \ldots, x_{m-1}), \tau)$, $x \in S_k$, but $x \notin S$. Intuitively, this means that some module produces a transition on one of its outputs before some module is prepared to receive it. These types of failures are called *safety failures*.

A *timing failure* occurs when some module does not receive an input in time. In other words, either some input fails to occur or occurs later than required. There are potentially several ways to characterize timing failures formally, with each choice having different effects on the difficulty of verification. In particular, for the most general definition, it is no longer possible to use mirrors without some extra complexity which is beyond the scope of this chapter.

To verify a timed system, we must use one of the timed state space exploration algorithms described in Chapter 7. Let's consider using the discrete-time analysis method to find the timed state space for the implementation and the mirror of the specification. We can again compose them and check if the failure state is reachable. The application of region- or zone-based analysis methods for timing verification are a bit more involved and beyond the scope of this chapter.

8.3 SOURCES

Logics to represent time have long been discussed in philosophy circles. The modern temporal logics have their origin in the work of Kripke [212]. In recent years, it has seen increasing use in verification of both hardware and software systems [123, 303, 316]. The application of temporal logic to the verification of asynchronous circuits was proposed by Browne et al. [50] and Dill and Clarke [116]. A similar approach is taken by Berthet and Cerny using characteristic functions rather than a temporal logic [41]. Weih and Greenstreet have combined model checking with symbolic trajectory evaluation to verify speed-independent datapath circuits [398]. Lee et al. utilized a formalism similar to temporal logic called synchronized transitions to verify asynchronous circuits [228]. Bailey et al. utilized the Circal language and its associated algebra to verify asynchronous designs [21]. Adaptations of temporal logic to include timing have been developed by numerous people. One of the more famous ones is due to Abadi and Lamport [1]. Techniques for timed model checking of asynchronous circuits are described by Hamaguchi et al. [159], Burch [61], Yoneda and Schlingloff [415], and Vakilotojar et al. [386].

Trace theory was first applied to circuits by Snepscheut [361] and Udding [381] for the specification of delay-insensitive circuits. The application of trace theory to the verification of speed-independent circuits was pioneered by Dill [113, 115]. Section 8.2 follows Dill's work closely. An improved verifier was developed by Ebergen and Berks [117]. The notion of strong conformance was introduced by Gopalakrishnan et al. [151].

Recently, several techniques have been proposed to avoid a complete enumeration of the state space. These techniques utilize *partial orders* [149, 317, 416], *stubborn sets* [388], *unfoldings* [264, 266], or *cube approximations* [26, 28] to greatly reduce the size of the representation needed for the state space. Yoneda and Schlingloff developed a partial order method for the verification of timed systems [415].

Timed trace theory and a verification method based on discrete time were described by Burch [62, 63]. Devadas developed a technique for verifying Huffman circuits using back-annotated bounded delay information [112]. Rokicki and Myers employed a zone-based technique to verify timed asynchronous circuits [287, 325, 326]. These techniques have since been improved and applied to numerous examples by Belluomini et al. [31, 32, 33, 34]. Semenov and Yakovlev have developed an unfolding technique to verify timed circuits modeled using time Petri nets [349]. Yoneda and Ryu extended their timed partial order method to timed trace theoretic verification [414]. The problems with conformance and mirroring in the verification of timed circuits are described in [425]. Recently, there has been some research that employs implicit or *relative timing* assumptions (i.e., the timing of some sequence of events is assumed to be larger than some other sequence) to verify timed circuits [292, 310, 366].

Problems

8.1 Linear-Time Temporal Logic
Consider the situation where there are two shops to which the winery can deliver wine. It communicates with the first using the wires *req_wine1* and *ack_wine1* and the second with *req_wine2* and *ack_wine2*. Write an LTL formula that says that the winery is *fair*. In other words, the winery will deliver wine to both shops.

8.2 Linear-Time Temporal Logic
Again consider the situation where there are two shops to which the winery can deliver wine. Write an LTL formula that says that the winery at some point decides to sell only to one shop in the future.

8.3 Linear-Time Temporal Logic
If either the winery can stop producing wine or the patron loses interest in buying wine, which of LTL formulas 8.1 to 8.8 would be violated? How could they be fixed?

8.4 Protocol Verification
Check if the SG in Figure 8.15(a) satisfies the following LTL formula:

$$\Box(y \Rightarrow (x \textbf{ U } z))$$

If not, indicate which states violate the formula.

8.5 Protocol Verification
Check if the SG in Figure 8.15(b) satisfies the following LTL formula:

$$\Box(y \Rightarrow (x \textbf{ U } z))$$

If not, indicate which states violate the formula.

8.6 Timed Linear-Time Temporal Logic
Again consider the situation where there are two shops to which the winery can deliver wine. Write a timed LTL formula that says that the winery is fair in bounded time. In other words, the winery will not stop delivering wine to one shop for more than 60 minutes.

8.7 Trace Structures
Give a receptive trace structure for a NAND gate.

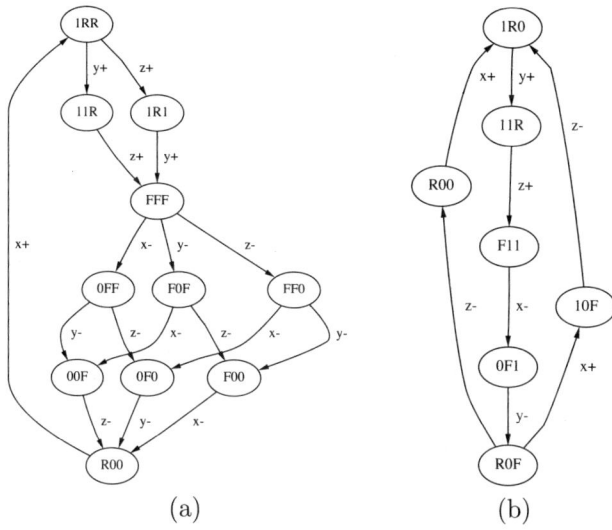

Fig. 8.15 SGs for Problems 8.4 and 8.5 (state vector $\langle x, y, z \rangle$).

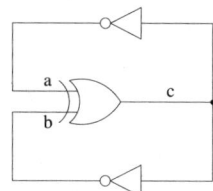

Fig. 8.16 Circuit for Problem 8.9.

8.8 Trace Structures
Give a receptive trace structure for an XOR gate.

8.9 Trace Structure Composition
Use composition of trace structures to determine whether or not the circuit shown in Figure 8.16 is failure-free.

8.10 Trace Structure Composition
Use composition of trace structures to determine whether or not the circuit shown in Figure 8.17 is failure-free.

8.11 Canonical Trace Structures
Transform the trace structure shown in Figure 8.18 to a canonical prefix-closed trace structure.

318   VERIFICATION

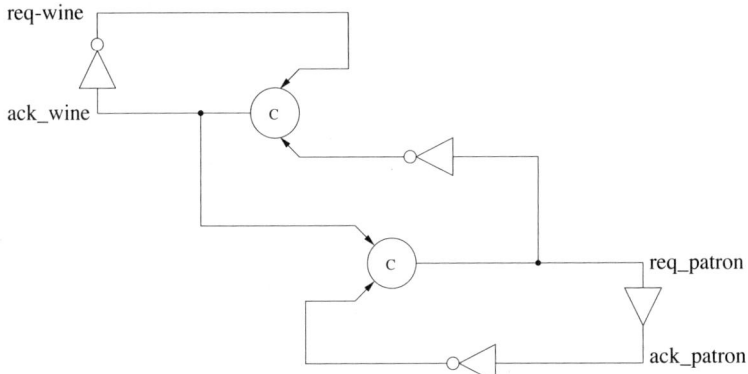

Fig. 8.17   Circuit for Problem 8.10.

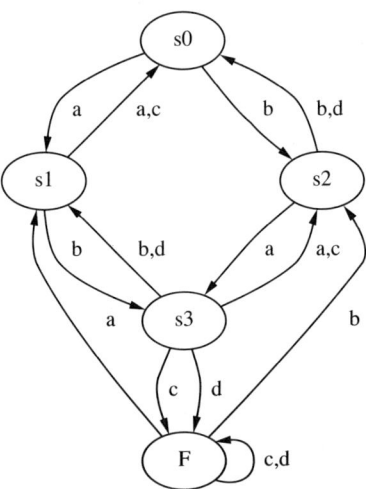

Fig. 8.18   Trace structure for Problem 8.11.

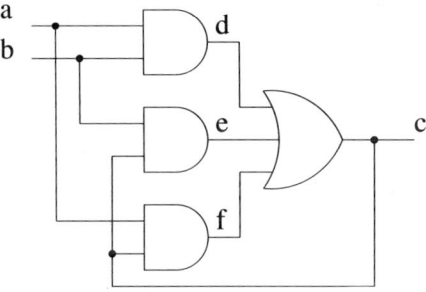

Fig. 8.19 Circuit for Problem 8.12.

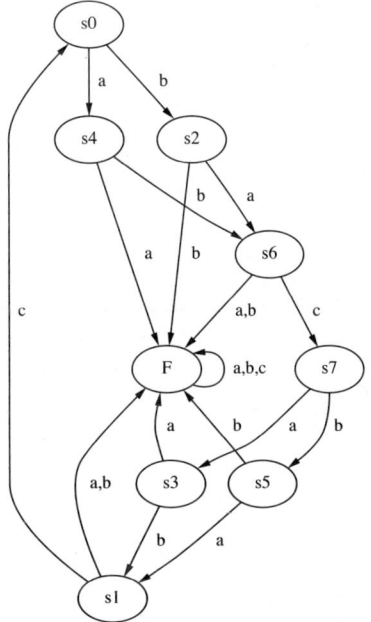

Fig. 8.20 Specification for Problems 8.12 and 8.13.

8.12 Mirrors and Verification

Use composition to create a trace structure for the circuit shown in Figure 8.19. Find the mirror of the specification for a C-element shown in Figure 8.20. Compose the trace structure for the circuit and the mirror of the specification to determine if the circuit conforms to the specification (i.e., the circuit is a correct implementation of a C-element). If the circuit does not conform, give a sequence of transitions that causes the circuit to fail.

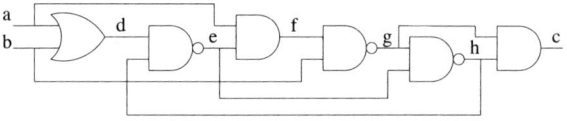

Fig. 8.21 Circuit for Problem 8.13.

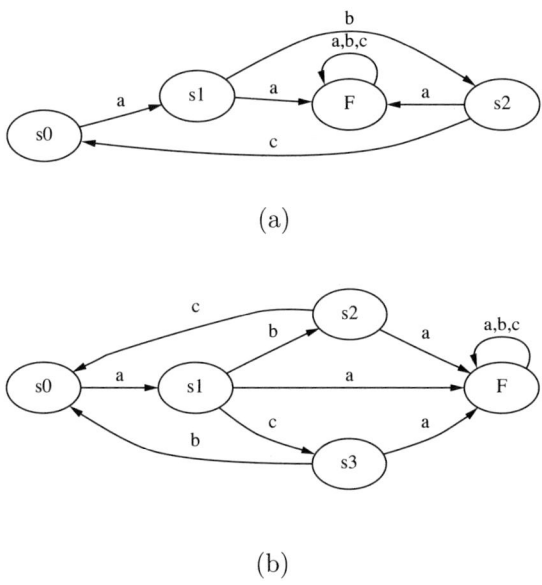

Fig. 8.22 (a) Sequencer and (b) fork.

8.13 Mirrors and Verification

Use composition to create a trace structure for the circuit shown in Figure 8.21. Find the mirror of the specification for a C-element shown in Figure 8.20. Compose the trace structure for the circuit and the mirror of the specification to determine if the circuit conforms to the specification (i.e., the circuit is a correct implementation of a C-element). If the circuit does not conform, give a sequence of transitions that causes the circuit to fail.

8.14 Strong Conformance

A sequencer receives an input a and generates an output b followed by another output c. The trace structure for a sequencer is shown in Figure 8.22(a). A fork receives an input a and generates outputs b and c in parallel. The trace structure for a fork is shown in Figure 8.22(b). Show that a sequencer conforms to a fork, but a fork does not conform to a sequencer. Does a sequencer strongly conform to a fork?

9
Applications

The memories of a man in his old age are the deeds of a man in his prime.
—Pink Floyd

In theory, there's no difference between theory and practice; in practice, there is.
—Chuck Reid

It's all very well in practice, but it will never work in theory.
—French management saying

An ounce of action is worth a ton of theory.
—Friedrich Engels

A theory must be tempered with reality.
—Jawaharlal Nehru

If the facts don't fit the theory, change the facts.
—Albert Einstein

In this chapter we give a brief history of the application of asynchronous design. An in-depth study is given for one recent successful asynchronous design, Intel's RAPPID instruction-length decoder. Issues in performance analysis and testing of asynchronous designs are also discussed. The chapter concludes with a discussion of the *synchronization problem*. Although this problem may have prevented the commercial application of RAPPID, it may ultimately be the problem that makes asynchronous design a necessity.

322   APPLICATIONS

9.1 BRIEF HISTORY OF ASYNCHRONOUS CIRCUIT DESIGN

Since the early days, asynchronous circuits have been used in many interesting applications. In the 1950s and 1960s, asynchronous design was used in many early mainframe computers, including the ILLIAC and ILLIAC II designed at the University of Illinois and the Atlas and MU-5 designed at the University of Manchester. The ILLIAC and ILLIAC II were designed using the speed-independent design techniques developed by Muller and his colleagues. The ILLIAC, completed in 1952, was 10 feet long, 2 feet wide, $8\frac{1}{2}$ feet high, contained 2800 vacuum tubes, and weighed 5 tons. ILLIAC II was completed in 1962, and it was 100 times faster than its predecessor. This computer contained 55,000 transistors, and it could perform a floating-point multiply in 6.3 μs. The ILLIAC II used three concurrently operating controls: an arithmetic control, interplay control for data transfers, and a supervisory control called Advanced Control. The Advanced Control is responsible for fetching and storing operands, address construction and indexing, partial decoding of orders for the other controls, etc. The three controls are largely asynchronous and speed-independent. The reasons cited for using speed-independent design were increased reliability and ease of maintenance. These controls collected *reply* signals to indicate that all operations for the current step are complete before going on to the next step. The arithmetic unit was not speed-independent, as they believed that would increase its complexity and cost while decreasing its speed. The electromechanical peripheral devices were also not speed-independent, since they were inherently synchronous. This computer was used until 1967, and among its accomplishments was the discovery of three Mersenne prime numbers, including the largest then known prime number, $2^{11213} - 1$, which is over 3000 digits.

In the 1960s and 1970s at Washington University in St. Louis, asynchronous *macromodules* were developed. Macromodules are "building blocks such as registers, adders, memories, control devices, etc., from which it is possible for the electronically-naive to construct arbitrarily large and complex computers that work" [302]. The set of macromodules was designed such that they were functionally large enough to be useful, easy to interconnect to form larger computing engines, and robust enough to allow designers to worry only about logical and not electrical problems. Asynchronous design was chosen for this reason. Macromodules were used to build macromodular computer systems by placing them in a rack and interconnecting them by wires. These wires carry data signals as well as a bundled data control signal that indicates the arrival of valid data. There were also wires for control to sequence operations. Macromodular systems were developed in such a way as to be almost directly realizably from a flowchart description of an algorithm. Through this procedure many special-purpose computing engines were designed using macromodules to solve numerous problems, many from the biomedical field. Also in the 1970s, asynchronous techniques were used at the University of

Utah in the design of the first operational dataflow computer, DDM-1, and at Evans and Sutherland in the design of the first commercial graphics system.

In the late 1980s, Matsushita, Sanyo, Sharp, and Mitsubishi developed *data-driven processors*. Rather than the traditional view of a single program counter controlling the timing of instruction execution, the flow of data controls the operation speed. In other words, when all needed data arrives, it is operated on. Several signal processors were designed using this idea. Most recently, a data-driven media processor (DDMP) has been designed at Sharp capable of 2400 million signal processing operations per second while consuming only 1.32 W. Videonics has utilized the DDMP design in a high-speed video DSP. The asynchronous design of the DDMP is cited to have simplified the board layout and reduced RF interference.

In 1989, researchers at Caltech designed the first fully asynchronous microprocessor. The processor has a 16-bit datapath with 16- and 32-bit instructions. It has twelve 16-bit registers, four buses, an ALU, and two adders. The chip design consisted of about 20,000 transistors, and it was fabricated in both 2 μm and 1.6 μm MOSIS SCMOS. The 2 μm version could perform 12 million ALU instructions per second, while the 1.6 μm version could perform 18 million. The chips were shown to operate with VDD ranges from 0.35 to 7 V, and they were shown to achieve almost double the performance when cooled in liquid nitrogen. While the architecture of this design is quite simple and the results are modest, this design is significant for several reasons. First, the design is entirely *quasi-delay insensitive* (QDI), which means that it will operate correctly regardless of delays, except on specific isochronic forks. Second, the design was derived from a high-level channel description much like that described in Chapter 2. Using program transformations, ideas like pipelining were introduced. Third, the entire design took five people only five months. The group at Caltech went on to design the first asynchronous microprocessor in gallium arsenide technology. This processor ran at 100 MIPS while consuming 2 W. The most recent design from this group is an asynchronous MIPS R3000 microprocessor. This design introduced new ways to reduce the overhead of completion detection through bit-level pipelining of functional units and pipelining global completion detection. The design was fabricated in 0.6 μm CMOS, and it uses 2 million transistors, of which 1.25 million are in its caches. The measured performance ranged from 60 MIPS and 220 mW at 1.5 V and 25°C to 180 MIPS and 4 W at 3.3 V and 25°C. Running Dhrystone, the chip achieved about 185 MHz at 3.3 V.

In 1994, the AMULET group at the University of Manchester completed the AMULET1, the first asynchronous processor to be code-compatible with an existing synchronous processor, the ARM microprocessor. Their design style followed Sutherland's two-phase micropipeline idea. Their chip was fabricated in a CMOS 1 μm process and a 0.7 μm process. The performance was measured for the 1 μm part from 3.5 to 6 V using the Dhrystone benchmark code. At these voltages, the processor could complete between 15 and 25 thousand Dhrystones per second. The MIPS/watt value was also measured to be from 175 down to 50. The chip was also shown to operate correctly between $-50°C$ and $120°C$. This project was followed up with the design of the AMULET2e, which targeted embedded system applications. In this design, an AMULET2 core is coupled with a cache/RAM, a memory interface, and other control functions on chip. One key difference is that the design used a four-phase bundled-data style rather than two-phase, as it was found to be simpler and more efficient. The design was fabricated in 0.5 μm CMOS, and its measured performance at 3.3 V was 74 kDhrystones, which is roughly equivalent to 42 MIPS. At this peak rate, it consumes 150 mW. One other interesting measurement they performed was the radio-frequency emission spectrum or EMC. Results show significant spread of the peaks as compared with a clocked system. One last interesting result was that the power consumed when the processor enters "halt" mode drops to under 0.1 mW. The most recent design by this group is the AMULET3, which has incorporated ARM thumb code and new architectural features to improve performance.

In 1994, a group at SUN suggested that replicating synchronous architecture may not be the best way to demonstrate the advantages of asynchronous design. Instead, they suggested a radically different architecture, the *counterflow pipeline* in which instructions are injected up the pipeline while contents of registers are injected down. When an instruction meets the register values it needs, it computes a result. Key to the performance of such a design are circuits to move data very quickly. This group has subsequently designed several very fast FIFO circuits. Their test chips fabricated in 0.6 μm CMOS have been shown to have a maximum throughput of between 1.1 and 1.7 Giga data items per second.

Philips Research Laboratories has designed numerous asynchronous designs targeting low power. This group developed a fully automated design procedure from a hardware specification in their language TANGRAM to a chip, and they have applied this procedure to rapidly design several commercially interesting designs. In 1994, this group produced an error corrector chip for the digital compact cassette (DCC) player that consumed only 10 mW at 5 V which is one-fifth of the power consumed by its synchronous counterpart. This design also required only 20 percent more area. In 1997, they designed asynchronous standby circuits for a pager decoder which dissipated four times less power and are only 40 percent larger than their comparable synchronous design. Since the standby circuit in modern pagers is responsible for a substantial portion of the power consumption while using only a small amount of silicon area, this result maps to a 37 percent decrease in power for the entire pager while showing only a negligible overall area increase. In 1998, this group designed an 80C51 microcontroller. A chip fabricated in 0.5 μm CMOS was shown to be three to four times more power efficient than its synchronous counterpart, consuming only 9 mW when operating at 4 MIPS. Perhaps the most notable accomplishment to come out of this group is a fully asynchronous pager being sold by Philips that uses the standby circuits and the 80C51 microcontroller just described. While the power savings reported was important, the major reason for Philips to use the asynchronous pager design is the fact that the asynchronous design had an emission spectrum more evenly spread over the frequency range. Due to interference produced at the clock frequency and its harmonics, the synchronous version shuts off its digital circuitry as a message is received so as not to interfere with the RF communication. The spread-spectrum emission pattern for the asynchronous design allowed the digital circuitry to remain active as the message is received. This permitted their pager to be capable of being universal in that it can accept all three of the international pager standards.

Another recent asynchronous application came out of the RAPPID project at Intel conducted between 1995 and 1999. RAPPID is a fully asynchronous instruction-length decoder for the PentiumII 32-bit MMX instruction set. The chip design in this project achieved a three fold improvement in speed and a two fold improvement in power compared with the existing synchronous design. This design is described in some detail in the following section.

9.2 AN ASYNCHRONOUS INSTRUCTION-LENGTH DECODER

RAPPID (Revolving Asynchronous Pentium Processor Instruction Decoder) is a fully asynchronous instruction-length decoder for the complete PentiumII 32-bit MMX instruction set. In this instruction set, each instruction can be from 1 to 15 bytes long, depending on a large number of factors. In order to allow concurrent execution of instructions, it is necessary to rapidly determine the positions of each instruction in a cache line. This was at the time a critical bottleneck in this architecture. A partial list of the rules that determine the length of an instruction is given below.

- Opcode can be 1 or 2 bytes.
- Opcode determines presence of the ModR/M byte.
- ModR/M determines presence of the SIB byte.
- ModR/M and SIB set length of displacement field.
- Opcode determines length of immediate field.
- Instructions may be preceded by as many as 15 prefix bytes.
- A prefix may change the length of an instruction.
- The maximum instruction length is 15 bytes.

For real applications, it turns out that there are only a few common instruction lengths. As shown in the top graph from Figure 9.1, 75 percent of instructions are 3 bytes or less in length. Nearly all instructions are 7 bytes or less. Furthermore, prefix bytes that modify instruction lengths are extremely rare. This presents an opportunity for an asynchronous design to optimize for the common case by optimizing for instructions of length 7 or less with no

326 APPLICATIONS

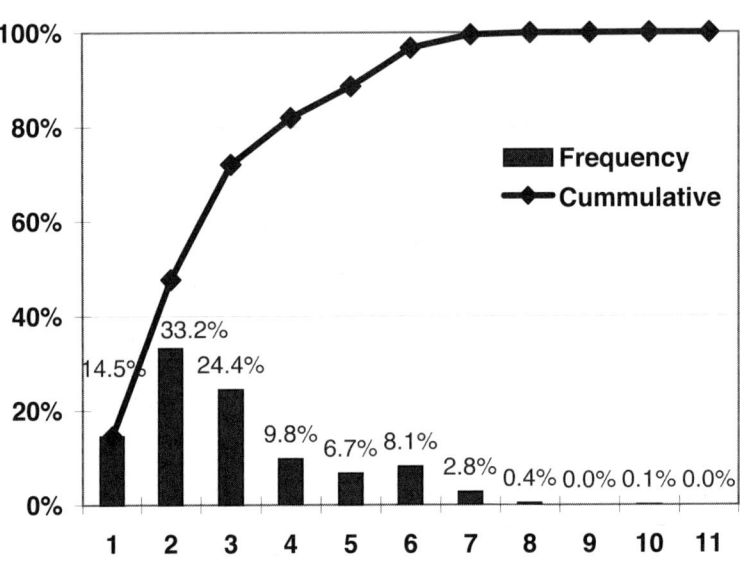

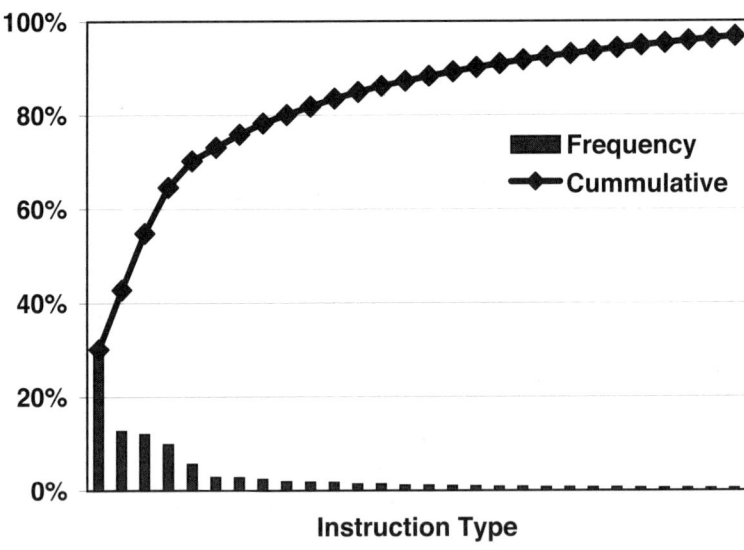

Fig. 9.1 Histogram for proportion of instruction lengths and cumulative length statistics, and histogram for proportion of instruction types and cumulative statistics.

AN ASYNCHRONOUS INSTRUCTION-LENGTH DECODER

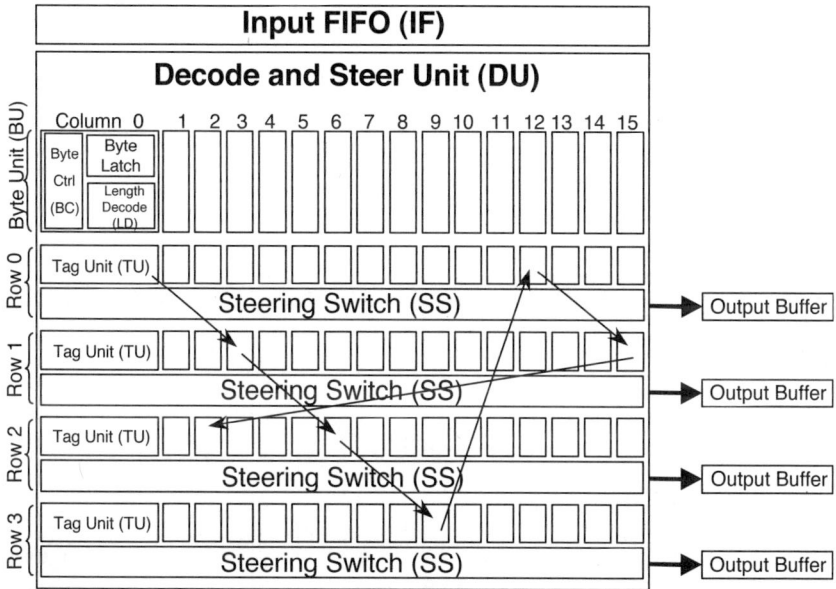

Fig. 9.2 RAPPID microarchitecture.

prefix bytes. Other less efficient methods are then used for longer instructions and instructions with prefix bytes.

The RAPPID microarchitecture is shown in Figure 9.2. The RAPPID decoder reads in a 16-byte cache line, and it decodes each byte as if it is the first byte of a new instruction. Each byte speculatively determines the length of an instruction beginning with this byte. It does this by looking at three additional downstream bytes. The actual first byte of the current instruction is marked with a tag. This byte uses the length that it determined to decide which byte is the first byte of the next instruction. It then signals that byte while notifying all bytes in between to cancel their length calculations and forwards the bytes of the current instruction to an output buffer. To improve performance, four rows of tag units and output buffers are used in a round-robin four-issue fashion.

As shown in benchmark analysis at the bottom of Figure 9.1, only a small number of instruction types are common. In fact, 15 percent of the opcode types occur 90 percent of the time. This provides another opportunity for an asynchronous design to be optimized for the common case. Each length decode unit is essentially a large, complex PLA structure. Using the statistics shown in Figure 9.1, it is possible to restructure the combinational logic to be faster for the common case. Consider the logic shown in Figure 9.3(a). Assume that gate A causes the OR gate to turn on 90 percent of the time, gate B 6 percent, and gate C 4 percent. If the delay of a two-input gate is

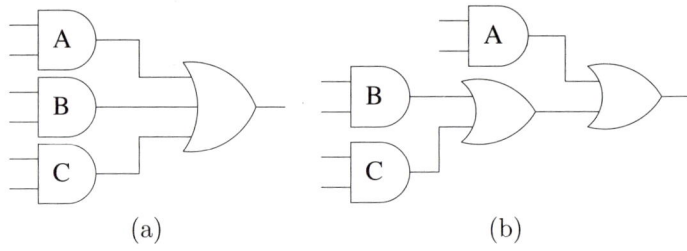

Fig. 9.3 (a) Balanced tree logic function. (b) Unbalanced tree logic function.

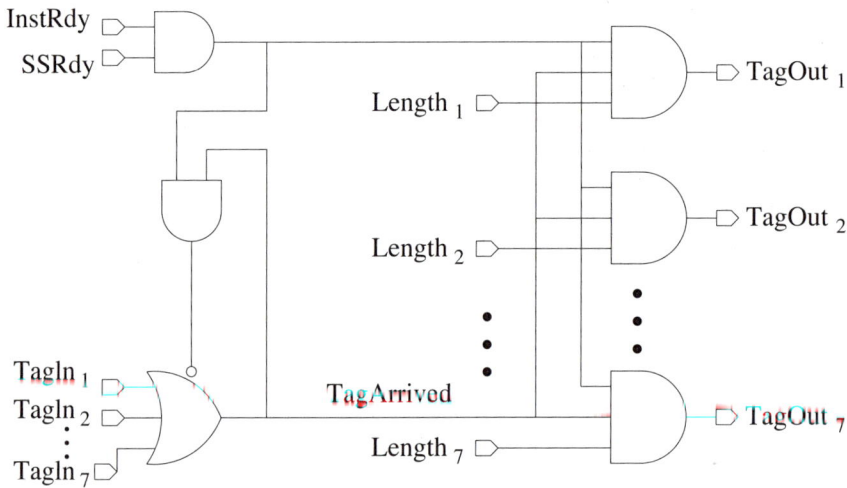

Fig. 9.4 Tag unit circuit.

1 and a three-input gate 1.5, the delay of the logic shown in Figure 9.3(a) is 2.5. However, the logic shown in Figure 9.3(b) has an average-case delay of only 2.1. The length decode logic takes advantage of this idea, and it is implemented using large unbalanced trees of domino logic that have been optimized for common instructions.

The key to achieving high performance is the tag unit, which must be able to rapidly tag instructions. The timed circuit for one tag unit is shown in Figure 9.4. Assuming that the instruction is ready (i.e., *InstRdy* is high, indicating that one $Length_i$ is high and all bytes of the instruction are available) and the steering switch is ready (i.e., *SSRdy* is high), then when a tag arrives (i.e., one of $TagIn_j$ is high), the first byte of the next instruction is tagged (i.e., $TagOut_i$ is set to high). In the case of a branch, the tag is forwarded to a branch control circuit which determines where to inject the tag back into the new cache line.

If the instruction and steering switch are ready when a column gets tagged, it takes only two gate delays from TagIn to TagOut. In other words, a synchronization signal can be created every two gate delays. It is difficult to imagine distributing a clock which has a period of only two gate delays. The tag unit in the chip is capable of tagging up to 4.5 instructions/ns.

This circuit, however, requires timing assumptions for correct operation. In typical asynchronous communication, a request is transmitted followed by an acknowledge being received to indicate that the circuit can reset. In this case, there is no explicit acknowledgment, but rather, acknowledgment comes by way of a timing assumption. Once a tag arrives (i.e., *TagArrived* is high), if the instruction and steering switch are ready, the course is set to begin to reset *TagArrived*. The result is that the signal produced on $TagOut_i$ is a pulse. Let us consider now the effect of receiving a pulse on a *TagIn* signal. If either the instruction or steering switch are not ready, then *TagArrived* gets set by the pulse, in effect latching the pulse. *TagArrived* will not get reset by the disappearance of the pulse but rather the arrival of a state in which both the instruction and steering switch are ready.

For this circuit to operate correctly, there are two critical timing assumptions. First, the pulse created must be long enough to be latched by the next tag unit. This can be satisfied by adding delay to the AND gate used to reset *TagArrived*. An arbitrary amount of delay, however, cannot be added since the pulse must not also be so long that another pulse could come before the circuit has reset. Therefore, we have a *two-sided timing constraint*. Analysis methods such as those described in Chapter 7 are needed to verify that this design operates correctly.

The RAPPID test chip was fabricated in May 1998 using a 0.25 μm CMOS process. The test chip was capable of decoding and steering instructions at a rate of 2.5 to 4.5 instructions per nanosecond. This is about three times faster than the peak performance of the fastest synchronous three-issue product in the same fabrication process clocked at 400 MHz which is capable of only 1.2 instructions per nanosecond. The chip operated correctly between 1.0 and 2.5 V while the synchronous design could only tolerate about 1.9 to 2.1 V. The RAPPID design also consumes only one-half of the energy of the clocked design. The RAPPID design was found to achieve these gains with only a 22 percent area penalty over the clocked design.

9.3 PERFORMANCE ANALYSIS

One major difficulty in designing an asynchronous circuit such as RAPPID is determining its performance. It is not simply a matter of finding the critical path delay or counting the number of clock cycles per operation. One of the major driving forces behind the design of RAPPID is optimizing for the common case. This means that worst-case analysis as done for synchronous design may actually be quite pessimistic, as our goal is to achieve high rates of performance on average.

To address this problem, one must take a probabilistic approach to performance analysis. Consider, for example, the TEL structure model described in Chapter 4. In a TEL structure, the delay between two events is modeled as a range $[l, u]$, where l is a lower bound and u is an upper bound of delay. For purposes of the performance analysis, it is necessary to extend this model to include a distribution function for this delay. One simple approach is to simply assume that the delay falls uniformly in this range. If this design is a timed circuit, though, where the correctness may depend on the delay never stepping out of this range, a uniform assumption is hopefully not realistic. Another possibility would be to use a more interesting distribution such as a truncated Gaussian.

Once a probabilistic delay model has been chosen, the next step is to use it to determine performance. The most direct approach is a Monte Carlo simulation.

> **Example 9.3.1** Consider the timed circuit shown in Figure 7.38 and the delay numbers given in Table 7.1. Let us also assume that although theoretically the patron may have infinite response time, in practice it rarely takes him more than 10 minutes to come fetch the wine. If we take the standard synchronous performance analysis approach of using just the worst-case delay, we find that the cycle time of this circuit is 18.3 minutes. On the other hand, if we consider each delay to be distributed within its range using a truncated Gaussian with a mean in the middle and a standard deviation of one-fourth of the range, we find the cycle time to be 14.2 minutes. Since an asynchronous circuit operates at its average rate, this is a more true reflection of the actual performance.

9.4 TESTING ASYNCHRONOUS CIRCUITS

Another major obstacle to the commercial acceptance of asynchronous circuits, such as RAPPID, is the perception that they are more difficult to test. Once a chip has been fabricated, it is necessary to test it to determine the presence of manufacturing defects, or *faults*, before delivering the chip to the consumer. For asynchronous circuits, this is complicated by the fact that there is no global clock which can be used to single step the design through a sequence of steps. Asynchronous circuits also tend to have more state holding elements, which increases the overhead needed to apply and examine test vectors. Huffman circuits employ redundant circuitry to remove hazards, and redundant circuitry tends to hide some faults, making them untestable. Finally, asynchronous circuits may fail due to glitches caused by *delay faults*, which are particularly difficult to detect.

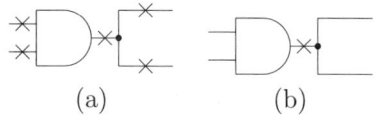

Fig. 9.5 (a) Potential fault locations. (b) Output stuck-at fault model.

However, it's not all bad news for asynchronous circuits. Since many asynchronous styles use handshakes to communicate data, for many possible faults, a defective circuit simply halts. In particular, in the *stuck-at fault model*, a defect is assumed to cause a wire to become permanently *stuck-at-0* or *stuck-at-1*. If an acknowledge wire is stuck-at-0, the corresponding request is never acknowledged, causing the circuit to stop and wait forever. This is obviously easy to detect with a timeout. In this case, the circuit is said to be *self-checking*. Since in a delay-insensitive circuit every transition must be acknowledged by the receiver of the transition, delay-insensitive circuits will halt in the presence of any stuck-at fault on any wire in the design.

As shown in Chapter 1, the class of circuits that can be designed using the delay-insensitive model is very limited. Unfortunately, for a more general class of designs such as Muller circuits designed under the speed-independent model, the circuit may not halt for all stuck-at faults. For the gate shown in Figure 9.5(a), a stuck-at fault can occur at any of the five locations marked with an ×. Under this model, not all faults cause a Muller circuit to halt. In the *output stuck-at fault model*, there is only one possible fault location shown in Figure 9.5(b). Since in a Muller circuit, a transition on an output must be acknowledged by one of the two branches of the isochronic fork, a fault at this location results in the circuit halting. Therefore, Muller circuits do halt for any output stuck-at fault.

Some Muller circuits such as the one shown in Figure 9.6(a) do halt for all stuck-at faults. Others, however, such as the one in Figure 9.6(b) do not. In these cases, the fault can cause a *premature firing*.

> **Example 9.4.1** Consider the fault where *r1* is stuck-at-0 in the circuit shown in Figure 9.6(b). Assume that all signals start out initially low. After *req_wine* goes high, *x* goes high, allowing *req_patron* to go high since *r1* is stuck-at-0. This is a premature firing since *req_patron* is not supposed to go high until after *ack_wine* goes high, followed by *req_wine* going low. Similarly, if *a2* is stuck-at-0, it can cause *ack_wine* to go low prematurely (i.e., before *ack_patron* goes low).

It is clear from these examples that the problem is due to isochronic forks. If we consider all wire forks and determine those which must be isochronic for correct circuit operation, we can determine a fault model which lies between the output stuck-at fault model and the more general stuck-at fault model. Namely, in the *isochronic fork fault model*, faults can be detected on

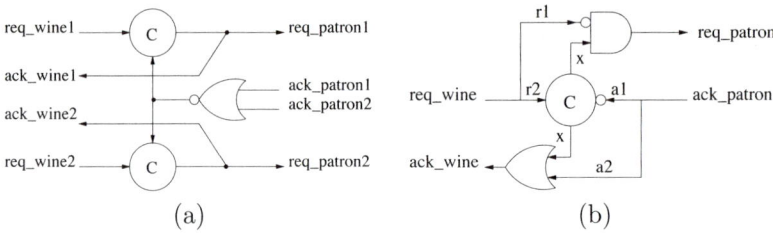

Fig. 9.6 (a) Circuit in which all stuck-at faults cause it to halt. (b) Circuit in which some stuck-at faults cause premature firings.

all branches of a nonisochronic fault by the circuit halting, but only on the input to forks that are isochronic.

Granted even the general stuck-at fault model is too simplistic to truly capture the symptoms of manufacturing defects. However, more complex *delay fault* and *bridging fault* models have been successfully adapted to asynchronous circuits. These topics, though, are beyond the scope of this short discussion.

Once we have decided on a fault model, the next step is to add any necessary circuitry to apply and analyze test vectors, such as *scan paths*. We must also generate a sufficient set of test vectors to guarantee a high degree of coverage of all possible faults in our model. As in models, the main hurdle to testing asynchronous circuits appears to be that the traditional synchronous testing methods do not work right off the shelf. Again, however, many of the popular methods of test have been adapted to the asynchronous design problem.

9.5 THE SYNCHRONIZATION PROBLEM

Despite the excellent results demonstrated by the RAPPID design, it has not been used in a commercial product. One important reason is that the asynchronous design must communicate with the rest of the microprocessor that operates synchronously. Unfortunately, this is difficult to do reliably without substantial latency penalties. When this latency penalty is taken into account, most, if not all, of the performance advantage gained by the RAPPID design is lost.

Consider again our wine patron shopping from two shops, one that sells chardonnay and another that sells merlot. When one shop calls him, he immediately heads off to that shop to buy the wine. What is he to do if they both notify him at nearly the same instant that they have a fresh bottle of wine to sell? He must make a choice. However, if both types of wine are equally appealing to him at that instant, he may sit there pondering it for some time. However, if the shopkeepers get impatient, they have a tendency to drink the wine themselves. Therefore, if he is really indecisive and cannot make up

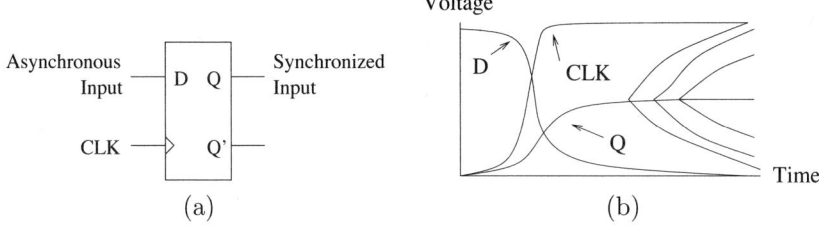

Fig. 9.7 (a) Simple, dangerous synchronizer. (b) Oscilloscope view of metastable behavior.

his mind for too long, he will not get either bottle of wine. This state in which the patron is stuck considering two equally appealing choices is called a *metastable state*. If this state persists so long that something bad happens (like the shopkeepers drinking his wine), this is called a *synchronization failure*.

In a circuit, this can happen when a synchronous circuit must synchronize an asynchronous input. This can be done using a single D-type flip-flop as shown in Figure 9.7(a). However, if the clock edge arrives too close in time to data arriving from an asynchronous circuit, the circuit may enter a metastable state in which its output is at neither a logic 0 or logic 1 level, but rather, lies somewhere in between. This behavior is depicted in Figure 9.7(b). Assume that Q is initially low and that D has recently gone high. If D goes low again at about the same time that CLK rises, the output Q may start to rise and then get stuck between the logic levels as it observes D falling. Should Q rise or fall? Actually, either answer would be okay, but the flip-flop becomes indecisive. At some point, Q may continue to a logic 1 level, or it may drop to the logic 0 level. When this happens, however, is theoretically unbounded. If during this period of indecision, a circuit downstream from this flip-flop looks at the synchronized input, it will see an indeterminate value. This value may be interpreted by different subsequent logic stages as either a logic 0 or a logic 1. This can lead the system into an illegal or incorrect state, causing the system to fail. Such a failure is traditionally called a synchronization failure. If care is not taken, the integration of asynchronous modules with synchronous modules can lead to an unacceptable probability of failure. Even if no asynchronous modules are used, synchronous modules operating at different clock rates or out of phase can have the same problem. The latter problem is becoming more significant as it becomes increasingly difficult to distribute a single global clock to all parts of the chip. Many designers today are considering the necessity of having multiple clock domains on a single chip, and they will need to face this problem.

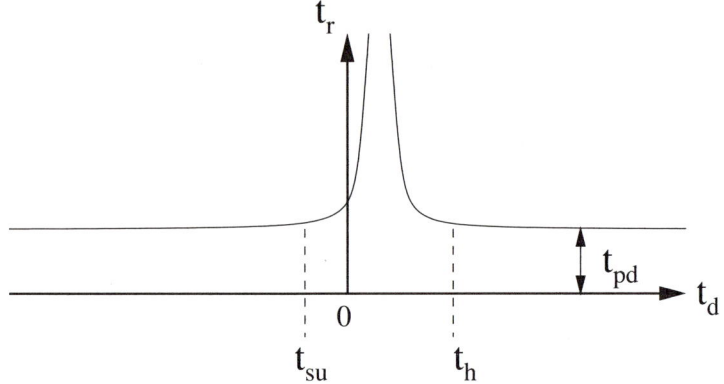

Fig. 9.8 Flip-flop response time as a function of input arrival time in relation to clock arrival time (clock arrives at time 0).

9.5.1 Probability of Synchronization Failure

Figure 9.8 shows a representative plot based on measured data for the response time of a flip-flop as a function of the arrival time, t_d, of data with respect to the clock. If data only changes before the *setup time*, t_{su}, and after the *hold time*, t_h, of a flip-flop, the *response time*, t_r, is roughly constant and equal to the *propagation delay* through the flip-flop, t_{pd}. If, on the other hand, the data arrives between the setup and hold times, the delay increases. In fact, if the data arrives at just the absolutely wrong time, the response time is unbounded.

If data can arrive asynchronously with respect to the clock, we can consider that it arrives at a time which is uniformly distributed within the clock cycle. Therefore, the probability that the data arrives at a time t_d which falls between t_{su} and t_h is given below.

$$P(t_d \in [t_{su}, t_h]) \;=\; \frac{t_h - t_{su}}{T} \tag{9.1}$$

where T is the length of the clock period. If we assume that the flip-flop is given some bounded amount of time, t_b, to decide whether or not to accept the newly arrived data, the probability of a synchronization failure is related to the probability that the response time, t_r, exceeds t_b. In the metastable region between t_{su} and t_h, it can be shown that the response time increases in an approximately exponential fashion. Therefore, if t_d falls in this range, the probability that $t_r > t_b$ can be expressed as follows:

$$P(t_r > t_b \mid t_d \in [t_{su}, t_h]) \;=\; \frac{1}{k + (1-k)e^{(t_b - t_{pd})/\tau}} \tag{9.2}$$

where k and τ are circuit parameters, with k being a positive fraction less than 1 and τ being a time constant with values on the order of a few picoseconds for modern technologies. Combine Equations 9.1 and 9.2 using Bayes rule:

$$P(t_r > t_b) = P(t_d \in [t_{su}, t_h]) \cdot P(t_r > t_b \mid t_d \in [t_{su}, t_h]) \quad (9.3)$$

$$= \frac{t_h - t_{su}}{T} \cdot \frac{1}{k + (1-k)e^{(t_b - t_{pd})/\tau}} \quad (9.4)$$

If $t_b - t_{pd} \geq 5\tau$, Equation 9.3 can be simplified as follows:

$$P(t_r > t_b) \approx \frac{t_h - t_{su}}{T} \cdot \frac{e^{-(t_b - t_{pd})/\tau}}{1 - k} \quad (9.5)$$

By combining constants, Equation 9.5 can be changed to

$$P(t_r > t_b) \approx \frac{T_0}{T} \cdot e^{-t_b/\tau} \quad (9.6)$$

Equation 9.6 is convenient since there is only two circuit-dependent parameters T_0 and τ that need to be determined experimentally. These parameters appear to scale linearly with feature size. Equation 9.6 has been verified experimentally and found to be a good estimate as long as t_b is not too close to t_{pd}. It is important to note that there is no finite value of t_b such that $P(t_r > t_b) = 0$. Therefore, the response time in the worst-case is unbounded.

A *synchronization error* occurs when t_r is greater than the time available to respond, t_a. A synchronization failure occurs when there is an inconsistency caused by the error. Failures occur less often than errors since a consistent interpretation still may be made even when there is an error. The expected number of errors is

$$E_e(t_a) = P(t_r > t_a) \cdot \lambda \cdot t \quad (9.7)$$

where λ is the average rate of change of the signal being sampled and t is the time over which the errors are counted. If we set $E_e(t_a)$ to 1, change t to MTBF (*mean time between failure*), substitute Equation 9.6 for $P(t_r > t_a)$, and rearrange Equation 9.7, we get

$$\text{MTBF} = \frac{T \cdot e^{t_a/\tau}}{T_0 \cdot \lambda} \quad (9.8)$$

This equation increases rapidly as t_a is increased. Therefore, even though there is no absolute bound in which no failure can ever occur, there does exist an engineering bound in which there is an acceptably low likelihood of error.

9.5.2 Reducing the Probability of Failure

Many techniques have been devised to address the metastability problem and reduce the probability of synchronization failure to an acceptable level when interfacing between synchronous and asynchronous modules. The goal of each

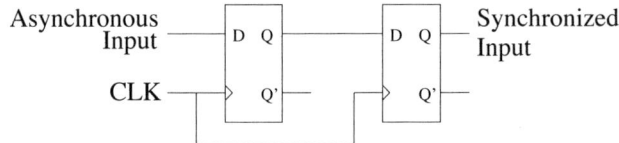

Fig. 9.9 Double latch solution to reduce synchronization failure.

of these techniques is to increase the amount of time to resolve the metastability (i.e., increase t_a). The simplest approach to achieve this is to use two (or more) latches in series as shown in Figure 9.9 to sample asynchronous signals arriving at a synchronous module. This increases the time allowed for a metastable condition to resolve. In other words, if n extra latches are added in series with an asynchronous input, the new value of t_a is given by

$$t'_a = t_a + n(T + t_{pd}) \qquad (9.9)$$

where T is the clock period and t_{pd} is the propagation delay through the added flip-flops. The cost, though, is an extra n cycles of delay when communicating data from an asynchronous module to a synchronous module, even when there is no metastability. This scheme also only minimizes the probability and does not eliminate the possibility of synchronization failure, as there is still some chance that a metastable condition could persist longer than n clock cycles.

Example 9.5.1 Assume that τ is measured to be about 20 ps and T_0 about 8 ns. If the clock frequency is 2 GHz, T is 500 ps. If asynchronous inputs are coming at an average rate of 1 GHz, λ is 10^9 samples per second. Let us also assume that we can tolerate a metastability for four-fifths of the clock period or $t_a = 400$ ps. Using Equation 9.8 we find the mean time between failures to be only 30 ms! If the propagation delay through a flip-flop is 120 ps and we add a second latch, then t_a becomes 780 ps, and the mean time between failure becomes about 63 days. If we add a third flip-flop, the mean time between failure increases to over 30 million years!

9.5.3 Eliminating the Probability of Failure

To eliminate synchronization failures completely, it is necessary to be able to force the synchronous system to wait an arbitrary amount of time for a metastable input to stabilize. In order for the synchronous circuit to wait, it is necessary for the asynchronous module to be able to cause the synchronous circuit's clock to stop when it is either not ready to communicate new data or not ready to receive new data. A *stoppable clock* can be constructed from a gated ring oscillator as shown in Figure 9.10. The basic operation is that when the *RUN* signal is activated, the clock operates at a nominal rate set by the number of inverters in the ring. To stop the clock, the *RUN* signal must be deactivated between two rising clock edges. The clock restarts as soon as

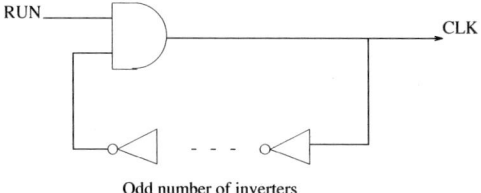

Fig. 9.10 Stoppable ring oscillator clock.

the *RUN* signal is reactivated. In other words, the clock should be stopped synchronously and is restarted asynchronously.

If the synchronous module decides when it needs data from the asynchronous module, the behavior is as follows. When the synchronous module needs data from an asynchronous module, it can request the data on the rising edge of the clock and in parallel set *RUN* low. If you want a guaranteed high pulse width, then *RUN* must be set low on the falling clock edge. When the data arrives from the asynchronous module, the acknowledgment from this module can be used to set *RUN* high. If *RUN* is set high before the end of the clock cycle, the next clock cycle can begin again without delay. If on the other hand, the asynchronous module is slow in providing data, the low phase of *CLK* will be stretched until the data arrives.

If the asynchronous module can decide when to send data, a *mutual exclusion* (ME) element is needed as shown in Figure 9.11 to guarantee that a synchronous module either receives data from an asynchronous unit or a pulse from the clock generator, but never both at the same time. If the asynchronous data arrives too close to the next clock pulse, both the data and the clock pulse may be delayed waiting for the metastability to resolve before determining which is to be handled first. An ME element has two inputs, *R1* and *R2*, and two outputs, *A1* and *A2*. It can receive rising transitions on both inputs concurrently, but it will respond with only a single rising transition on one of the corresponding outputs. There are three possible situations. The first is that the asynchronous module does not request to send data during this clock cycle. In this case, the ME simply acts as a buffer and the next rising clock edge is produced. The second case is the asynchronous request comes before the next rising clock edge is needed. In this case, the ME issues an *ACK* to the asynchronous module, and it prevents the next clock cycle from starting until *REQ* goes low. The third case is that *REQ* goes high just as *CLK* is about to rise again. This causes a metastable state to occur, but the ME is guaranteed by design to either allow the asynchronous module to communicate by setting *ACK* high and stopping the clock or by refusing to acknowledge the asynchronous module this cycle and allowing *CLK* to rise. Note that theoretically it may do neither of these things for an unbounded amount of time.

338 APPLICATIONS

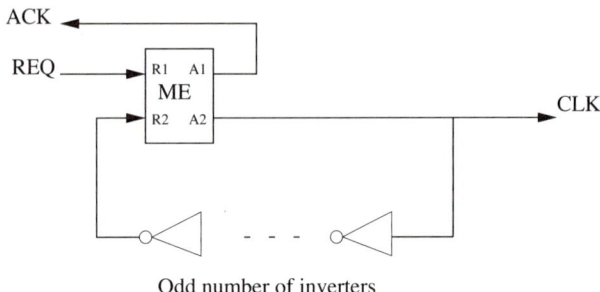

Fig. 9.11 Stoppable ring oscillator clock with ME.

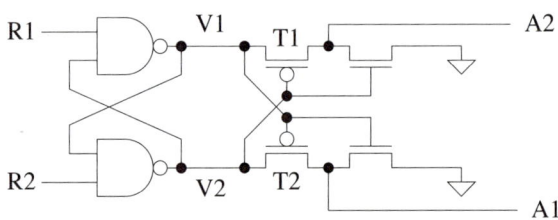

Fig. 9.12 Circuit for mutual exclusion.

A circuit diagram for a CMOS ME element is shown in Figure 9.12. When this circuit goes metastable, *V1* and *V2* differ by less than a threshold voltage, so *T1* and *T2* are off. Therefore, both *A1* and *A2* remain low. Once *V1* and *V2* differ by more than a threshold voltage, either *T1* or *T2* will turn on, pulling up its corresponding output.

A stoppable clock can be used to design a *globally asynchronous locally synchronous* (GALS) architecture. Communication between modules is done asynchronously using request/acknowledge protocols while computation is done synchronously within the modules using a locally generated clock. The basic structure of such a module is shown in Figure 9.13. The module's internal clock is stopped when it must wait for data to arrive from, or to be accepted by, other modules. If an asynchronous module can request to communicate data to a synchronous module at arbitrary times as discussed above, the stoppable clock shown in Figure 9.11 is needed. If the synchronous unit determines when data is to be transferred to/from the asynchronous modules, there is no need for a ME element, since the decision to wait on asynchronous communication is synchronized to the internal clock. In this case, the stoppable clock shown in Figure 9.10 can be used.

A *globally synchronous locally asynchronous* architecture is shown in Figure 9.14. One possible approach to increasing a synchronous, pipelined microprocessor's speed is to replace the slowest pipeline stages with asynchronous modules that have a better average-case performance, RAPPID for example. If the interfacing problem can be addressed, this allows a performance gain

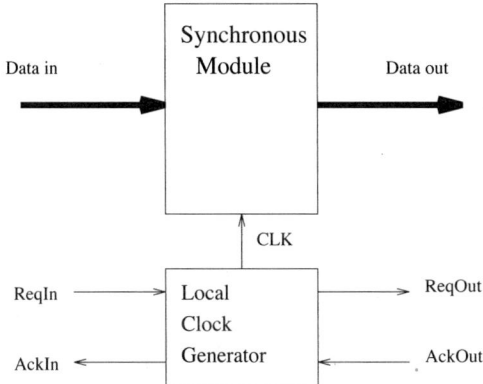

Fig. 9.13 Basic module of a GALS architecture.

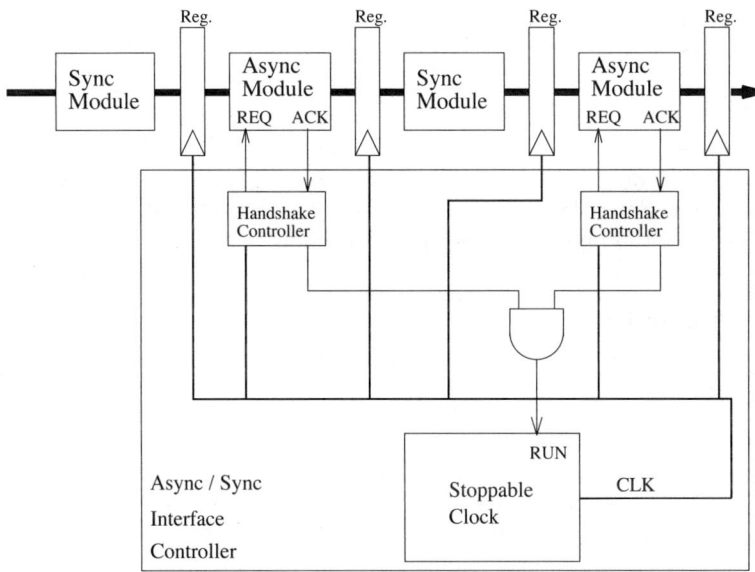

Fig. 9.14 Architecture for a globally synchronous locally asynchronous system.

without redesigning the entire chip. While the entire system communicates synchronously, one or more local modules may compute asynchronously. In other words, the system is globally synchronous locally asynchronous.

This interface methodology, while similar to the GALS architecture, allows for stages in high-speed pipelines to be either synchronous or asynchronous. The architecture shown in Figure 9.14 assumes true single-phase clocking configured in such a way that data is latched into the next stage on the rising edge of the clock. The CLK signal is generated using a stoppable ring

340   APPLICATIONS

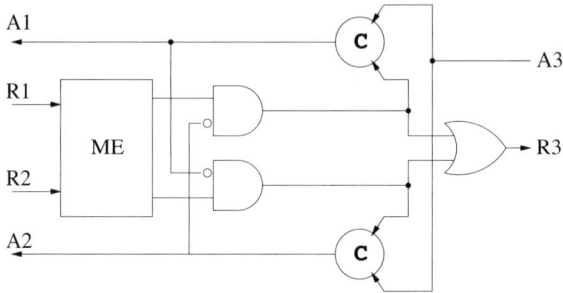

Fig. 9.15 Circuit for arbitration.

oscillator. Besides being used to sequence data between pipeline stages, the *CLK* signal is also used to generate the handshake protocol that controls the asynchronous modules. The interface controller is composed of the stoppable clock generator, one handshake control circuit for each asynchronous module, and an *AND* gate to collect the *ACK* signals to generate the *RUN* signal. The circuit behavior of the interface controller is as follows. Shortly after the rising edge of the *CLK* signal, the *RUN* signal is set low. The *RUN* signal is set high again only after all the asynchronous modules have completed their computation. Since data moves in and out of each asynchronous module with every cycle in the pipeline, no mutual exclusion elements are necessary.

9.5.4 Arbitration

A closely related problem to synchronization is *arbitration*. Arbitration is necessary when two or more modules would like mutually exclusive access to a shared resource. If the modules requesting the resource are asynchronous with respect to each other, arbitration also cannot be guaranteed to be accomplished in a bounded amount of time. If all modules are asynchronous, however, it is possible to build an *arbiter* with zero probability of failure.

An arbiter should be capable of receiving requests from multiple sources, select one to service, and forward the request to the resource requiring mutually exclusive access. An arbiter circuit for two requesting modules is shown in Figure 9.15. If this circuit receives a request on both *R1* and *R2* concurrently, the ME element selects one to service by asserting one of its two output wires. This in turn causes the resource to be requested by asserting *R3*. When the resource has completed its operation, it asserts *A3* and the acknowledgment is forwarded to the requesting module that was chosen.

Larger arbiters can be built from two-input arbiters by connecting them up in the form of a tree. These trees can be either balanced or unbalanced, as shown in Figure 9.16 (note that the symbol that looks like either a sideways "A" or an eye watching its inputs means an arbiter). If it is balanced as shown in Figure 9.16(a) for a four-way arbiter, all modules have equal priority access

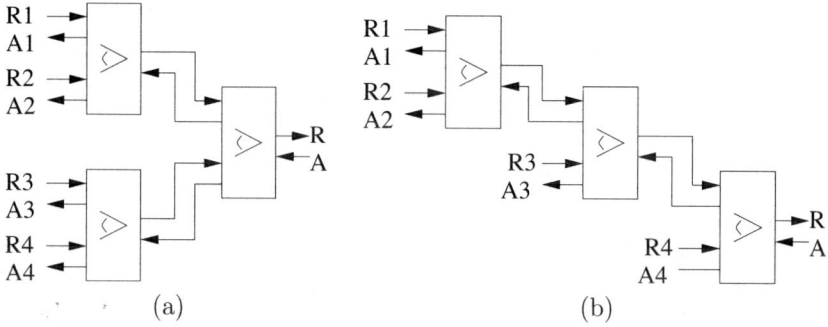

Fig. 9.16 (a) Balanced four-way arbiter. (b) Unbalanced four-way arbiter.

to the resource. If it is unbalanced as shown in Figure 9.16(b), modules closest to the end get a higher priority. In this example, if all four requests are high, $R4$ has a 50 percent chance of obtaining the resource, $R3$ has a 25 percent chance, and $R2$ and $R1$ each have only a 12.5 percent chance.

9.6 THE FUTURE OF ASYNCHRONOUS CIRCUIT DESIGN

Attempting to predict the future is an exercise that should be avoided. In reviewing the literature during the preparation of this book, papers in the 1960s and 1970s espouse many of the same advantages for asynchronous design that show up in the papers published today. Why after all these years has asynchronous circuit design not taken over the world as it has been widely predicted? The trouble is there are a lot of very smart people who have been capable of overcoming each hurdle that synchronous design has faced.

If asynchronous design is so much better, why do designers continue to struggle with synchronous design? The reason is that the synchronous design paradigm is so simple to understand and use. The trouble with this reasoning, though, is that in high-performance circuit design, the simple synchronous design paradigm does not truly exist anymore. Carl Anderson, an IBM fellow who led the design of a recent 170 million-transistor PowerPC design, stated in a talk at Tau2000, "Clocking is a major pain!" He went on to state that all future high-performance microprocessors must use multiple clocking domains. In modern technologies, the time of flight of data (not to mention the clock signal) between areas of the chip can take multiple clock periods.

So when is the asynchronous revolution going to take place? Actually, I do not believe it will. Rather, what we see happening is that as global synchrony is becoming ever more difficult, people are rediscovering ideas like GALS as evidenced by the recent IPCMOS project at IBM. Once designers start to think of communication being asynchronous between the synchronous modules they are comfortable with, it is only a matter of time before some asynchronous modules infiltrate the ranks. This would allow designs such as RAPPID where high performance is achieved using asynchronous design on only a single module to be incorporated in a larger system. There are, however, already some domains, such as those requiring low EMI, where asynchronous design has shown its merit as evidenced by Philip's asynchronous pager.

What will we do if asynchronous design never takes over the world? We can be comforted by the fact that the issues that we face in learning how asynchronous design is to be done are important even in today's synchronous design problems. For example, the elimination of hazards has shown to be worthwhile for reducing power consumption, and it is also a necessity in several synchronous design styles. Also, the two-sided timing issues faced in the design of timed circuits are also encountered by the designers of aggressive circuit styles such as *delayed-reset* and *self-resetting* domino circuits being experimented with today.

That being said, we would still like to see it take over the world, so what is the main obstacle to this? The major hurdle that remains, in this author's opinion, is fear. Designers do not fully understand the asynchronous design problem. In the past, this has led designers to attempt asynchronous design without all the tools and background, only to become disillusioned when unforeseen problems lead to the design's failure. This can be overcome through education, and it is the hope of this author that this book will play an important role in bringing this about.

9.7 SOURCES

There is little material available about the design of the ILLIAC computer. A useful reference on the design of the ILLIAC II is an article by Brearley [46] which serves as an annotated bibliography listing 40 papers and technical reports describing the ILLIAC II design and the speed-independent design methodology used in its design. We have been unable to find any references that discuss the circuit design of the Atlas and MU-5 mainframe computers which were completed and used at the University of Manchester in 1962 and 1966, respectively. The macromodule project at Washington University in St. Louis is described in some detail in papers by Clark and others [87, 88, 302, 369]. The DDM-1 is described by Davis in [102, 103]. The only documentation of the LDS-1, the graphics system designed at Evans and Sutherland, appears to be in internal technical reports. A micropipelined dataflow processor is described in [76]. The work on data-driven processors is described in [207, 208, 373, 376, 410]. Other design work in the DSP and communications domain includes that in [6, 182, 269, 295].

The first fully asynchronous microprocessor designed at Caltech is described in [251, 257, 258]. The subsequent design of a GaAs version of this microprocessor is described in [379], and the asynchronous MIPS R3000 is described in [259]. The Amulet microprocessors are described in [134, 135, 138, 406]. Numerous other researchers have also designed asynchronous microprocessors and special-purpose processors [10, 13, 76, 191, 288, 291, 324]. SUN's counterflow pipeline architecture is described in [364]. Experimental work from this group appears in [275, 372]. The low-power asynchronous designs from Phillips Research Laboratories are described in [38, 136, 192, 193]. The RAPPID project is described in [330, 367]. This design is also described in some detail in four patents [141, 142, 143, 144].

There have been numerous recent experiments with using asynchronous design to build functional units either to take advantage of delay variations due to data dependencies or operating conditions or to allow iterative algorithms to be carried out using a local clock. These designs include adders [80, 148, 199, 245, 256, 319, 331, 419], multipliers [74, 196, 293, 336, 419], and dividers [227, 321, 404]. Some of the most successful work has been applying self-timing to iterative division in that such a divider has been used in a SPARC microprocessor [400]. There have been several asynchronous memories designed as well [79, 127, 206, 356].

There has been some interesting work in applying novel circuit structures to asynchronous design. This includes using pass-transistor gates [399], bipolar circuit elements [403], dynamic logic [260], self-reseting gates [178, 231], threshold logic [226, 304], phased logic [236], and superconducting devices such as rapid single flux quantum (RSFQ) devices [108, 109, 110, 111, 216, 240]. An interesting technique for improving power consumption was proposed by Nielsen et al. in which the power supply is dynamically adjusted to balance power consumption and speed requirements [294].

In the domain of performance analysis, Burns uses a single delay parameter and presents techniques for determining the cycle period [65]. This performance estimate is used to guide transistor sizing. Williams developed techniques for analyzing throughput and latency in micropipelines and rings [402]. Ebergen and Berks analyzed the response time of linear pipelines [118]. Xie and Beerel present a performance analysis approach based on probabilistic delay models [408]. The stochastic cycle period analysis in Section 9.3 is from Mercer and Myers [270].

The self-checking property of asynchronous circuits has been investigated by numerous researchers [27, 162, 393]. These researchers proved that speed-independent circuits always halt under the output stuck-at fault model. Liebelt and Burgess extended these results to *excitory* output stuck-at faults (i.e., faults that take the circuit out of the valid set of states) [232]. Hazewindus demonstrated the potential for premature firing under the more general stuck-at fault model [162]. An early paper on test pattern generation is due to Putzolu [318]. Techniques for testing macromodule designs are presented by Khoche [195]. A partial-scan technique to test for delay faults is given by Kishinevsky et al. [201]. Design techniques to produce asynchronous circuits that are robust path delay fault testable are given in [194, 299]. Roncken has developed techniques to enable I_{DDQ} testing of asynchronous circuits. An excellent survey of testing issues as they relate to asynchronous design is given in [176] and forms the basis for Section 9.4.

The synchronization problem has been known for some time. The earliest paper we found on asynchronous design addresses it [239]. In Lubkin's 1952 paper, he makes the following comment about the synchronization problem:

> If an entity similar to one of Maxwell's famous demons were available to convert such "maybe's" to either "yes" or "no," even in arbitrary fashion, the problem would be solved.

He goes on to discuss how the designers of the ENIAC used additional flip-flops to allow more time to synchronize asynchronous inputs. This paper states that this technique does not eliminate the chance of error, but rather, it simply reduces its probability. This paper also presents a method of determining the probability of error. The problem appears to be largely ignored until 1966 when Catt rediscovered it and presents a different formulation of this error probability [70]. Again, it appears that the synchronization problem was not widely known or understood. Evidence of this is that several asynchronous arbiters designed in the early 1970s suffered from metastability problems if arrival times of signals are not carefully controlled [90, 91, 307, 315].

Finally, in 1973 experimental evidence of the synchronization problem presented by Chaney and Molnar appears to have awakened the community to the problem [75]. After this paper, a number of papers were published that provided experimental evidence of metastability due to asynchronous inputs, and mathematical models were developed to explain the experimental results [95, 125, 169, 200, 217, 308, 327, 394]. Pechoucek's paper also shows that the only way to reduce the probability to zero is to generate the clock locally and be able to stop the clock when metastability occurs.

One of the earliest proofs to show that metastability in a bistable is unavoidable is by Hurtado, who applies dynamical systems theory [177]. Marino demonstrates that a perfect inertial delay can be used to implement a perfect synchronizer, and he then goes on to show that several possible inertial delay designs are not perfect [246]. Thus, this casts further doubt on the existence of a perfect synchronizer. Marino later generalized and extended Hurtado's theory to arbitrary sequential systems [247]. He proved that if activity occurs asynchronously with respect to the activity of some system, metastabilibity cannot be avoided. Kleeman and Cantoni extended Marino's theory by removing the restriction that inputs have a bounded first derivative [205]. Barros and Johnson use an axiomatic method to show the equivalence of a bounded-time arbiter, synchronizer, and inertial delay [23]. Unger has recently shown a relationship between hazards and metastable states [385]. An excellent discussion of the synchronization problem, ways to reduce the probability of error, and the use of stoppable clocks is given by Stucki and Cox [368]. The mathematical treatment in Section 9.5 follows this paper.

Many designs have been proposed for synchronizers. Veendrick shows that the probability of metastability is independent of circuit noise in the synchronizer and that it could be reduced somewhat through careful design and layout [394]. Kleeman and Cantoni show that using redundancy and masking does not eliminate the probability of synchronization failure [204]. Manner describes how *quantum synchronizers* can solve the problem in principle, but not in practice [244]. Sakurai shows how careful sizing can reduce the probability of failure in synchronizers and arbiters [335]. Walker and Cantoni recently published a synchronizer design which uses a bank of parallel rather than serial registers operating using a clock period of nT [397]. Another interesting design is published in [78] in which EMI caused by the clock is exploited to build a better synchronizer. One of the most interesting synchronizers was due to Seizovic, in which he proposes to pipeline the synchronization process [348]. *Pipeline synchronization* essentially breaks up the synchronization into a series of asynchronous pipeline stages which each attempt to synchronize their request signal to the clock. When metastability occurs in one stage, its request to the next stage is delayed. When the next stage sees the request, it will see it at a somewhat different time, and it will hopefully not enter the metastability region. As the length of the pipeline is increased, the likelihood that metastability persists until the last stage is greatly reduced. This scheme is used in the Myranet local area network. This network at the time had about 80 million asynchronous events per second with $\tau = 230$ ps. If the synchronous clock rate being synchronized to is at 80 MHz, the MTBF is on the order of 2 hours. Using an eight-stage pipeline for synchronization and a two-phase clock, a latency of 50 ns is incurred in which 28 ns is available for synchronization reducing the MTBF to about 10^{37} years.

Stoppable clocks date back to the 1960s with work done by Chuck Seitz which was used in early display systems and other products of the Evans and Sutherland company [347, 362]. Numerous researchers have developed GALS architectures based on the idea of a stoppable clock [77, 163, 328, 347, 368, 392, 423]. Some of the schemes such as those proposed in [233, 328, 368, 423] allow an asynchronous module to request to communicate data to a synchronous module at arbitrary times. The approach in [328] is based on an asynchronous synchronizer called a *Q-flop*. This synchronizer receives a potentially unstable input and a clock, and it produces a latched value and an acknowledgement when it has latched the value successfully. Q-flops are used in *Q-modules* which are essentially synchronous modules clocked using a locally generated stoppable clock. These Q-modules are then interconnected asynchronously. The schemes proposed in [77, 347] assume that the synchronous unit determines when data is to be transferred to/from the asynchronous modules. Recently, a group from IBM introduced a new GALS approach of the second kind called interlocked pipelined CMOS (IPCMOS) [341]. They implemented a test chip in a 0.18 μm 1.5 V CMOS process which consisted of the critical path from a pipelined floating-point multiplier. Their experimental results showed a typical performance of 3.3 GHz with a best-case performance of 4.5 GHz. An analysis comparing synchronous versus GALS is given in [2]. The globally synchronous locally asynchronous architecture is due to Sjogren [357].

Numerous early arbiter designs were prone to failure due to metastability. A safe arbiter is proposed by Seitz in [346]. The first published MOS design of an ME was also due to Seitz [346]. Arbiters with improved response times are presented in [183, 409]. Modular arbiter design are presented in [69, 387], which can be extended to an arbitrary number of requesters. A silicon compiler that produced circuits with arbiters is presented in [22]. In [35] it is shown that extra care must be taken when designing a three-way arbiter as a straightforward design leads to additional metastability problems.

Problems

9.1 Probability of Synchronization Failure
Assume that $\tau = 20$ ps, $T_0 = 8$ ns, the clock frequency is 4 GHz, asynchronous inputs arrive at an average rate of 2 GHz, and $t_a = 200$ ps.

9.1.1. What is the MTBF?

9.1.2. How many latches does it take to reduce MTBF to 10,000 years?

9.1.3. If no extra latchs are used, at what rate can asynchronous inputs arrive such that the MTBF is over 1 year?

9.1.4. For what value of τ does the MTBF exceed 1 year?

9.2 Probability of Synchronization Failure
Assume that $\tau = 10$ ps, $T_0 = 4$ ns, the clock frequency is 5 GHz, asynchronous inputs arrive at an average rate of 100 MHz, and $t_a = 20$ ps

9.2.1. What is the MTBF?

9.2.2. How many latches does it take to reduce MTBF to 100,000 years?

9.2.3. If no extra latchs are used, at what clock frequency does the MTBF exceed one year? (Assume that t_a is 10 percent of the cycle time.)

9.2.4. What clock frequency would produce an MTBF of more than 1 year if t_a is 50 percent of the cycle time?

Appendix A
VHDL Packages

I have always imagined that Paradise will be a kind of library.
—Jorge Luis Borges

A.1 NONDETERMINISM.VHD

This package defines functions to model random selection and random delays.

```
library ieee;
use ieee.math_real.all;
use ieee.std_logic_1164.all;
use ieee.std_logic_arith.all;
package nondeterminism is
   shared variable s1:integer:=844396720;
   shared variable s2:integer:=821616997;
   -- Returns a number between 1 and num.
   impure function selection(constant num:in integer) return integer;
   -- Returns a std_logic_vector of size bits between 1 and num.
   impure function selection(constant num:in integer;
                   constant size:in integer) return std_logic_vector;
   -- Returns random delay between lower and upper.
   impure function delay(constant l:in integer;
              constant u:in integer) return time;
```

```vhdl
end nondeterminism;
package body nondeterminism is
  impure function selection(constant num:in integer) return integer is
    variable result:integer;
    variable tmp_real:real;
  begin
    uniform(s1,s2,tmp_real);
    result := 1 + integer(trunc(tmp_real * real(num)));
    return(result);
  end selection;
  impure function selection(constant num:in integer;
                            constant size:in integer)
    return std_logic_vector is
    variable result:std_logic_vector(size-1 downto 0);
    variable tmp_real:real;
  begin
    uniform(s1,s2,tmp_real);
    result := conv_std_logic_vector(integer(trunc(tmp_real * real(num)))
                                    +1,size);
    return(result);
  end selection;
  impure function delay(constant l:in integer;
                        constant u:in integer) return time is
    variable result:time;
    variable tmp:real;
  begin
    uniform(s1,s2,tmp);
    result:=(((tmp * real(u - 1)) + real(l)) * 1 ns);
    return result;
  end delay;
end nondeterminism;
```

A.2 CHANNEL.VHD

This package defines a channel data type and send and receive functions.

```vhdl
library IEEE;
use IEEE.std_logic_1164.all;
package channel is
  constant MAX_BIT_WIDTH:natural:=32;
  subtype datatype is std_logic_vector((MAX_BIT_WIDTH-1) downto 0 );
  constant dataZ:datatype:=datatype'(others => 'Z');
  constant data0:datatype:=datatype'(others => '0');
  constant dataACK:dataType:=dataType'(others => '1');
  type channel is record
    dataright,dataleft:datatype;
    pending_send,pending_recv,sync:std_logic;
  end record;
  type bools is array (natural range <>) of boolean;
  -- Used to send data on a channel
  procedure send(signal c1:inout channel);
  procedure send(signal c1:inout channel;signal d1:inout std_logic);
  procedure send(signal c1:inout channel;signal d1:inout std_logic;
                 signal c2:inout channel;signal d2:inout std_logic);
```

```vhdl
    procedure send(signal c1:inout channel;signal d1:inout std_logic;
                   signal c2:inout channel;signal d2:inout std_logic;
                   signal c3:inout channel;signal d3:inout std_logic);
    procedure send(signal c1:inout channel;signal d1:inout std_logic_vector);
    procedure send(signal c1:inout channel;signal d1:inout std_logic_vector;
                   signal c2:inout channel;signal d2:inout std_logic_vector);
    procedure send(signal c1:inout channel;signal d1:inout std_logic_vector;
                   signal c2:inout channel;signal d2:inout std_logic_vector;
                   signal c3:inout channel;signal d3:inout std_logic_vector);
    -- Used to receive data on a channel
    procedure receive(signal c1:inout channel);
    procedure receive(signal c1:inout channel;signal d1:inout std_logic);
    procedure receive(signal c1:inout channel;signal d1:inout std_logic;
                      signal c2:inout channel;signal d2:inout std_logic);
    procedure receive(signal c1:inout channel;signal d1:inout std_logic;
                      signal c2:inout channel;signal d2:inout std_logic;
                      signal c3:inout channel;signal d3:inout std_logic);
    procedure receive(signal c1:inout channel;signal d1:inout std_logic_vector);
    procedure receive(signal c1:inout channel;signal d1:inout std_logic_vector;
                      signal c2:inout channel;signal d2:inout std_logic_vector);
    procedure receive(signal c1:inout channel;signal d1:inout std_logic_vector;
                      signal c2:inout channel;signal d2:inout std_logic_vector;
                      signal c3:inout channel;signal d3:inout std_logic_vector);
    -- Initialization function called in a port declaration
    -- as a default value to initialize a channel.
    function init_channel return channel;
    function active return channel;
    function passive return channel;
    -- Test for pending communication on a channel
    function probe(signal chan:in channel) return boolean;
end channel;
package body channel is
    procedure validate(signal data:in std_logic_vector) is
    begin
        assert( data'LENGTH <= MAX_BIT_WIDTH )
            report "Bit width is too wide" severity failure;
    end validate;
    function init_channel return channel is
    begin
        return(dataright=>dataZ,dataleft=>dataZ,
               pending_send=>'Z',pending_recv=>'Z',sync=>'Z');
    end init_channel;
    function active return channel is
    begin
        return(dataright=>dataZ,dataleft=>dataZ,
               pending_send=>'Z',pending_recv=>'Z',sync=>'Z');
    end active;
    function passive return channel is
    begin
        return(dataright=>dataZ,dataleft=>dataZ,
               pending_send=>'Z',pending_recv=>'Z',sync=>'Z');
    end passive;
    procedure send_handshake(variable done:inout boolean;
                             variable reset:inout boolean;
                             signal chan:inout channel) is
    begin
```

```vhdl
      variable reset:bools(1 to 3) := (others => false);
    begin
      c1.dataright <= zero_extend(d1);
      c2.dataright <= zero_extend(d2);
      c3.dataright <= zero_extend(d3);
      loop
        send_handshake(done(1),reset(1),c1);
        send_handshake(done(2),reset(2),c2);
        send_handshake(done(3),reset(3),c3);
        exit when ((done(1)=true) and (done(2)=true) and (done(3)=true));
        wait on c1.sync,c2.sync,c3.sync;
      end loop;
    end send;
    procedure recv_hse(variable done:inout boolean;
                      variable reset:inout boolean;
                      signal chan:inout channel) is
    begin
      if (done = false) then
        if (reset = false) then
          if ((chan.pending_recv='Z') and
              (chan.sync='Z')) then
            chan.pending_recv<='1';
          end if;
          if (chan.sync='1') then
            chan.pending_recv<='Z';
          end if;
          if ((chan.pending_send='1') and
              (chan.pending_recv='1') and
              (chan.sync='Z')) then
            chan.sync<='1';
          end if;
          if ((chan.pending_send='Z') and
              (chan.pending_recv='Z') and
              (chan.sync='1')) then
            chan.sync <= 'Z';
            reset:=true;
          end if;
        elsif (chan.sync='Z') then
          done:=true;
        end if;
      end if;
    end recv_hse;
    procedure recv_hse(variable done:inout boolean;
                      variable reset:inout boolean;
                      signal chan:inout channel;signal data:out std_logic) is
    begin
      if (done=false) then
        if (reset=false) then
          if ((chan.pending_recv='Z') and
              (chan.sync= 'Z')) then
            chan.pending_recv<='1';
          end if;
          if (chan.sync='1') then
            chan.pending_recv<='Z';
          end if;
          if ((chan.pending_send='1') and
```

```vhdl
              (chan.pending_recv='1') and
              (chan.sync='Z')) then
          chan.sync<='1';
          data<=chan.dataright(0);
        end if;
        if ((chan.pending_send='Z') and
            (chan.pending_recv='Z') and
            (chan.sync='1')) then
          chan.sync<='Z';
          reset:=true;
        end if;
      elsif (chan.sync='Z') then
        done:=true;
      end if;
    end if;
end recv_hse;
procedure recv_hse(variable done:inout boolean;
                   variable reset:inout boolean;
                   signal chan:inout channel;
                   signal data:out std_logic_vector) is
begin
  if (done=false) then
    if (reset=false) then
      if ((chan.pending_recv='Z') and
          (chan.sync='Z')) then
        chan.pending_recv<='1';
      end if;
      if (chan.sync='1') then
        chan.pending_recv<='Z';
      end if;
      if ((chan.pending_send='1') and
          (chan.pending_recv='1') and
          (chan.sync='Z')) then
        chan.sync<='1';
        data<=chan.dataright(data'length - 1 downto 0);
      end if;
      if ((chan.pending_send='Z') and
          (chan.pending_recv='Z') and
          (chan.sync='1')) then
        chan.sync<='Z';
        reset:=true;
      end if;
    elsif (chan.sync='Z') then
      done:=true;
    end if;
  end if;
end recv_hse;
procedure receive (signal c1:inout channel) is
  variable done:bools(1 to 1) := (others => false);
  variable reset:bools(1 to 1) := (others => false);
begin
  loop
    recv_hse(done(1),reset(1),c1);
    exit when (done(1)=true);
    wait on c1.pending_send,c1.sync,c1.pending_recv;
  end loop;
```

```
                             signal sig3:inout std_logic;constant val3:std_logic;
                             constant l3:integer;constant u3:integer);
  procedure vassign(signal sig:inout std_logic;constant val:std_logic;
                    constant l:integer;constant u:integer);
  procedure vassign(signal sig1:inout std_logic;constant val1:std_logic;
                    constant l1:integer;constant u1:integer;
                    signal sig2:inout std_logic;constant val2:std_logic;
                    constant l2:integer;constant u2:integer);
  procedure vassign(signal sig1:inout std_logic;constant val1:std_logic;
                    constant l1:integer;constant u1:integer;
                    signal sig2:inout std_logic;constant val2:std_logic;
                    constant l2:integer;constant u2:integer;
                    signal sig3:inout std_logic;constant val3:std_logic;
                    constant l3:integer;constant u3:integer);
  procedure guard(signal sig:in std_logic;constant val:std_logic);
  procedure guard_or(signal sig1:in std_logic;constant val1:std_logic;
                    signal sig2:in std_logic;constant val2:std_logic);
  procedure guard_or(signal sig1:in std_logic;constant val1:std_logic;
                    signal sig2:in std_logic;constant val2:std_logic;
                    signal sig3:in std_logic;constant val3:std_logic);
  procedure guard_and(signal sig1:in std_logic;constant val1:std_logic;
                    signal sig2:in std_logic;constant val2:std_logic);
  procedure guard_and(signal sig1:in std_logic;constant val1:std_logic;
                    signal sig2:in std_logic;constant val2:std_logic;
                    signal sig3:in std_logic;constant val3:std_logic);
end handshake;
package body handshake is
  procedure assign(signal sig:inout std_logic;constant val:std_logic;
                    constant l:integer;constant u:integer) is
  begin
    assert (sig /= val) report "Vacuous assignment" severity failure;
    sig <= val after delay(l,u);
    wait until sig = val;
  end assign;
  procedure assign(signal sig1:inout std_logic;constant val1:std_logic;
                    constant l1:integer;constant u1:integer;
                    signal sig2:inout std_logic;constant val2:std_logic;
                    constant l2:integer;constant u2:integer) is
  begin
    assert (sig1 /= val1 or sig2 /= val2)
      report "Vacuous assignment" severity failure;
    sig1 <= val1 after delay(l1,u1);
    sig2 <= val2 after delay(l2,u2);
    wait until sig1 = val1 and sig2 = val2;
  end assign;
  procedure assign(signal sig1:inout std_logic;constant val1:std_logic;
                    constant l1:integer;constant u1:integer;
                    signal sig2:inout std_logic;constant val2:std_logic;
                    constant l2:integer;constant u2:integer;
                    signal sig3:inout std_logic;constant val3:std_logic;
                    constant l3:integer;constant u3:integer) is
  begin
    assert (sig1 /= val1 or sig2 /= val2 or sig3 /= val3)
      report "Vacuous assignment" severity failure;
    sig1 <= val1 after delay(l1,u1);
    sig2 <= val2 after delay(l2,u2);
```

```vhdl
      sig3 <= val3 after delay(l3,u3);
      wait until sig1 = val1 and sig2 = val2 and sig3 = val3;
    end assign;
    procedure vassign(signal sig:inout std_logic;constant val:std_logic;
                     constant l:integer;constant u:integer) is
    begin
      if (sig /= val) then
        sig <= val after delay(l,u);
        wait until sig = val;
      end if;
    end vassign;
    procedure vassign(signal sig1:inout std_logic;constant val1:std_logic;
                     constant l1:integer;constant u1:integer;
                     signal sig2:inout std_logic;constant val2:std_logic;
                     constant l2:integer;constant u2:integer) is
    begin
      if (sig1 /= val1) then
        sig1 <= val1 after delay(l1,u1);
      end if;
      if (sig2 /= val2) then
        sig2 <= val2 after delay(l2,u2);
      end if;
      if (sig1 /= val1 or sig2 /= val2) then
        wait until sig1 = val1 and sig2 = val2;
      end if;
    end vassign;
    procedure vassign(signal sig1:inout std_logic;constant val1:std_logic;
                     constant l1:integer;constant u1:integer;
                     signal sig2:inout std_logic;constant val2:std_logic;
                     constant l2:integer;constant u2:integer;
                     signal sig3:inout std_logic;constant val3:std_logic;
                     constant l3:integer;constant u3:integer) is
    begin
      if (sig1 /= val1) then
        sig1 <= val1 after delay(l1,u1);
      end if;
      if (sig2 /= val2) then
        sig2 <= val2 after delay(l2,u2);
      end if;
      if (sig3 /= val3) then
        sig3 <= val3 after delay(l3,u3);
      end if;
      if (sig1 /= val1 or sig2 /= val2 or sig3 /= val3) then
        wait until sig1 = val1 and sig2 = val2 and sig3 = val3;
      end if;
    end vassign;
    procedure guard(signal sig:in std_logic;constant val:std_logic) is
    begin
      if (sig /= val) then
        wait until sig = val;
      end if;
    end guard;
    procedure guard_or(signal sig1:in std_logic;constant val1:std_logic;
                      signal sig2:in std_logic;constant val2:std_logic) is
    begin
      if (sig1 /= val1 and sig2 /= val2) then
```

```
      wait until sig1 = val1 or sig2 = val2;
    end if;
  end guard_or;
  procedure guard_or(signal sig1:in std_logic;constant val1:std_logic;
                    signal sig2:in std_logic;constant val2:std_logic;
                    signal sig3:in std_logic;constant val3:std_logic) is
  begin
    if (sig1 /= val1 and sig2 /= val2 and sig3 /= val3) then
      wait until sig1 = val1 or sig2 = val2 or sig3 = val3;
    end if;
  end guard_or;
  procedure guard_and(signal sig1:in std_logic;constant val1:std_logic;
                    signal sig2:in std_logic;constant val2:std_logic) is
  begin
    if ( sig1 /= val1 or sig2 /= val2 ) then
      wait until sig1 = val1 and sig2 = val2;
    end if;
  end guard_and;
  procedure guard_and(signal sig1:in std_logic;constant val1:std_logic;
                    signal sig2:in std_logic;constant val2:std_logic;
                    signal sig3:in std_logic;constant val3:std_logic) is
  begin
    if (sig1 /= val1 or sig2 /= val2 or sig3 /= val3) then
      wait until sig1 = val1 and sig2 = val2 and sig3 = val3;
    end if;
  end guard_and;
end handshake;
```

Appendix B
Sets and Relations

Your theory is crazy, but it's not crazy enough to be true.
—Niels Bohr, to a young physicist

First things first, but not necessarily in that order.
—Doctor Who

What we imagine is order is merely the prevailing form of chaos.
—Kerry Thornley, *Principia Discordia*, 5th edition

The art of progress is to preserve order amid change.
—A. N. Whitehead

Confusion is a word we have invented for an order which is not understood.
—Henry Miller (1891 – 1980)

Not till we are lost, in other words, not till we have lost the world, do we begin to find ourselves, and realize the infinite extent of our relations.
—Henry David Thoreau (1817 – 1862)

Throughout the book we have assumed a basic knowledge of set theory. This appendix provides a brief review of some of the basic concepts of set theory used in this book.

B.1 BASIC SET THEORY

A *set* S is any collection of objects that can be distinguished. Each object x which is in S is called a *member* of S (denoted $x \in S$). When an object x is not a member of S, it is denoted by $x \notin S$. A set is determined by its members. Therefore, two sets X and Y are equal when they consist of the same members (denoted $X = Y$). This means that if $X = Y$ and $a \in X$, then $a \in Y$. This is known as the *principle of extension*. If two sets are not equal, it is denoted $X \neq Y$. There are three basic properties of equality:

1. $X = X$ (*reflexive*)
2. $X = Y$ implies $Y = X$ (*symmetric*)
3. $X = Y$ and $Y = Z$ then $X = Z$ (*transitive*)

   Example B.1.1 The set $\{1, 2, 3, 5, 6, 10, 15, 30\}$ is the set whose members are the divisors of 30. The sets $\{30, 15, 10, 6, 5, 3, 2, 1\}$ and $\{1, 1, 2, 3, 5, 5, 6, 10, 15, 30\}$ are equal to the set $\{1, 2, 3, 5, 6, 10, 15, 30\}$ since they have the same members.

Large or infinite sets are described using the help of *predicates*. A predicate $P(x)$ takes an object and returns true or false. When a set S is defined using a predicate $P(x)$, the set S contains those objects a such that $P(a)$ is true. This is known as the *principle of abstraction*. This is denoted using *set builder notation* as follows:

$$S = \{x \mid P(x)\}$$

This is read as "the set of all objects x such that $P(x)$ is true." There are some useful variants. For example, the following sets can be used interchangeably:

$$\{x \mid x \in A \text{ and } P(x)\} = \{x \in A \mid P(x)\}$$
$$\{y \mid y = f(x) \text{ and } P(x)\} = \{f(x) \mid P(x)\}$$

Example B.1.2 The set $\{x \in \mathcal{N} \mid x \text{ divides } 30\}$, where $\mathcal{N}$ the set of natural numbers, is equivalent to the set $\{1, 2, 3, 5, 6, 10, 15, 30\}$.

Another useful relation on sets is *subset*. If X and Y are sets such that every member of X is also a member of Y, then X is a subset of Y (denoted $X \subseteq Y$). If, on the other hand, every member of Y is a member of X, then X is a superset of Y (denoted $X \supseteq Y$). If $X \subseteq Y$ and $X \neq Y$, then X is a proper subset of Y (denoted $X \subset Y$). Proper superset is similarly defined (denoted $X \supset Y$). The subset relation has the following three basic properties:

1. $X \subseteq X$ (*reflexive*)
2. $X \subseteq Y$ and $Y \subseteq X$ implies that $X = Y$ (*antisymmetric*)
3. $X \subseteq Y$ and $Y \subseteq Z$, then $X \subseteq Z$ (*transitive*)

The set which includes no elements is called the *empty set* (denoted $\emptyset$). For any set X, the empty set is a subset of it (i.e., $\emptyset \subseteq X$). Each set $X \neq \emptyset$ has at least two subsets X and $\emptyset$. Each member of a set $x \in X$ also determines a subset of X (i.e., $\{x\} \subseteq X$). Similarly, if a set has at least two members, each pair of objects makes up a subset. The *power set* of a set X is all subsets of X (denoted 2^X). The number of members of a set X is denoted $|X|$. The number of members of 2^X is equal to $2^{|X|}$.

Example B.1.3 If $X = \{2, 3, 5\}$ and $Y = \{1, 2, 3, 5, 6, 10, 15, 30\}$, then

$$X \subseteq Y$$
$$2^X = \{\emptyset, \{2\}, \{3\}, \{5\}, \{2,3\}, \{2,5\}, \{3,5\}, X\}$$

The *union* of two sets X and Y (denoted $X \cup Y$) is the set composed of all objects that are a member of either X or Y (i.e., $X \cup Y = \{x \mid x \in X \text{ or } x \in Y\}$). The *intersection* of two sets X and Y (denoted $X \cap Y$) is the set composed of all objects that are a member of both X and Y (i.e., $X \cap Y = \{x \mid x \in X \text{ and } x \in Y\}$).

Example B.1.4 If $X = \{2, 3\}$ and $Y = \{2, 5\}$, then

$$X \cup Y = \{2, 3, 5\}$$
$$X \cap Y = \{2\}$$

Two sets X and Y are *disjoint* if their intersection contains no members (i.e., $X \cap Y = \emptyset$). Otherwise, the sets *intersect* (i.e., $X \cap Y \neq \emptyset$). A *disjoint collection* is a set of sets in which each pair of member sets is disjoint. A *partition* of a set X is a disjoint collection π of nonempty and disjoint subsets of X such that each member of X is contained within some set in π.

Example B.1.5 The set $\{\{1\}, \{2, 3, 5\}, \{6, 10, 15\}, \{30\}\}$ is a partition of the set $\{1, 2, 3, 5, 6, 10, 15, 30\}$.

The set U is called the *universal set*, and it is composed of all objects in some domain being discussed. The *absolute complement* of a set X (denoted $\overline{X}$) are those elements in U which are not in X (i.e., $\{x \in U \mid x \notin X\}$). The *relative complement* of a set X with respect to a set Y (denoted $Y - X$) are those elements in Y which are not in X (i.e., $Y - X = Y \cap \overline{X} = \{x \in Y \mid x \notin X\}$). The *symmetric difference* of two sets X and Y (denoted $X + Y$) are those objects in exactly one of the two sets [i.e., $(A - B) \cup (B - A)$].

Example B.1.6 If $U = \{1, 2, 3, 5, 6, 10, 15, 30\}$, $X = \{2, 3\}$, and $Y = \{2, 5\}$, then

$$\overline{X} = \{1, 5, 6, 10, 15, 30\}$$
$$X - Y = \{3\}$$
$$X + Y = \{3, 5\}$$

There are numerous useful *identities* (equations that are true regardless of what U and the subset letters represent) shown in Table B.1. Note that each

Table B.1 Useful identities of set theory.

Law	Union	Intersection
Associative	$A \cup (B \cup C) = (A \cup B) \cup C$	$A \cap (B \cap C) = (A \cap B) \cap C$
Communitive	$A \cup B = B \cup A$	$A \cap B = B \cap A$
Distributive	$A \cup (B \cap C) = (A \cup B) \cap (A \cup C)$	$A \cap (B \cup C) = (A \cap B) \cup (A \cap C)$
Identity	$A \cup \emptyset = A$	$A \cap U = A$
Inverse	$A \cup \overline{A} = U$	$A \cap \overline{A} = \emptyset$
Idempotent	$A \cup A = A$	$A \cap A = A$
Absorption	$A \cup (A \cap B) = A$	$A \cap (A \cup B) = A$
DeMorgan	$\overline{A \cup B} = \overline{A} \cap \overline{B}$	$\overline{A \cap B} = \overline{A} \cup \overline{B}$

identity appears twice and that the second can be found by interchanging $\cup$ and $\cap$ as well as $\emptyset$ and U. The identities in the third column are called *duals* of those in the second, and vice versa. In general, we can use this *principle of duality* to translate any theorem in terms of $\cup$, $\cap$, and complement to a dual theorem. One last useful property is the following equivalence:

$$A \subseteq B \Leftrightarrow A \cap B = A \Leftrightarrow A \cup B = B$$

B.2 RELATIONS

Using set theory, we can define *binary relations* to show relationships between two items. Examples include things like "a is less than b" or "Chris is husband to Ching and father of John." First, we define an *ordered pair* as a set of two objects which have a specified order. An ordered pair of x and y is denoted by $\langle x, y \rangle$ and is equivalent to the set $\{\{x\}, \{x, y\}\}$. A binary relation is simply a set of ordered pairs. We say that x is ρ-*related to* y (denoted $x\rho y$) when ρ is a binary relation and $\langle x, y \rangle \in \rho$. The *domain* and *range* of ρ are

$$\begin{aligned} D_\rho &= \{x \mid \exists y \,.\, \langle x, y \rangle \in \rho\} \\ R_\rho &= \{y \mid \exists x \,.\, \langle x, y \rangle \in \rho\} \end{aligned}$$

Example B.2.1 A binary relation ρ that says that x times y equals 30 is defined as follows:

$$\rho = \{\langle 1, 30\rangle, \langle 2, 15\rangle, \langle 3, 10\rangle, \langle 5, 6\rangle, \langle 6, 5\rangle, \langle 10, 3\rangle, \langle 15, 2\rangle, \langle 30, 1\rangle\}$$

Using ordered pairs, we can recursively define an *ordered triple* $\langle x, y, z \rangle$ as being equivalent to the ordered pair $\langle \langle x, y \rangle, z \rangle$. A *ternary relation* is simply a set of ordered triples. We can further define for any size n an *ordered n-tuple* and use them to define *n-ary relations*.

One of the simplest binary relations is the *cartesian product*, which is the set of all pairs $\langle x, y \rangle$, where x is a member of some set X and y is a member of some set Y. It is defined formally as follows:

$$X \times Y = \{\langle x, y \rangle \mid x \in X \wedge y \in Y\}$$

If $X \supseteq D_\rho$ and $Y \supseteq R_\rho$, then $\rho \subseteq X \times Y$ and ρ is a *relation from X to Y*.

Example B.2.2 The cartesian product of $X = \{2, 3, 5\}$ and $Y = \{6, 10\}$ is defined as follows:

$$X \times Y = \{\langle 2, 6 \rangle, \langle 2, 10 \rangle, \langle 3, 6 \rangle, \langle 3, 10 \rangle, \langle 5, 6 \rangle, \langle 5, 10 \rangle\}$$

A relation ρ in a set X is an *equivalence relation* iff it is reflexive (i.e., $x\rho x$ for all $x \in X$), symmetric (i.e., $x\rho y$ implies $y\rho x$), and transitive (i.e., $x\rho y$ and $y\rho z$ imply $x\rho z$). A set $A \subseteq X$ is an *equivalence class* iff there exists an $x \in A$ such that A is equal to the set of all y for which $x\rho y$. The equivalence class implied by x is denoted $[x]$. Using ρ, we can partition a set X into a set of equivalence classes called a *quotient set*, which is denoted by X/ρ.

Example B.2.3 The binary relation ρ on the set $X = \{1,2,3,5,6,10, 15,30\}$ defined below is an equivalence relation.

$$\rho = \{\langle 1, 1 \rangle, \langle 2, 2 \rangle, \langle 2, 3 \rangle, \langle 2, 5 \rangle, \langle 3, 2 \rangle, \langle 3, 3 \rangle, \langle 3, 5 \rangle, \langle 5, 2 \rangle, \langle 5, 5 \rangle,$$
$$\langle 6, 6 \rangle, \langle 6, 10 \rangle, \langle 6, 15 \rangle, \langle 10, 6 \rangle, \langle 10, 10 \rangle, \langle 10, 15 \rangle, \langle 15, 6 \rangle,$$
$$\langle 15, 10 \rangle, \langle 15, 15 \rangle, \langle 30, 30 \rangle\}$$
$$X/\rho = \{\{1\}, \{2, 3, 5\}, \{6, 10, 15\}, \{30\}\}$$

A *function* is a binary relation in which no two members have the same first element. More formally, a binary relation f is a function if $\langle x, y \rangle$ and $\langle x, z \rangle$ are members of f, then $y = z$. If f is a function and $\langle x, y \rangle \in f$ (i.e., xfy), then x is an argument of f and y is the *image* of x under f. A function f is *into* Y if $R_f \subseteq Y$. A function f is *onto* Y if $R_f = Y$. A function f is *one-to-one* if $f(x) = f(y)$ implies that $x = y$. Functions can be extended to more variables by using arguments that are ordered n-tuples.

Example B.2.4 The function f on the set $X = \{1, 2, 3, 5, 6, 10, 15, 30\}$ is defined as the result of dividing 30 by x. It is onto X and one-to-one.

$$f = \{\langle 1, 30 \rangle, \langle 2, 15 \rangle, \langle 3, 10 \rangle, \langle 5, 6 \rangle, \langle 6, 5 \rangle, \langle 10, 3 \rangle, \langle 15, 2 \rangle, \langle 30, 1 \rangle\}$$

A binary relation ρ is called a *partial order* if it is reflexive, antisymmetric (i.e., $x\rho y$ and $y\rho x$ implies that $x = y$), and transitive. A *partially ordered set (poset)* is a pair $\langle X, \leq \rangle$, where $\leq$ partially orders X. A partial order is a *simple* (or *linear*) ordering if for every pair of elements from the domain x and y either $x\rho y$ or $y\rho x$. An example of a simple ordering is $\leq$ on the real numbers. A *simply ordered set* is also called a *chain*. Two posets $\langle X, \leq \rangle$ and $\langle X', \leq' \rangle$ are isomorphic if there exists a one-to-one mapping between X and X' that preserves the ordering.

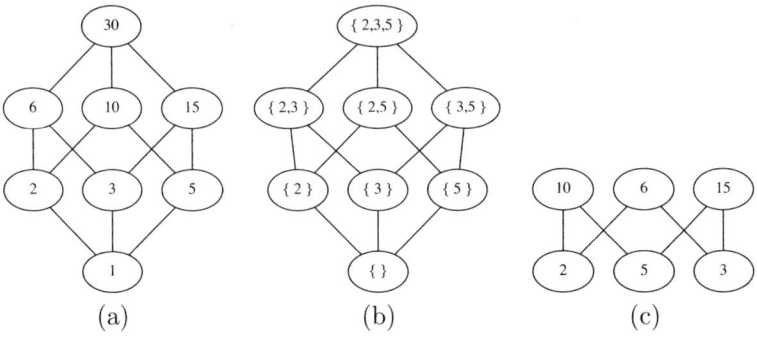

Fig. B.1 (a) Poset composed of the divisors of 30. (b) Isomorphic poset for the power set of {2, 3, 5}. (c) Poset with no least or greatest member.

Example B.2.5 Consider the set $\{1, 2, 3, 5, 6, 10, 15, 30\}$. It can be partially ordered using the relation $\leq$, which is defined to be x divides y. A diagram for this poset is shown in Figure B.1(a), where y is placed above x if $x \leq y$. Reflexive and transitive edges are omitted. A poset created from the power set of $\{2, 3, 5\}$ with $\subseteq$ as the ordering relation is shown in Figure B.1(b). These two posets are isomorphic.

A *least member* of X with respect to $\leq$ is a x in X such that $x \leq y$ for all y in X. A least member is unique. A *minimal member* is a x in X such that there does not exist a y in X such that $y < x$. A minimal member need not be unique. Similarly, a *greatest member* is a x in X such that $y \leq x$ for all y in X. A *maximal member* is a x in X such that there does not exist a y in X such that $y > x$. A poset $\langle X, \leq \rangle$ is *well-ordered* when each nonempty subset of X has a least member. Any well-ordered set must be a chain.

Example B.2.6 The least and minimal member of the poset shown in Figure B.1(a) is 1. The greatest and maximal member is 30. The poset shown in Figure B.1(c) has no least or greatest member. Its minimal members are 2, 3, and 5. Its maximal members are 6, 10, and 15.

For a poset $\langle X, \leq \rangle$ and $A \subseteq X$, an element $x \in X$ is an *upper bound* for A if for all $a \in A$, $a \leq x$. It is a *least upper bound* for A [denoted $lub(A)$] if x is an upper bound and $x \leq y$ for all y which are upper bounds of A. Similarly, an element $x \in X$ is a *lower bound* for A if for all $a \in A$, $x \leq a$. It is a *greatest lower bound* for A [denoted $glb(A)$] if x is a lower bound and $y \leq x$ for all y which are lower bounds of A. If A has a least upper bound, it is unique, and similarly for the greatest lower bound.

Example B.2.7 Consider the poset shown in Figure B.1(a). The upper bounds for $A = \{2, 5\}$ are 10 and 30, while $lub(2,5)$ is 10. The lower bounds for $A = \{6, 15\}$ are 1 and 3, while the $glb(6,15)$ is 3. For the poset shown in Figure B.1(c), the $lub(3,10)$ does not exist.

References

1. M. Abadi and L. Lamport. An old-fashioned recipe for real time, 1993.

2. M. Afghahi and C. Svensson. Performance of synchronous and asynchronous schemes for VLSI systems. *IEEE Transactions on Computers*, 41(7):858–872, July 1992.

3. V. Akella. *An Integrated Framework for the Automatic Synthesis of Efficient Self-Timed Circuits from Behavioral Specifications*. PhD thesis, University of Utah, 1992.

4. V. Akella and G. Gopalakrishnan. SHILPA: A high-level synthesis system for self-timed circuits. In *Proc. International Conference on Computer-Aided Design (ICCAD)*, pages 587–591. IEEE Computer Society Press, Los Alamitos, CA, November 1992.

5. V. Akella and G. Gopalakrishnan. Specification and validation of control-intensive IC's in hopCP. *IEEE Transactions on Software Engineering*, 20(6):405–423, 1994.

6. V. Akella, N. H. Vaidya, and G. R. Redinbo. Asynchronous comparison-based decoders for delay-insensitive codes. *IEEE Transactions on Computers*, 47(7):802–811, July 1998.

7. R. Alur. *Techniques for Automatic Verification of Real-Time Systems*. PhD thesis, Stanford University, August 1991.

8. R. Alur and D. L. Dill. A theory of timed automata. *Theoretical Computer Science*, 126:183–235, 1994.

9. R. Alur and R. P. Kurshan. Timing analysis in cospan. In *Hybrid Systems III*. Springer-Verlag, New York, 1996.

10. S. S. Appleton, S. V. Morton, and M. J. Liebelt. The design of a fast asynchronous microprocessor. *IEEE Technical Committee on Computer Architecture Newsletter*, October 1995.

11. D. B. Armstrong, A. D. Friedman, and P. R. Menon. Realization of asynchronous sequential circuits without inserted delay elements. *IEEE Transactions on Computers*, C-17(2):129–134, February 1968.

12. D. B. Armstrong, A. D. Friedman, and P. R. Menon. Design of asynchronous circuits assuming unbounded gate delays. *IEEE Transactions on Computers*, C-18(12):1110–1120, December 1969.

13. F. Asai, S. Komori, and T. Tamura. Self-timed design for a data-driven microprocessor. *IEICE Transactions*, E 74(11):3757–3765, November 1991.

14. *Proc. 6th International Symposium on Advanced Research in Asynchronous Circuits and Systems (ASYNC 2000)*, Eilat, Israel, April 2000. IEEE Computer Society Press, Los Alamitos, CA.

15. *Proc. 7th International Symposium on Asynchronous Circuits and Systems (ASYNC 2001)*, Salt Lake City, UT, March 2001. IEEE Computer Society Press, Los Alamitos, CA.

16. *Proc. International Symposium on Advanced Research in Asynchronous Circuits and Systems (ASYNC'94)*, Salt Lake City, UT, November 1994. IEEE Computer Society Press, Los Alamitos, CA.

17. *Proc. 2nd International Symposium on Advanced Research in Asynchronous Circuits and Systems (ASYNC'96)*, Aizu-Wakamatsu, Fukushima, Japan, March 1996. IEEE Computer Society Press, Los Alamitos, CA.

18. *Proc. 3rd International Symposium on Advanced Research in Asynchronous Circuits and Systems (ASYNC'97)*, Eindhoven, The Netherlands, April 1997. IEEE Computer Society Press, Los Alamitos, CA.

19. *Proc. 4th International Symposium on Advanced Research in Asynchronous Circuits and Systems (ASYNC'98)*, San Diego, CA, April 1998. IEEE Computer Society Press, Los Alamitos, CA.

20. *Proc. 5th International Symposium on Advanced Research in Asynchronous Circuits and Systems (ASYNC'99)*, Barcelona, Spain, April 1999. IEEE Computer Society Press, Los Alamitos, CA.

21. A. Bailey, G. A. McCaskill, and G. J. Milne. An exercise in the automatic verification of asynchronous designs. *Formal Methods in System Design*, 4:213–242, 1994.

22. T. S. Balraj and M. J. Foster. Miss Manners: A specialized silicon compiler for synchronizers. In Charles E. Leierson, editor, *Advanced Research in VLSI*, pages 3–20. MIT Press, Cambridge, MA, April 1986.

23. J. C. Barros and B. W. Johnson. Equivalence of the arbiter, the synchronizer, the latch, and the inertial delay. *IEEE Transactions on Computers*, 32(7):603–614, July 1983.

24. W. S. Bartky. A theory of asynchronous circuits III. Technical Report 96, University of Illinois, Urbana, IL, 1960.

25. P. Beerel and T. H.-Y. Meng. Automatic gate-level synthesis of speed-independent circuits. In *Proc. International Conference on Computer-Aided Design (ICCAD)*, pages 581–587. IEEE Computer Society Press, Los Alamitos, CA, November 1992.

26. P. A. Beerel, J. R. Burch, and T. H.-Y. Meng. Checking combinational equivalence of speed-independent circuits. *Formal Methods in System Design*, March 1998.

27. P. A. Beerel and T. H.-Y. Meng. Semi-modularity and testability of speed-independent circuits. *Integration, the VLSI Journal*, 13(3):301–322, September 1992.

28. P. A. Beerel, T. H.-Y. Meng, and J. Burch. Efficient verification of determinate speed-independent circuits. In *Proc. International Conference on Computer-Aided Design (ICCAD)*, pages 261–267. IEEE Computer Society Press, Los Alamitos, CA, November 1993.

29. P. A. Beerel, C. J. Myers, and T. H.-Y. Meng. Covering conditions and algorithms for the synthesis of speed-independent circuits. *IEEE Transactions on Computer-Aided Design*, March 1998.

30. J. Beister. A unified approach to combinational hazards. *IEEE Transactions on Computers*, C-23(6), 1974.

31. W. Belluomini. *Algorithms for the Synthesis and Verification of Timed Circuits and Systems*. PhD thesis, University of Utah, 1999.

32. W. Belluomini and C. J. Myers. Verification of timed systems using posets. In *Proc. International Conference on Computer Aided Verification*. Springer-Verlag, New York, 1998.

33. W. Belluomini and C. J. Myers. Timed state space exploration using posets. *IEEE Transactions on Computer-Aided Design*, 19(5):501–520, May 2000.

34. W. Belluomini and C. J. Myers. Timed circuit verification using tel structures. *IEEE Transactions on Computer-Aided Design*, 20(1):129–146, January 2001.

35. C. H. van Berkel and C. E. Molnar. Beware the three-way arbiter. *IEEE Journal of Solid-State Circuits*, 34(6):840–848, June 1999.

36. K. v. Berkel. *Handshake Circuits: An Intermediary Between Communicating Processes and VLSI*. PhD thesis, Eindhoven University of Technology, 1992.

37. K. v. Berkel. *Handshake Circuits: An Asynchronous Architecture for VLSI Programming*, volume 5 of *International Series on Parallel Computation*. Cambridge University Press, New York, 1993.

38. K. v. Berkel, R. Burgess, J. Kessels, A. Peeters, M. Roncken, and F. Schalij. A fully-asynchronous low-power error corrector for the DCC player. *IEEE Journal of Solid-State Circuits*, 29(12):1429–1439, December 1994.

39. K. v. Berkel, J. Kessels, M. Roncken, R. Saeijs, and F. Schalij. The VLSI-programming language Tangram and its translation into handshake circuits. In *Proc. European Conference on Design Automation (EDAC)*, pages 384–389, 1991.

40. K. v. Berkel and M. Rem. VLSI programming of asynchronous circuits for low power. In G. Birtwistle and A. Davis, editors, *Asynchronous Digital Circuit*

Design, Workshops in Computing, pages 152–210. Springer-Verlag, New York, 1995.

41. C. Berthet and E. Cerny. An algebraic model for asynchronous circuits verification. *IEEE Transactions on Computers*, 37(7):835–847, July 1988.

42. A. Blake. *Canonical Expressions in Boolean Algebra*. PhD thesis, University of Chicago, 1937.

43. M. Blaum and J. Bruck. Coding for skew-tolerant parallel asynchronous communications. *IEEE Transactions on Information Theory*, 39(2):379–388, March 1993.

44. M. Blaum and J. Bruck. Delay-insensitive pipelined communication on parallel buses. *IEEE Transactions on Computers*, 44(5):660–668, May 1995.

45. M. Bozga, O. Maler, A. Pnueli, and S. Yovine. Some progress in the symbolic verification of timed automata. In *Proc. International Conference on Computer Aided Verification*, 1997.

46. H. C. Brearley. ILLIAC II: A short description and annotated bibliography. *IEEE Transactions on Computers*, 14(6):399–403, June 1965.

47. J. G. Bredeson. Synthesis of multiple-input change hazard-free combinational switching circuits without feedback. *International Journal of Electronics (GB)*, 39(6):615–624, December 1975.

48. J. G. Bredeson and P. T. Hulina. Elimination of static and dynamic hazards for multiple input changes in combinational switching circuits. *Information and Control*, 20:114–224, 1972.

49. F. Brown. *Boolean Reasoning: The Logic of Boolean Equations*. Kluwer Academic Publishers, Boston, 1990.

50. M. C. Browne, E. M. Clarke, D. L. Dill, and B. Mishra. Automatic verification of sequential circuits using temporal logic. *IEEE Transactions on Computers*, 35(12):1035–1044, December 1986.

51. J. Bruno and S. M. Altman. A theory of asynchronous control networks. *IEEE Transactions on Computers*, 20(6):629–638, June 1971.

52. E. Brunvand. *Translating Concurrent Communicating Programs into Asynchronous Circuits*. PhD thesis, Carnegie Mellon University, 1991.

53. E. Brunvand. Designing self-timed systems using concurrent programs. *Journal of VLSI Signal Processing*, 7(1/2):47–59, February 1994.

54. J. A. Brzozowski and J. C. Ebergen. On the delay-sensitivity of gate networks. *IEEE Transactions on Computers*, 41(11):1349–1360, November 1992.

55. J. A. Brzozowski and C.-J. Seger. A characterization of ternary simulation of gate networks. *IEEE Transactions on Computers*, C-36(11):1318–1327, November 1987.

56. J. A. Brzozowski and C.-J. Seger. A unified framework for race analysis of asynchronous networks. *Journal of the ACM*, 36(1):20–45, January 1989.

57. J. A. Brzozowski and C.-J. H. Seger. *Asynchronous Circuits*. Springer-Verlag, New York, 1995.

58. J. A. Brzozowski and S. Singh. Definite asynchronous sequential circuits. *IEEE Transactions on Computers*, C-17(1):18–26, January 1968.
59. J. A. Brzozowski and M. Yoeli. Practical approach to asynchronous gate networks. *Proceedings of the IEE*, 123(6):495–498, June 1976.
60. J. A. Brzozowski and M. Yoeli. On a ternary model of gate networks. *IEEE Transactions on Computers*, C-28(3):178–184, March 1979.
61. J. R. Burch. Combining ctl, trace theory and timing models. In *Proc. First Workshop on Automatic Verification Methods for Finite State Systems*, 1989.
62. J. R. Burch. Modeling timing assumptions with trace theory. In *Proc. International Conference on Computer Design (ICCD)*, 1989.
63. J. R. Burch. *Trace Algebra for Automatic Verification of Real-Time Concurrent Systems*. PhD thesis, Carnegie Mellon University, 1992.
64. S. M. Burns. Automated compilation of concurrent programs into self-timed circuits. Master's thesis, California Institute of Technology, 1988.
65. S. M. Burns. *Performance Analysis and Optimization of Asynchronous Circuits*. PhD thesis, California Institute of Technology, 1991.
66. S. M. Burns. General condition for the decomposition of state holding elements. In *Proc. International Symposium on Advanced Research in Asynchronous Circuits and Systems*. IEEE Computer Society Press, Los Alamitos, CA, March 1996.
67. S. M. Burns and A. J. Martin. Syntax-directed translation of concurrent programs into self-timed circuits. In J. Allen and F. T. Leighton, editors, *Advanced Research in VLSI*, pages 35–50. MIT Press, Cambridge, MA, 1988.
68. S. M. Burns and A. J. Martin. Synthesis of self-timed circuits by program transformation. In G. J. Milne, editor, *The Fusion of Hardware Design and Verification*, pages 99–116. Elsevier Science Publishers, New York, 1988.
69. J. Calvo, J. I. Acha, and M. Valencia. Asynchronous modular arbiter. *IEEE Transactions on Computers*, 35(1):67–70, January 1986.
70. I. Catt. Time loss through gating of asynchronous logic signal pulses. *IEEE Transactions on Electronic Computers*, EC-15:108–111, February 1966.
71. S. Chakraborty. *Polynomial-Time Techniques for Approximate Timing Analysis of Asynchronous Systems*. PhD thesis, Stanford University, August 1998.
72. S. Chakraborty, D. L. Dill, and K. Y. Yun. Min-max timing analysis and an application to asynchronous circuits. *Proceedings of the IEEE*, 87(2):332–346, February 1999.
73. S. T. Chakradhar, S. Banerjee, R. K. Roy, and D. K. Pradhan. Synthesis of initializable asynchronous circuits. *IEEE Transactions on VLSI Systems*, 4(2):254–263, June 1996.
74. V. Chandramouli, E. Brunvand, and K. F. Smith. Self-timed design in GaAs: Case study of a high-speed, parallel multiplier. *IEEE Transactions on VLSI Systems*, 4(1):146–149, March 1996.
75. T. J. Chaney and C. E. Molnar. Anomalous behavior of synchronizer and arbiter circuits. *IEEE Transactions on Computers*, C-22(4):421–422, April 1973.

REFERENCES

76. C.-M. Chang and S.-L. Lu. Design of a static MIMD data flow processor using micropipelines. *IEEE Transactions on VLSI Systems*, 3(3):370–378, September 1995.

77. D. M. Chapiro. *Globally-Asynchronous Locally-Synchronous Systems*. PhD thesis, Stanford University, October 1984.

78. J. F. Chappel and S. G. Zaky. EMI effects and timing design for increased reliability in digital systems. *IEEE Transactions on Circuits and Systems I: Fundamental Theory and Applications*, 44(2):130–142, February 1997.

79. T. I. Chappell, B. A. Chappell, S. E. Schuster, J. W. Allan, S. P. Klepner, R. V. Joshi, and R. L. Franch. A 2-ns cycle, 3.8-ns access 512-kb CMOS ECL SRAM with a fully pipelined architecture. *IEEE Journal of Solid-State Circuits*, 26(11):1577–1585, November 1991.

80. F.-C. Cheng, S. H. Unger, and M. Theobald. Self-timed carry-lookahead adders. *IEEE Transactions on Computers*, 49(7):659–672, July 2000.

81. W.-C. Chou, P. A. Beerel, and K. Y. Yun. Average-case technology mapping of asynchronous burst-mode circuits. *IEEE Transactions on Computer-Aided Design*, 18(10):1418–1434, October 1999.

82. T.-A. Chu. On the models for designing VLSI asynchronous digital circuits. *Integration, the VLSI Journal*, 4(2):99–113, June 1986.

83. T.-A. Chu. *Synthesis of Self-Timed VLSI Circuits from Graph-Theoretic Specifications*. PhD thesis, MIT Laboratory for Computer Science, June 1987.

84. T.-A. Chu. Synthesis of hazard-free control circuits from asynchronous finite state machine specifications. *Journal of VLSI Signal Processing*, 7(1/2):61–84, February 1994.

85. H. Y. H. Chuang and S. Das. Synthesis of multiple-input change asynchronous machines using controlled excitation and flip-flops. *IEEE Transactions on Computers*, C-22(12):1103–1109, December 1973.

86. Y. H. Chuang. Transition logic circuits and a synthesis method. *IEEE Transactions on Computers*, C-18(2):154–168, February 1969.

87. W. A. Clark. Macromodular computer systems. In *AFIPS Conference Proc.: 1967 Spring Joint Computer Conference*, volume 30, pages 335–336. Academic Press, New York, 1967.

88. W. A. Clark and C. E. Molnar. Macromodular computer systems. In R. W. Stacy and B. D. Waxman, editors, *Computers in Biomedical Research*, volume IV, chapter 3, pages 45–85. Academic Press, New York, 1974.

89. B. Coates, A. Davis, and K. Stevens. The Post Office experience: Designing a large asynchronous chip. *Integration, the VLSI Journal*, 15(3):341–366, October 1993.

90. P. Corsini. Self-synchronizing asynchronous arbiter. *Digital Processes*, 1:67–73, 1975.

91. P. Corsini. Speed-independent asynchronous arbiter. *IEE Journal on Computers and Digital Techniques*, 2(5):221–222, October 1979.

92. J. Cortadella, M. Kishinevsky, A. Kondratyev, L. Lavagno, E. Pastor, and A. Yakovlev. Decomposition and technology mapping of speed-independent

circuits using Boolean relations. *IEEE Transactions on Computer-Aided Design*, 18(9), September 1999.

93. J. Cortadella, M. Kishinevsky, A. Kondratyev, L. Lavagno, and A. Yakovlev. A region-based theory for state assignment in speed-independent circuits. *IEEE Transactions on Computer-Aided Design*, 16(8):793–812, August 1997.

94. O. Coudert and J. C. Madre. New ideas for solving covering problems. In *Proc. ACM/IEEE Design Automation Conference*, pages 641–646, June 1995.

95. G. R. Couranz and D. F. Wann. Theoretical and experimental behavior of synchronizers operating in the metastable region. *IEEE Transactions on Computers*, 24(6):604–616, June 1975.

96. I. David, R. Ginosar, and M. Yoeli. An efficient implementation of boolean functions as self-timed circuits. *IEEE Transactions on Computers*, 41(1):2–11, January 1992.

97. I. David, R. Ginosar, and M. Yoeli. Implementing sequential machines as self-timed circuits. *IEEE Transactions on Computers*, 41(1):12–17, January 1992.

98. R. David. Modular design of asynchronous circuits defined by graphs. *IEEE Transactions on Computers*, 26(8):727–737, August 1977.

99. A. Davis. Synthesizing asynchronous circuits: Practice and experience. In G. Birtwistle and A. Davis, editors, *Asynchronous Digital Circuit Design*, Workshops in Computing, pages 104–150. Springer-Verlag, New York, 1995.

100. A. Davis, B. Coates, and K. Stevens. Automatic synthesis of fast compact asynchronous control circuits. In S. Furber and M. Edwards, editors, *Asynchronous Design Methodologies*, volume A-28 of *IFIP Transactions*, pages 193–207. Elsevier Science Publishers, New York, 1993.

101. A. Davis, B. Coates, and K. Stevens. The Post Office experience: Designing a large asynchronous chip. In *Proc. Hawaii International Conference on System Sciences*, volume I, pages 409–418. IEEE Computer Society Press, Los Alamitos, CA, January 1993.

102. A. L. Davis. The architecture and system method of ddm-1: A recursive-structured data driven machine. In *Proc. 5th Annual Symposium on Computer Architecture*, 1978.

103. A. L. Davis. A data-driven machine architecture suitable for VLSI implementation. In C. L. Seitz, editor, *Proc. Caltech Conference on Very Large Scale Integration*, pages 479–494, January 1979.

104. P. Day and J. V. Woods. Investigation into micropipeline latch design styles. *IEEE Transactions on VLSI Systems*, 3(2):264–272, June 1995.

105. M. Dean, T. Williams, and D. Dill. Efficient self-timing with level-encoded 2-phase dual-rail (LEDR). In C. H. Séquin, editor, *Advanced Research in VLSI*, pages 55–70. MIT Press, Cambridge, MA, 1991.

106. M. E. Dean. *STRiP: A Self-Timed RISC Processor Architecture*. PhD thesis, Stanford University, 1992.

107. M. E. Dean, D. L. Dill, and M. Horowitz. Self-timed logic using current-sensing completion detection (CSCD). *Journal of VLSI Signal Processing*, 7(1/2):7–16, February 1994.

108. Z. J. Deng, N. Yoshikawa, S. R. Whiteley, and T. Van Duzer. Data-driven self-timed RSFQ digital integrated circuit and system. *IEEE Transactions on Applied Superconductivity*, 7(2):3634–3637, June 1997.

109. Z. J. Deng, N. Yoshikawa, S. R. Whiteley, and T. Van Duzer. Data-driven self-timed RSFQ high-speed test system. *IEEE Transactions on Applied Superconductivity*, 7(4):3830–3833, December 1997.

110. Z. J. Deng, N. Yoshikawa, S. R. Whiteley, and T. Van Duzer. Self-timing and vector processing in RSFQ digital circuit technology. *IEEE Transactions on Applied Superconductivity*, 9(1):7–17, March 1999.

111. Z. J. Deng, N. Yoshikawa, S. R. Whiteley, and T. Van Duzer. Simulation and 18 Gb/s testing of a data-driven self-timed RSFQ demultiplexer. *IEEE Transactions on Applied Superconductivity*, 9(2):4349–4352, June 1999.

112. S. Devadas, K. Keutzer, S. Malik, and A. Wang. Verification of asynchronous interface circuits with bounded wire delays. *Journal of VLSI Signal Processing*, 7(1/2):161–182, February 1994.

113. D. L. Dill. Trace theory for automatic hierarchical verification of speed-independent circuits. In J. Allen and F. T. Leighton, editors, *Advanced Research in VLSI*, pages 51–65. MIT Press, Cambridge, MA, 1988.

114. D. L. Dill. Timing assumptions and verification of finite-state concurrent systems. In *Proc. Workshop on Automatic Verification Methods for Finite-State Systems*, 1989.

115. D. L. Dill. *Trace Theory for Automatic Hierarchical Verification of Speed-Independent Circuits*. ACM Distinguished Dissertations. MIT Press, Cambridge, MA, 1989.

116. D. L. Dill and E. M. Clarke. Automatic verification of asynchronous circuits using temporal logic. *IEE Proceedings, Computers and Digital Techniques*, 133:272–282, September 1986.

117. J. Ebergen and R. Berks. VERDECT: A verifier for asynchronous circuits. *IEEE Technical Committee on Computer Architecture Newsletter*, October 1995.

118. J. Ebergen and R. Berks. Response time properties of linear asynchronous pipelines. *Proceedings of the IEEE*, 87(2):308–318, February 1999.

119. J. C. Ebergen. *Translating Programs into Delay-Insensitive Circuits*. PhD thesis, Eindhoven University of Technology, 1987.

120. J. C. Ebergen. *Translating Programs into Delay-Insensitive Circuits*, volume 56 of *CWI Tract*. Centre for Mathematics and Computer Science, 1989.

121. J. C. Ebergen. A formal approach to designing delay-insensitive circuits. *Distributed Computing*, 5(3):107–119, 1991.

122. E. B. Eichelberger. Hazard detection in combinational and sequential switching circuits. *IBM Journal of Research and Development*, 9:90–99, March 1965.

123. E. A. Emerson. Temporal and modal logic. In J. v. Leeuwen, editor, *Handbook of Theoretical Computer Science*, pages 995–1072. North-Holland, Amsterdam, 1989.

124. P. D. Fisher and S.-F. Wu. Race-free state assignments for synthesizing large-scale asynchronous sequential logic circuits. *IEEE Transactions on Computers*, 42(9):1025–1034, September 1993.

125. W. Fleischhammer and O. Dortok. The anomalous behavior of flip-flops in synchronizer circuits. *IEEE Transactions on Computers*, 28(3):273–276, March 1979.

126. J. Frackowiak. Methoden der analyse und synthese von hasardarmen schaltnetzen mit minimalen kosten i. *Elektronische Informationsverarbeitung und Kybernetik*, 10(2/3):149–187, 1974.

127. E. H. Frank and R. F. Sproull. A self-timed static RAM. In R. Bryant, editor, *Proc. 3rd Caltech Conference on VLSI*, pages 275–285. Computer Science Press, Rockville, MD, 1983.

128. A. D. Friedman, R. L. Graham, and J. D. Ullman. Universal single transition time asynchronous state assignments. *IEEE Transactions on Computers*, C-18:541–547, June 1969.

129. A. D. Friedman and P. R. Menon. Synthesis of asynchronous sequential circuits with multiple-input changes. *IEEE Transactions on Computers*, C-17(6):559–566, June 1968.

130. A. D. Friedman and P. R. Menon. Systems of asynchronously operating modules. *IEEE Transactions on Computers*, 20:100–104, 1971.

131. R. M. Fuhrer, B. Lin, and S. M. Nowick. Symbolic hazard-free minimization and encoding of asynchronous finite state machines. In *Proc. International Conference on Computer-Aided Design (ICCAD)*. IEEE Computer Society Press, Los Alamitos, CA, 1995.

132. R. M. Fuhrer and S. M. Nowick. OPTIMISTA: State minimization of asynchronous FSMs for optimum output logic. In *Proc. International Conference on Computer-Aided Design (ICCAD)*, pages 7–13. IEEE Computer Society Press, Los Alamitos, CA, November 1999.

133. S. B. Furber and P. Day. Four-phase micropipeline latch control circuits. *IEEE Transactions on VLSI Systems*, 4(2):247–253, June 1996.

134. S. B. Furber, J. D. Garside, and D. A. Gilbert. AMULET3: A high-performance self-timed ARM microprocessor. In *Proc. International Conference on Computer Design (ICCD)*, October 1998.

135. S. B. Furber, J. D. Garside, P. Riocreux, S. Temple, P. Day, J. Liu, and N. C. Paver. AMULET2e: An asynchronous embedded controller. *Proceedings of the IEEE*, 87(2):243–256, February 1999.

136. H. v. Gageldonk, D. Baumann, K. v. Berkel, D. Gloor, A. Peeters, and G. Stegmann. An asynchronous low-power 80c51 microcontroller. In *Proc. International Symposium on Advanced Research in Asynchronous Circuits and Systems*, pages 96–107, 1998.

137. M. J. Gamble. A novel current-sensing completion-detection circuit adapted to the micropipeline methodology. Master's thesis, University of Manitoba, Canada, 1994.

138. J. D. Garside, S. B. Furber, and S.-H. Chung. AMULET3 revealed. In *Proc. International Symposium on Advanced Research in Asynchronous Circuits and Systems*, pages 51–59, April 1999.

139. B. Gilchrist, J. H. Pomerene, and S. Y. Wong. Fast carry logic for digital computers. *IRE Transactions on Electronic Computers*, EC-4(4):133–136, December 1955.

140. J. Gimpel. A reduction technique for prime implicant tables. *IEEE Transactions on Electronic Computers*, EC-14:535–541, August 1965.

141. R. Ginosar, R. Kol, K. Stevens, P. Beerel, K. Yun, C. Myers, and S. Rotem. Apparatus and method for parallel processing and self-timed serial marking of variable length instructions. U.S. patent 5,978,899 granted November 2, 1999.

142. R. Ginosar, R. Kol, K. Stevens, P. Beerel, K. Yun, C. Myers, and S. Rotem. Apparatus and method for self-timed marking of variable length instructions having length-affecting prefix bytes. U.S. patent 5,948,096 granted September 7, 1999.

143. R. Ginosar, R. Kol, K. Stevens, P. Beerel, K. Yun, C. Myers, and S. Rotem. Branch instruction handling in a self-timed marking system. U.S. patent 5,931,944 granted August 3, 1999.

144. R. Ginosar, R. Kol, K. Stevens, P. Beerel, K. Yun, C. Myers, and S. Rotem. Efficient self-timed marking of lengthy variable length instructions. U.S. patent 5,941,982 granted August 24, 1999.

145. S. Ginsburg. On the reduction of superfluous states in sequential machines. *Journal of the ACM*, 6:252–282, April 1959.

146. S. Ginsburg. A synthesis technique for minimal state sequential machines. *IRE Transactions on Electronic Computers*, EC-8:13–24, March 1959.

147. A. De Gloria and M. Olivieri. Efficient semicustom micropipeline design. *IEEE Transactions on VLSI Systems*, 3(3):464–469, September 1995.

148. A. De Gloria and M. Olivieri. Statistical carry lookahead adders. *IEEE Transactions on Computers*, 45(3):340–347, March 1996.

149. P. Godefroid. Using partial orders to improve automatic verification methods. In *Proc. International Workshop on Computer Aided Verification*, pages 176–185, June 1990.

150. G. Gopalakrishnan and V. Akella. High level optimizations in compiling process descriptions to asynchronous circuits. *Journal of VLSI Signal Processing*, 7(1/2):33–45, February 1994.

151. G. Gopalakrishnan, E. Brunvand, N. Michell, and S. Nowick. A correctness criterion for asynchronous circuit validation and optimization. *IEEE Transactions on Computer-Aided Design*, 13(11):1309–1318, November 1994.

152. G. Gopalakrishnan, P. Kudva, and E. Brunvand. Peephole optimization of asynchronous macromodule networks. *IEEE Transactions on VLSI Systems*, 7(1):30–37, March 1999.

153. E. Grass, V. Bartlett, and I. Kale. Completion-detection techniques for asynchronous circuits. *IEICE Transactions on Information and Systems*, E80-D(3):344–350, March 1997.

154. E. Grass and S. Jones. Activity monitoring completion detection (AMCD): A new approach to achieve self-timing. *Electronics Letters*, 32(2):86–88, 1996.

155. A. Grasselli and F. Luccio. A method for minimizing the number of internal states in incompletely specified sequential networks. *IEEE Transactions on Electronic Computers*, pages 350–359, June 1965.

156. A. Grasselli and F. Luccio. Some covering problems in switching theory. In G. Biorci, editor, *Network and Switching Theory*. Academic Press, 1966.

157. J. Gu and R. Puri. Asynchronous circuit synthesis with boolean satisfiability. *IEEE Transactions on Computer-Aided Design*, 14(8):961–973, August 1995.

158. G. D. Hachtel and F. Somenzi. *Logic Synthesis and Verification Algorithms*. Kluwer Academic Publishers, Boston, 1996.

159. K. Hamaguchi, H. Hiraishi, and S. Yajima. Formal verification of speed-dependent asynchronous circuits using symbolic model checking of branching time regular temporal logic. In K. G. Larsen and A. Skou, editors, *Proc. International Workshop on Computer Aided Verification*, volume 575 of *Lecture Notes in Computer Science*, pages 410–420. Springer-Verlag, New York, 1991.

160. A. B. Hayes. Stored state asynchronous sequential circuits. *IEEE Transactions on Computers*, C-30(8):596–600, August 1981.

161. A. B. Hayes. Self-timed IC design with PPL's. In R. Bryant, editor, *Proc. 3rd Caltech Conference on VLSI*, pages 257–274. Computer Science Press, Rockville, MD, 1983.

162. P. J. Hazewindus. *Testing Delay-Insensitive Circuits*. PhD thesis, California Institute of Technology, 1992.

163. A. Hemani, T. Meincke, S. Kumar, A. Postula, T. Olsson, P. Nilsson, J. Öberg, P. Ellervee, and D. Lundqvist. Lowering power consumption in clock by using globally asynchronous, locally synchronous design style. In *Proc. ACM/IEEE Design Automation Conference*, 1999.

164. J. Hennessy and D. Patterson. *Computer Organization & Design: The Hardware/Software Interface*. Morgan Kaufmann, San Francisco, 1998.

165. Masaharu Hirayama. A silicon compiler system based on asynchronous architecture. *IEEE Transactions on Computer-Aided Design*, 6(3):297–304, May 1987.

166. C. A. R. Hoare. Communicating sequential processes. *Communications of the ACM*, 21(8):666–677, August 1978.

167. C. A. R. Hoare. *Communicating Sequential Processes*. Prentice-Hall, Englewood Cliffs, NJ, 1985.

168. L. A. Hollaar. Direct implementation of asynchronous control units. *IEEE Transactions on Computers*, C-31(12):1133–1141, December 1982.

169. J. U. Horstmann, H. W. Eichel, and R. L. Coates. Metastability behavior of CMOS ASIC flip-flops in theory and test. *IEEE Journal of Solid-State Circuits*, 24(1):146–157, February 1989.

170. D. A. Huffman. The synthesis of sequential switching circuits. *Journal of the Franklin Institute*, March/April 1954.

171. D. A. Huffman. Design of hazard-free switching circuits. *Journal of the ACM*, 4:47–62, January 1957.

172. D. A. Huffman. The synthesis of sequential switching circuits. In E. F. Moore, editor, *Sequential Machines: Selected Papers*. Addison-Wesley, Reading, MA, 1964.

173. H. Hulgaard. *Timing Analysis and Verification of Timed Asynchronous Circuits*. PhD thesis, University of Washington, 1995.

174. H. Hulgaard and S. M. Burns. Bounded delay timing analysis of a class of CSP programs. *Formal Methods in System Design*, 11(3):265–294, October 1997.

175. H. Hulgaard, S. M. Burns, T. Amon, and G. Borriello. An algorithm for exact bounds on the time separation of events in concurrent systems. *IEEE Transactions on Computers*, 44(11):1306–1317, November 1995.

176. H. Hulgaard, S. M. Burns, and G. Borriello. Testing asynchronous circuits: A survey. *Integration, the VLSI Journal*, 19(3):111–131, November 1995.

177. M. Hurtado. *Structure and Performance of Asymptotically Bistable Dynamical Systems*. PhD thesis, Sever Institute of Technology, Washington University, 1975.

178. W. Hwang, R. V. Joshi, and G. D. Gristede. A scannable pulse-to-static conversion register array for self-timed circuits. *IEEE Journal of Solid-State Circuits*, 35(1):125–128, January 2000.

179. S. Robinson III and R. House. Gimpel's reduction technique extended to the covering problem with costs. *IEEE Transactions on Electronic Computers*, EC-16:509–514, February 1967.

180. H. Jacobson, C. J. Myers, and G. Gopalakrishnan. Achieving fast and exact hazard-free logic minimization of extended burst-mode gC finite state machines. In *Proc. International Conference on Computer-Aided Design (ICCAD)*, November 2000.

181. K. W. James and K. Y. Yun. Average-case optimized transistor-level technology mapping of extended burst-mode circuits. In *Proc. International Symposium on Advanced Research in Asynchronous Circuits and Systems*, pages 70–79, 1998.

182. D. Johnson, V. Akella, and B. Stott. Micropipelined asynchronous discrete cosine transform (DCT/IDTC) processor. *IEEE Transactions on VLSI Systems*, 6(4):731–740, December 1998.

183. M. B. Josephs and J. T. Yantchev. CMOS design of the tree arbiter element. *IEEE Transactions on VLSI Systems*, 4(4):472–476, December 1996.

184. J. R. Jump and P. S. Thiagarajan. On the interconnection of asynchronous control structures. *Journal of the ACM*, 22:596–612, October 1975.

185. S. T. Jung and C. S. Jhon. Direct synthesis of efficient speed-independent circuits from deterministic signal transition graphs. In *Proc. International Symposium on Circuits and Systems*, pages 307–310, June 1994.

186. S. T. Jung and C. J. Myers. Direct synthesis of timed asynchronous circuits. In *Proc. International Conference on Computer-Aided Design (ICCAD)*, pages 332–337, November 1999.

187. S. T. Jung, U. S. Park, J. S. Kim, and C. S. Jhon. Automatic synthesis of gate-level speed-independent control circuits from signal transition graphs. In *Proc. International Symposium on Circuits and Systems*, pages 1411–1414, 1995.

188. J.-W. Kang, C.-L. Wey, and P. D. Fisher. Application of bipartite graphs for achieving race-free state assignments. *IEEE Transactions on Computers*, 44(8):1002–1011, August 1995.

189. V. Kantabutra and A. G. Andreou. A state assignment approach to asynchronous CMOS circuit design. *IEEE Transactions on Computers*, 43(4):460–469, April 1994.

190. R. M. Keller. Towards a theory of universal speed-independent modules. *IEEE Transactions on Computers*, C-23(1):21–33, January 1974.

191. R. Kelly and L. E. M. Brackenbury. Design and modelling of a high performance differential bipolar self-timed microprocessor. *IEE Proceedings, Computers and Digital Techniques*, 144(6):371–380, November 1997.

192. J. Kessels, T. Kramer, G. d. Besten, A. Peeters, and V. Timm. Applying asynchronous circuits in contactless smart cards. In *Proc. International Symposium on Advanced Research in Asynchronous Circuits and Systems*, pages 36–44. IEEE Computer Society Press, Los Alamitos, CA, April 2000.

193. J. Kessels and P. Marston. Designing asynchronous standby circuits for a low-power pager. *Proceedings of the IEEE*, 87(2):257–267, February 1999.

194. K. Keutzer, L. Luciano, and A. Sangiovanni-Vincentelli. Synthesis for testability techniques for asynchronous circuits. *IEEE Transactions on Computer-Aided Design*, 14(12):1569–1577, December 1995.

195. A. Khoche. *Testing Macro-Module Based Self-Timed Circuits*. PhD thesis, University of Utah, 1996.

196. K. Killpack, E. Mercer, and C. J. Myers. A standard-cell self-timed multiplier for energy and area critical synchronous systems. In E. Brunvand and C. Myers, editors, *Advanced Research in VLSI*, pages 188–201. IEEE Computer Society Press, Los Alamitos, CA, March 2001.

197. I. Kimura. Extensions of asynchronous circuits and the delay problem I: Good extensions and the delay problem of the first kind. *Journal of Computer and System Sciences*, 2(3):251–287, October 1968.

198. I. Kimura. Extensions of asynchronous circuits and the delay problem II: Spike-free extensions and the delay problem of the second kind. *Journal of Computer and System Sciences*, 5(2):129–162, April 1971.

199. D. J. Kinniment. An evaluation of asynchronous addition. *IEEE Transactions on VLSI Systems*, 4(1):137–140, March 1996.

200. D. J. Kinniment and J. V. Woods. Synchronization and arbitration circuits in digital systems. *Proceedings of the IEE*, 123(10):961–966, October 1976.

201. M. Kishinevsky, A. Kondratyev, L. Lavagno, A. Saldanha, and A. Taubin. Partial-scan delay fault testing of asynchronous circuits. *IEEE Transactions on Computer-Aided Design*, 17(11):1184–1199, November 1998.

202. M. Kishinevsky, A. Kondratyev, A. Taubin, and V. Varshavsky. Analysis and identification of speed-independent circuits on an event model. *Formal Methods in System Design*, 4(1):33–75, 1994.

203. M. Kishinevsky, A. Kondratyev, A. Taubin, and V. Varshavsky. *Concurrent Hardware: The Theory and Practice of Self-Timed Design*. Series in Parallel Computing. Wiley, New York, 1994.

204. L. Kleeman and A. Cantoni. Can redundancy and masking improve the performance of synchronizers. *IEEE Transactions on Computers*, 35:643–646, July 1986.

205. L. Kleeman and A. Cantoni. On the unavoidability of metastable behavior in digital systems. *IEEE Transactions on Computers*, C-36(1):109–112, January 1987.

206. T. Kobayashi, K. Arimoto, Y. Ikeda, M. Hatanaka, K. Mashiko, M. Yamada, and T. Nakano. A high-speed 46K×4 CMOS DRAM using on-chip self-timing techniques. *IEEE Journal of Solid-State Circuits*, 21(5):655–661, October 1986.

207. S. Komori, H. Takata, T. Tamura, F. Asai, T. Ohno, O. Tomisawa, T. Yamasaki, K. Shima, K. Asada, and H. Terada. An elastic pipeline mechanism by self-timed circuits. *IEEE Journal of Solid-State Circuits*, 23(1):111–117, February 1988.

208. S. Komori, H. Takata, T. Tamura, F. Asai, T. Ohno, O. Tomisawa, T. Yamasaki, K. Shima, H. Nishikawa, and H. Terada. A 40-MFLOPS 32-bit floating-point processor with elastic pipeline scheme. *IEEE Journal of Solid-State Circuits*, 24(5):1341–1347, October 1989.

209. A. Kondratyev, J. Cortadella, M. Kishinevsky, L. Lavagno, and A. Yakovlev. Logic decomposition of speed-independent circuits. *Proceedings of the IEEE*, 87(2):347–362, February 1999.

210. A. Kondratyev, M. Kishinevsky, and A. Yakovlev. Hazard-free implementation of speed-independent circuits. *IEEE Transactions on Computer-Aided Design*, 17(9):749–771, September 1998.

211. C. Krieger. Solving state coding problems in timed asynchronous circuits. Master's thesis, University of Utah, 1999.

212. S. A. Kripke. Semantical analysis of modal logic I: Normal propositional calculi. *Zeitschrift fuer Mathematsche Logik und Grundlagen der Matheematik*, 9:67–96, 1963.

213. P. Kudva, G. Gopalakrishnan, H. Jacobson, and S. M. Nowick. Synthesis of hazard-free customized CMOS complex-gate networks under multiple-input changes. In *Proc. ACM/IEEE Design Automation Conference*, 1996.

214. J. G. Kuhl and S. M. Reddy. A multicode single transition-time state assignment for asynchronous sequential machines. *IEEE Transactions on Computers*, 27:927–934, October 1978.

215. D.S. Kung. Hazard-non-increasing gate-level optimization algorithms. In *Proc. International Conference on Computer-Aided Design (ICCAD)*, pages 631–634. IEEE Computer Society Press, Los Alamitos, CA, November 1992.

216. I. Kurosawa, H. Nakagawa, M. Aoyagi, M. Maezawa, Y. Kameda, and T. Nanya. A basic circuit for asynchronous superconductive logic using RSFQ gates. *Superconductor-Science-Technology*, 9(4A):A46–49, April 1996.

217. G. Lacroix, P. Marchegay, and G. Piel. Comments on "the anomalous behavior of flip-flops in synchronizer circuits". *IEEE Transactions on Computers*, 31(1):77–78, January 1982.

218. H. Lampinen and O. Vainio. Circuit design for current-sensing completion detection. In *Proc. International Symposium on Circuits and Systems*, volume 2, pages 185–188, June 1998.

219. C. G. Langdon. *Analysis and Synthesis of Asynchronous Circuits Under Different Delay Assumptions*. PhD thesis, Syracuse University, October 1967.

220. C. G. Langdon. Delay-free asynchronous circuits with constrained line delays. *IEEE Transactions on Computers*, 17:1131–1143, December 1968.

221. L. Lavagno. *Synthesis and Testing of Bounded Wire Delay Asynchronous Circuits from Signal Transition Graphs*. PhD thesis, University of California, Berkeley, November 1992.

222. L. Lavagno, K. Keutzer, and A. Sangiovanni-Vincentelli. Synthesis of hazard-free asynchronous circuits with bounded wire delays. *IEEE Transactions on Computer-Aided Design*, 14(1):61–86, January 1995.

223. L. Lavagno, C. W. Moon, R. K. Brayton, and A. Sangiovanni-Vincentelli. An efficient heuristic procedure for solving the state assignment problem for event-based specifications. *IEEE Transactions on Computer-Aided Design*, 14(1):45–60, January 1995.

224. L. Lavagno and A. Sangiovanni-Vincentelli. *Algorithms for Synthesis and Testing of Asynchronous Circuits*. Kluwer Academic Publishers, Boston, 1993.

225. L. Lavagno, N. Shenoy, and A. Sangiovanni-Vincentelli. Linear programming for hazard elimination in asynchronous circuits. *Journal of VLSI Signal Processing*, 7(1/2):137–160, February 1994.

226. Y. Leblebici, H. Özdemir, A. Kepkep, and U. Çilingiroğlu. A compact high-speed (31,5) parallel counter circuit based on capacitive threshold-logic gates. *IEEE Journal of Solid-State Circuits*, 31(8):1177–1183, August 1996.

227. K. Lee and K. Choi. Self-timed divider based on RSD number system. *IEEE Transactions on VLSI Systems*, 4(2):292–295, June 1996.

228. T. W. S. Lee, M. R. Greenstreet, and C.-J. Seger. Automatic verification of asynchronous circuits. *IEEE Design and Test of Computers*, 12(1):24–31, Spring 1995.

229. S. C. Leung and H. F. Li. A syntax-directed translation for the synthesis of delay-insensitive circuits. *IEEE Transactions on VLSI Systems*, 2(2):196–210, June 1994.

230. S. C. Leung and H. F. Li. On the realizability and synthesis of delay-insensitive behaviors. *IEEE Transactions on Computer-Aided Design*, 14(7):833–848, July 1995.

231. L. A. Lev et al. A 64-b microprocessor with multimedia support. *IEEE Journal of Solid-State Circuits*, 30(11):1227–1238, November 1995.

232. M. J. Liebelt and N. Burgess. Detecting exitory stuck-at faults in semimodular asynchronous circuits. *IEEE Transactions on Computers*, 48(4):442–448, April 1999.

233. W. Lim. Design methodology for stoppable clock systems. *IEE Proceedings, Computers and Digital Techniques*, 133(1):65–69, January 1986.

234. B. Lin and S. Devadas. Synthesis of hazard-free multilevel logic under multi-input changes from binary decision diagrams. *IEEE Transactions on Computer-Aided Design*, 14(8):974–985, August 1995.

235. K.-J. Lin, C.-W. Kuo, and C.-S. Lin. Synthesis of hazard-free asynchronous circuits based on characteristic graph. *IEEE Transactions on Computers*, 46(11):1246–1263, November 1997.

236. D. H. Linder and J. C. Harden. Phased logic: Supporting the synchronous design paradigm with delay-insensitive circuitry. *IEEE Transactions on Computers*, 45(9):1031–1044, September 1996.

237. P. F. Lister. Design methodology for self-timed VLSI systems. *IEE Proceedings, Computers and Digital Techniques*, 132(1):25–32, January 1985.

238. C. N. Liu. A state variable assignment method for asynchronous sequential switching circuits. *Journal of the ACM*, 10:209–216, 1963.

239. S. Lubkin. Asynchronous circuits in digital computers. *Mathematical Tables and Other Aids to Computation*, pages 238–241, October 1952.

240. M. Maezawa, I. Kurosawa, M. Aoyagi, H. Nakagawa, Y. Kameda, and T. Nanya. Rapid single-flux-quantum dual-rail logic for asynchronous circuits. *IEEE Transactions on Applied Superconductivity*, 7(2):2705–2708, June 1997.

241. G. Magó. Realization methods for asynchronous sequential circuits. *IEEE Transactions on Computers*, C-20(3):290–297, March 1971.

242. G. K. Maki and J. H. Tracey. A state assignment procedure for asynchronous sequential circuits. *IEEE Transactions on Computers*, 20:666–668, June 1971.

243. G. K. Maki, J. H. Tracey, and R. J. Smith. Generation of design equations in asynchronous sequential circuits. *IEEE Transactions on Computers*, 18:467–472, May 1969.

244. R. Männer. Metastable states in asynchronous digital systems - avoidable or unavoidable. *Microelectronics and Reliability*, 28(2):295–307, 1988.

245. R. Manohar and J. A. Tierno. Asynchronous parallel prefix computation. *IEEE Transactions on Computer-Aided Design*, 47(11):1244–1252, November 1998.

246. L. R. Marino. The effect of asynchronous inputs on sequential network reliability. *IEEE Transactions on Computers*, 26:1082–1090, 1977.

247. L. R. Marino. General theory of metastable operation. *IEEE Transactions on Computers*, C-30(2):107–115, February 1981.

248. A. J. Martin. The probe: An addition to communication primitives. *Information Processing Letters*, 20(3):125–130, 1985. Erratum: *IPL* 21(2):107, 1985.

249. A. J. Martin. Compiling communicating processes into delay-insensitive VLSI circuits. *Distributed Computing*, 1(4):226–234, 1986.

250. A. J. Martin. A synthesis method for self-timed VLSI circuits. In *Proc. International Conference on Computer Design (ICCD)*, pages 224–229, Rye Brook, NY, 1987. IEEE Computer Society Press, Los Alamitos, CA.

251. A. J. Martin. The design of a delay-insensitive microprocessor: An example of circuit synthesis by program transformation. In M. Leeser and G. Brown, editors, *Hardware Specification, Verification and Synthesis: Mathematical Aspects*, volume 408 of *Lecture Notes in Computer Science*, pages 244–259. Springer-Verlag, New York, 1989.

252. A. J. Martin. Formal program transformations for VLSI circuit synthesis. In E. W. Dijkstra, editor, *Formal Development of Programs and Proofs*, UT Year of Programming Series, pages 59–80. Addison-Wesley, Reading, MA, 1989.

253. A. J. Martin. The limitations to delay-insensitivity in asynchronous circuits. In W. J. Dally, editor, *Advanced Research in VLSI*, pages 263–278. MIT Press, Cambridge, MA, 1990.

254. A. J. Martin. Programming in VLSI: From communicating processes to delay-insensitive circuits. In C. A. R. Hoare, editor, *Developments in Concurrency and Communication*, UT Year of Programming Series, pages 1–64. Addison-Wesley, Reading, MA, 1990.

255. A. J. Martin. Synthesis of asynchronous VLSI circuits. In J. Straunstrup, editor, *Formal Methods for VLSI Design*, chapter 6, pages 237–283. North-Holland, Amsterdam, 1990.

256. A. J. Martin. Asynchronous datapaths and the design of an asynchronous adder. *Formal Methods in System Design*, 1(1):119–137, July 1992.

257. A. J. Martin, S. M. Burns, T. K. Lee, D. Borkovic, and P. J. Hazewindus. The design of an asynchronous microprocessor. In C. L. Seitz, editor, *Advanced Research in VLSI*, pages 351–373. MIT Press, Cambridge, MA, 1989.

258. A. J. Martin, S. M. Burns, T. K. Lee, D. Borkovic, and P. J. Hazewindus. The first asynchronous microprocessor: The test results. *Computer Architecture News*, 17(4):95–110, June 1989.

259. A. J. Martin, A. Lines, R. Manohar, M. Nystroem, P. Penzes, R. Southworth, and U. Cummings. The design of an asynchronous MIPS R3000 microprocessor. In *Advanced Research in VLSI*, pages 164–181. IEEE Computer Society Press, Los Alamitos, CA, September 1997.

260. A. J. McAuley. Dynamic asynchronous logic for high-speed CMOS systems. *IEEE Journal of Solid-State Circuits*, 27(3):382–388, March 1992.

261. E. J. McCluskey. Minimization of boolean functions. *Bell System Technical Journal*, 35:1417–1444, November 1956.

262. E. J. McCluskey. Fundamental mode and pulse mode sequential circuits. In *Proc. IFIP Congress 1962, Munich, Germany, Information Processing*, pages 725–730. Holland Publishing Co., 1962.

263. E. J. McCluskey. Transient behavior of combinational logic networks. *Redundancy Techniques for Computing Systems*, pages 9–46, 1962.

264. K. McMillan. Using unfoldings to avoid the state explosion problem in the verification of asynchronous circuits. In G. v. Bochman and D. K. Probst, editors, *Proc. International Workshop on Computer Aided Verification*, volume 663 of *Lecture Notes in Computer Science*, pages 164–177. Springer-Verlag, New York, 1992.

265. K. McMillan and D. L. Dill. Algorithms for interface timing verification. In *Proc. International Conference on Computer Design (ICCD)*. IEEE Computer Society Press, Los Alamitos, CA, 1992.

266. K. L. McMillan. Trace theoretic verification of asynchronous circuits using unfoldings. In *Proc. International Workshop on Computer Aided Verification*, 1995.

267. T. H.-Y. Meng. *Synchronization Design for Digital Systems*. Kluwer Academic Publishers, Boston, 1991. Contributions by D. Messerschmitt, S. Nowick, and D. Dill.

268. T. H.-Y. Meng, R. W. Brodersen, and D. G. Messerschmitt. Automatic synthesis of asynchronous circuits from high-level specifications. *IEEE Transactions on Computer-Aided Design*, 8(11):1185–1205, November 1989.

269. T. H.-Y. Meng, R. W. Brodersen, and D. G. Messerschmitt. Asynchronous design for programmable digital signal processors. *IEEE Transactions on Signal Processing*, 39(4):939–952, April 1991.

270. E. Mercer and C. J. Myers. Stochastic cycle period analysis in timed circuits. In *Proc. International Symposium on Circuits and Systems*, May 2000.

271. G. De Micheli. *Synthesis and Optimization of Digital Circuits*. McGraw-Hill, New York, 1994.

272. R. E. Miller. *Sequential Circuits and Machines*, volume 2 of *Switching Theory*. Wiley, New York, 1965.

273. D. Misunas. Petri nets and speed independent design. *Communications of the ACM*, 16(8):474–481, August 1973.

274. C. E. Molnar, T.-P. Fang, and F. U. Rosenberger. Synthesis of delay-insensitive modules. In H. Fuchs, editor, *Proc. 1985 Chapel Hill Conference on Very Large Scale Integration*, pages 67–86. Computer Science Press, Rockville, MD, 1985.

275. C. E. Molnar, I. W. Jones, W. S. Coates, J. K. Lexau, S. M. Fairbanks, and I. E. Sutherland. Two FIFO ring performance experiments. *Proceedings of the IEEE*, 87(2):297–307, February 1999.

276. Y. Mukai and Y. Tohma. A method for the realization of fail-safe asynchronous sequential circuits. *IEEE Transactions on Computers*, 23(7):736–739, July 1974.

277. D. E. Muller. Asynchronous logics and application to information processing. In *Proc. Symposium on the Application of Switching Theory to Space Technology*, pages 289–297, Stanford, CA, 1962. Stanford University Press.

278. D. E. Muller. The general synthesis problem for asynchronous digital networks. In *Annual Symposium on Switching and Automata Theory*, New York, 1967.

279. D. E. Muller and W. S. Bartky. A theory of asynchronous circuits. In *Proc. International Symposium on the Theory of Switching*, pages 204–243, Cambridge, MA, April 1959. Harvard University Press.

280. T. Murata. Petri nets: Properties, analysis, applications. *Proceedings of the IEEE*, 77(4):541–580, April 1989.

281. C. Myers and H. Jacobson. Efficient exact two-level hazard-free logic minimization. In *Proc. International Symposium on Advanced Research in Asynchronous Circuits and Systems*, March 2001.

282. C. Myers and T. H.-Y. Meng. Synthesis of timed asynchronous circuits. In *Proc. International Conference on Computer Design (ICCD)*, pages 279–282. IEEE Computer Society Press, Los Alamitos, CA, October 1992.

283. C. J. Myers. *Computer-Aided Synthesis and Verification of Gate-Level Timed Circuits*. PhD thesis, Stanford University, October 1995.

284. C. J. Myers, P. A. Beerel, and T. H.-Y. Meng. Technology mapping of timed circuits. In *Asynchronous Design Methodologies*, IFIP Transactions, pages 138–147. Elsevier Science Publishers, New York, May 1995.

285. C. J. Myers and T. H.-Y. Meng. Synthesis of timed asynchronous circuits. *IEEE Transactions on VLSI Systems*, 1(2):106–119, June 1993.

286. C. J. Myers, T. G. Rokicki, and T. H.-Y. Meng. Automatic synthesis of gate-level timed circuits with choice. In *Advanced Research in VLSI*, pages 42–58. IEEE Computer Society Press, Los Alamitos, CA, 1995.

287. C. J. Myers, T. G. Rokicki, and T. H.-Y. Meng. POSET timing and its application to the synthesis and verification of gate-level timed circuits. *IEEE Transactions on Computer-Aided Design*, 18(6):769–786, June 1999.

288. T. Nanya, A. Takamura, M. Kuwako, M. Imai, T. Fujii, M. Ozawa, I. Fukasaku, Y. Ueno, F. Okamoto, H. Fujimoto, O. Fujita, M. Yamashina, and M. Fukuma. TITAC-2: A 32-bit scalable-delay-insensitive microprocessor. In *Symposium Record of HOT Chips IX*, pages 19–32, August 1997.

289. T. Nanya and Y. Tohma. On universal single transition time asynchronous state assignments. *IEEE Transactions on Computers*, 27(8):781–782, August 1978.

290. T. Nanya and Y. Tohma. Universal multicode STT state assignments for asynchronous sequential machines. *IEEE Transactions on Computers*, 28(11):811–818, November 1979.

291. T. Nanya, Y. Ueno, H. Kagotani, M. Kuwako, and A. Takamura. TITAC: Design of a quasi-delay-insensitive microprocessor. *IEEE Design and Test of Computers*, 11(2):50–63, 1994.

292. R. Negulescu and A. Peeters. Verification of speed-dependences in single-rail handshake circuits. In *Proc. International Symposium on Advanced Research in Asynchronous Circuits and Systems*, pages 159–170, 1998.

293. C. D. Nielsen and A. J. Martin. Design of a delay-insensitive multiply-accumulate unit. *Integration, the VLSI Journal*, 15(3):291–311, October 1993.

294. L. S. Nielsen, C. Niessen, J. Sparsø, and C. H. van Berkel. Low-power operation using self-timed and adaptive scaling of the supply voltage. *IEEE Transactions on VLSI Systems*, 2(4):391–397, December 1994.

295. L. S. Nielsen and J. Sparsø. Designing asynchronous circuits for low-power: An IFIR filter bank for a digital hearing aid. *Proceedings of the IEEE*, 87(2):268–281, February 1999.

296. B. J. Nordmann. Modular asynchronous control design. *IEEE Transactions on Computers*, 26(3):196–207, March 1977.

297. S. M. Nowick. Design of a low-latency asynchronous adder using speculative completion. *IEE Proceedings, Computers and Digital Techniques*, 143(5):301–307, September 1996.

298. S. M. Nowick and D. L. Dill. Exact two-level minimization of hazard-free logic with multiple-input changes. *IEEE Transactions on Computer-Aided Design*, 14(8):986–997, August 1995.

299. S. M. Nowick, N. K. Jha, and F.-C. Cheng. Synthesis of asynchronous circuits for stuck-at and robust path delay fault testability. *IEEE Transactions on Computer-Aided Design*, 16(12):1514–1521, December 1997.

300. S. M. Nowick, K. Y. Yun, and P. A. Beerel. Speculative completion for the design of high-performance asynchronous dynamic adders. In *Proc. International Symposium on Advanced Research in Asynchronous Circuits and Systems*, pages 210–223. IEEE Computer Society Press, Los Alamitos, CA, April 1997.

301. S.M. Nowick. *Automatic Synthesis of Burst-Mode Asynchronous Controllers*. PhD thesis, March 1993.

302. S. M. Ornstein, M. J. Stucki, and W. A. Clark. A functional description of macromodules. In *AFIPS Conference Proc.: 1967 Spring Joint Computer Conference*, volume 30, pages 337–355. Academic Press, New York, 1967.

303. J. S. Ostroff. *Temporal Logic for Real-Time Systems*. Wiley, New York, Taunton, England, 1989.

304. H. Özdemir, A. Kepkep, B. Pamir, Y. Leblebici, and U. Çilingiroğlu. A capacitive threshold-logic gate. *IEEE Journal of Solid-State Circuits*, 31(8):1141–1150, August 1996.

305. E. Pastor, J. Cortadella, A. Kondratyev, and O. Roig. Structural methods for the synthesis of speed-independent circuits. *IEEE Transactions on Computer-Aided Design*, 17(11):1108–1129, November 1998.

306. M. C. Paull and S. H. Unger. Minimizing the number of states in incompletely specified sequential switching functions. *IRE Transactions on Electronic Computers*, EC-8:356–367, September 1959.

307. R. C. Pearce, J. A. Field, and W. D. Little. Asynchronous arbiter module. *IEEE Transactions on Computers*, 24:931–932, September 1975.

308. M. Pechoucek. Anomalous response times of input synchronizers. *IEEE Transactions on Computers*, 25(2):133–139, February 1976.

309. A. Peeters. The "Asynchronous" Bibliography (BibTeX) database file async.bib. http://www.win.tue.nl/cs/pa/wsinap/doc/async.bib. Corresponding e-mail address: async-bib@win.tue.nl.

310. M. A. Peña, J. Cortadella, A. Kondratyev, and E. Pastor. Formal verification of safety properties in timed circuits. In *Proc. International Symposium on Advanced Research in Asynchronous Circuits and Systems*, pages 2–11. IEEE Computer Society Press, Los Alamitos, CA, April 2000.

311. J. L. Peterson. *Petri Net Theory and the Modeling of Systems*. Prentice Hall, 1981.

312. C. A. Petri. Communication with automata. Technical Report RADC-TR-65-377, Vol. 1, Suppl. 1, Applied Data Research, Princeton, NJ, 1966.

313. C. Piguet. Logic synthesis of race-free asynchronous CMOS circuits. *IEEE Journal of Solid-State Circuits*, 26(3):371–380, March 1991.

314. M. Pipponzi and F. Somenzi. An iterative approach to the binate covering problem. In *Proc. European Conference on Design Automation (EDAC)*, pages 208–211, March 1990.

315. W. W. Plummer. Asynchronous arbiters. *IEEE Transactions on Computers*, 21(1):37–42, January 1972.

316. A. Pnueli. The temporal logic of programs. In *Proc. 18th IEEE Symposium on Foundations of Computer Science*, pages 46–57, 1977.

317. D. K. Probst and H. F. Li. Using partial-order semantics to avoid the state explosion problem in asynchronous systems. In R. P. Kurshan and E. M. Clarke, editors, *Proc. International Workshop on Computer Aided Verification*, volume 531 of *Lecture Notes in Computer Science*, pages 146–155. Springer-Verlag, New York, 1990.

318. G. R. Putzolu. A heuristic algorithm for the testing of asynchronous circuits. *IEEE Transactions on Computers*, 20(6):639–647, June 1970.

319. R. Ramachandran and S.-L. Lu. Efficient arithmetic using self-timing. *IEEE Transactions on VLSI Systems*, 4(4):445–454, December 1996.

320. C. Ramchandani. Analysis of asynchronous concurrent systems by timed Petri nets. Technical Report 120, Project MAC, MIT, Cambridge, MA, February 1974.

321. M. Renaudin, B. El Hassan, and A. Guyot. New asynchronous pipeline scheme: Application to the design of a self-timed ring divider. *IEEE Journal of Solid-State Circuits*, 31(7):1001–1013, July 1996.

322. C. A. Rey and J. Vaucher. Self-synchronized asynchronous sequential machines. *IEEE Transactions on Computers*, 23(12):1306–1311, December 1974.

323. J. Rho, G. Hachtel, F. Somenzi, and R. Jacoby. Exact and heuristic algorithms for the minimization of incompletely specified state machines. *IEEE Transactions on Computer-Aided Design*, pages 167–177, February 1994.

324. W. F. Richardson and E. Brunvand. Architectural considerations for a self-timed decoupled processor. *IEE Proceedings, Computers and Digital Techniques*, 143(5):251–257, September 1996.

325. T. G. Rokicki. *Representing and Modeling Circuits*. PhD thesis, Stanford University, 1993.

326. T. G. Rokicki and C. J. Myers. Automatic verificaton of timed circuits. In *Proc. International Conference on Computer Aided Verification*, pages 468–480. Springer-Verlag, New York, 1994.

327. F. U. Rosenberger and C. E. Molnar. Comments on "metastability of CMOS latch/flip-flop". *IEEE Journal of Solid-State Circuits*, 27(1):128–130, January 1992. Reply by Robert W. Dutton, pages 131–132 of same issue.

328. F. U. Rosenberger, C. E. Molnar, T. J. Chaney, and T.-P. Fang. Q-modules: Internally clocked delay-insensitive modules. *IEEE Transactions on Computers*, C-37(9):1005–1018, September 1988.

329. L. Y. Rosenblum and A. V. Yakovlev. Signal graphs: From self-timed to timed ones. In *Proc. of International Workshop on Timed Petri Nets*, pages 199–207, Torino, Italy, July 1985. IEEE Computer Society Press, Los Alamitos, CA.

330. S. Rotem, K. Stevens, R. Ginosar, P. Beerel, C. Myers, K. Yun, R. Kol, C. Dike, M. Roncken, and B. Agapiev. RAPPID: An asynchronous instruction length decoder. In *Proc. International Symposium on Advanced Research in Asynchronous Circuits and Systems*, pages 60–70, April 1999.

331. G. A. Ruiz. Evaluation of three 32-bit CMOS adders in DCVS logic for self-timed circuits. *IEEE Journal of Solid-State Circuits*, 33(4):604–613, April 1998.

332. J. Rutten and M. Berkelaar. Efficient exact and heuristic minimization of hazard-free logic. In *Proc. International Conference on Computer Design (ICCD)*, pages 152–159, October 1998.

333. J. W. J. M. Rutten and M. R. C. M. Berkelaar. Improved state assignments for burst mode finite state machines. In *Proc. International Symposium on Advanced Research in Asynchronous Circuits and Systems*, pages 228–239. IEEE Computer Society Press, Los Alamitos, CA, April 1997.

334. J. W. J. M. Rutten, M. R. C. M. Berkelaar, C. A. J. van Eijk, and M. A. J. Kolsteren. An efficient divide and conquer algorithm for exact hazard free logic minimization. In *Proc. Design, Automation and Test in Europe (DATE)*, pages 749–754, April 1998.

335. T. Sakurai. Optimization of CMOS arbiter and synchronizer circuits with submicron MOSFETs. *IEEE Journal of Solid-State Circuits*, 23(4):901–906, August 1988.

336. M. Santoro and M. A. Horowitz. SPIM: A pipelined 64 × 64-bit iterative multiplier. *IEEE Journal of Solid-State Circuits*, 24(2):487–493, April 1989.

337. G. Saucier. Encoding of asynchronous sequential networks. *IEEE Transactions on Electronic Computers*, EC-16:365–369, June 1967.

338. G. Saucier. State assignment of asynchronous sequential machines using graph techniques. *IEEE Transactions on Computers*, 21:282–288, March 1972.

339. M. H. Sawasaki, C. Ykman-Couvreur, and B. Lin. Externally hazard-free implementations of asynchronous control circuits. *IEEE Transactions on Computer-Aided Design*, 16(8):835–848, August 1997.

340. D. H. Sawin and G. K. Maki. Asynchronous sequential machines designed for fault detection. *IEEE Transactions on Computers*, C-23(3):239–249, March 1974.

341. S. Schuster, W. Reohr, P. Cook, D. Heidel, M. Immediato, and K. Jenkins. Asynchronous interlocked pipelined CMOS circuits operating at 3.3-4.5 GHz. In *Proc. International Solid State Circuits Conference*, February 2000.

342. C-J. Seger and J.A. Brzozowski. An optimistic ternary simulation of gate races. *Theoretical Computer Science*, 61(1):49–66, October 1988.

343. C. Seitz. *Graph Representations for Logical Machines*. PhD thesis, Massachusetts Institute of Technology, 1971.

344. C. L. Seitz. Asynchronous machines exhibiting concurrency, 1970. *Record of the Project MAC Concurrent Parallel Computation*.

345. C. L. Seitz. Self-timed VLSI systems. In C. L. Seitz, editor, *Proc. First Caltech Conference on Very Large Scale Integration*, pages 345–355, Pasadena, CA, January 1979.

346. C. L. Seitz. Ideas about arbiters. *Lambda*, 1(1, First Quarter):10–14, 1980.

347. C. L. Seitz. System timing. In C. A. Mead and L. A. Conway, editors, *Introduction to VLSI Systems*, chapter 7. Addison-Wesley, Reading, MA, 1980.

348. J. N. Seizovic. Pipeline synchronization. In *Proc. International Symposium on Advanced Research in Asynchronous Circuits and Systems*, pages 87–96, November 1994.

349. A. Semenov and A. Yakovlev. Verification of asynchronous circuits using time Petri-net unfolding. In *Proc. ACM/IEEE Design Automation Conference*, pages 59–63, 1996.

350. A. Semenov, A. Yakovlev, E. Pastor, M. Peña, and J. Cortadella. Synthesis of speed-independent circuits from STG-unfolding segment. In *Proc. ACM/IEEE Design Automation Conference*, pages 16–21, 1997.

351. P. Siegel and G. De Micheli. Decomposition methods for library binding of speed-independent asynchronous designs. In *Proc. International Conference on Computer-Aided Design (ICCAD)*, pages 558–565, November 1994.

352. P. Siegel, G. De Micheli, and D. Dill. Automatic technology mapping for generalized fundamental-mode asynchronous designs. In *Proc. ACM/IEEE Design Automation Conference*, pages 61–67, June 1993.

353. P. S. K. Siegel. *Automatic Technology Mapping for Asynchronous Designs*. PhD thesis, Stanford University, February 1995.

354. M. Singh and S. M. Nowick. High-throughput asynchronous pipelines for fine-grain dynamic datapaths. In *Proc. International Symposium on Advanced Research in Asynchronous Circuits and Systems*, pages 198–209. IEEE Computer Society Press, Los Alamitos, CA, April 2000.

355. V. W.-Y. Sit, C.-S. Choy, and C.-F. Chan. Use of current sensing technique in designing asynchronous static RAM for self-timed systems. *Electronics Letters*, 33(8):667–668, 1997.

356. V. W.-Y. Sit, C.-S. Choy, and C.-F. Chan. A four-phase handshaking asynchronous static RAM design for self-timed systems. *IEEE Journal of Solid-State Circuits*, 34(1):90–96, January 1999.

357. A. E. Sjogren and C. J. Myers. Interfacing synchronous and asynchronous modules within a high-speed pipeline. 8(5):573–583, October 2000.

358. J. R. Smith and C. H. Roth. Analysis and synthesis of asynchronous sequential network using edge-sensitive flip-flops. *IEEE Transactions on Computers*, 20:847–855, 1971.

359. R. J. Smith. Generation of internal state assignments for large asynchronous sequential machines. *IEEE Transactions on Computers*, 23:924–932, September 1974.

360. J. Snepscheut. Deriving circuits from programs. In R. Bryant, editor, *Proc. 3rd Caltech Conference on VLSI*, pages 241–256. Computer Science Press, Rockville, MD, 1983.

361. J. L. A. van de Snepscheut. *Trace Theory and VLSI Design*, volume 200 of *Lecture Notes in Computer Science*. Springer-Verlag, New York, 1985.

362. R. F. Sproull and I. E. Sutherland. Stoppable clock. *Technical Memo 3438*, Sutherland, Sproull, and Associates, January, 1985.

363. R. F. Sproull and I. E. Sutherland. *Asynchronous Systems*. Sutherland, Sproull, and Associates, Palo Alto, CA, 1986. Vol. I: *Introduction*, Vol. II: *Logical Effort and Asynchronous Modules*, Vol. III: *Case Studies*.

364. R. F. Sproull, I. E. Sutherland, and C. E. Molnar. The counterflow pipeline processor architecture. *IEEE Design and Test of Computers*, 11(3):48–59, Fall 1994.

365. K. Stevens. Private communication, September 2000. Ken Stevens is with Intel Corporation.

366. K. Stevens, R. Ginosar, and S. Rotem. Relative timing. In *Proc. International Symposium on Advanced Research in Asynchronous Circuits and Systems*, pages 208–218, April 1999.

367. K. Stevens, S. Rotem, R. Ginosar, P. Beerel, C. Myers, K. Yun, R. Kol, C. Dike, and M. Roncken. An asynchronous instruction length decoder. *IEEE Journal of Solid-State Circuits*, 35(2):217–228, February 2001.

368. M. J. Stucki and J. R. Cox. Synchronization strategies. In C. L. Seitz, editor, *Proc. First Caltech Conference on Very Large Scale Integration*, pages 375–393, 1979.

369. M. J. Stucki, S. M. Ornstein, and W. A. Clark. Logical design of macromodules. In *AFIPS Conference Proc.: 1967 Spring Joint Computer Conference*, volume 30, pages 357–364. Academic Press, New York, 1967.

370. I. E. Sutherland. Micropipelines. *Communications of the ACM*, 32(6):720–738, June 1989.

371. I. E. Sutherland, C. E. Molnar, R. F. Sproull, and J. C. Mudge. The trimosbus. In C. L. Seitz, editor, *Proc. of the First Caltech Conference on Very Large Scale Integration*, pages 395–427, 1979.

372. Ivan Sutherland and Scott Fairbanks. GasP: A minimal FIFO control. In *Proc. International Symposium on Advanced Research in Asynchronous Circuits and Systems*, pages 46–53. IEEE Computer Society Press, Los Alamitos, CA, March 2001.

373. H. Takata, S. Komori, T. Tamura, F. Asai, H. Satoh, T. Ohno, T. Tokuda, H. Nishikawa, and H. Terada. A 100-mega-access per second matching memory for a data-driven microprocessor. *IEEE Journal of Solid-State Circuits*, 25(1):95–99, February 1990.

374. C. J. Tan. State assignments for asynchronous sequential machines. *IEEE Transactions on Computers*, 20(4):382–391, April 1971.

375. G. S. Taylor and G. M. Blair. Reduced complexity two-phase micropipeline latch controller. *IEEE Journal of Solid-State Circuits*, 33(10):1590–1593, October 1998.

376. H. Terada, S. Miyata, and M. Iwata. DDMP's: Self-timed super-pipelined data-driven multimedia processors. *Proceedings of the IEEE*, 87(2):282–296, February 1999.

377. R. A. Thacker, W. Belluomini, and C. J. Myers. Timed circuit synthesis using implicit methods. In *Proc. International Conference on VLSI Design*, pages 181–188, 1999.

378. M. Theobald and S. M. Nowick. Fast heuristic and exact algorithms for two-level hazard-free logic minimization. *IEEE Transactions on Computer-Aided Design*, 17(11):1130–1147, November 1998.

379. J. A. Tierno, A. J. Martin, D. Borkovic, and T. K. Lee. A 100-MIPS GaAs asynchronous microprocessor. *IEEE Design and Test of Computers*, 11(2):43–49, 1994.

380. J. H. Tracey. Internal state assignments for asynchronous sequential machines. *IEEE Transactions on Electronic Computers*, EC-15:551–560, August 1966.

381. J. T. Udding. A formal model for defining and classifying delay-insensitive circuits. *Distributed Computing*, 1(4):197–204, 1986.

382. S. H. Unger. *Asynchronous Sequential Switching Circuits*. Wiley-Interscience, New York, 1969.

383. S. H. Unger. Asynchronous sequential switching circuits with unrestricted input changes. *IEEE Transactions on Computers*, 20(12):1437–1444, December 1971.

384. S. H. Unger. Self-synchronizing circuits and nonfundamental mode operation. *IEEE Transactions on Computers*, 26(3):278–281, March 1977.

385. S. H. Unger. Hazards, critical races, and metastability. *IEEE Transactions on Computers*, 44(6):754–768, June 1995.

386. V. Vakilotojar and P. A. Beerel. RTL verification of asynchronous and heterogeneous systems using symbolic model checking. *Integration, the VLSI Journal*, 24(1):19–36, December 1997.

387. M. Valencia, M. J. Bellido, J. L. Huertas, A. J. Acosta, and S. Sanchez-Solano. Modular asynchronous arbiter insensitive to metastability. *IEEE Transactions on Computers*, 44(12):1456–1461, December 1995.

388. A. Valmari. A stubborn attack on state explosion. In *Proc. International Workshop on Computer Aided Verification*, pages 176–185, June 1990.

389. P. Vanbekbergen. *Synthesis of Asynchronous Control Circuits from Graph-Theoretic Specifications*. PhD thesis, Catholic University of Leuven, September 1993.

390. P. Vanbekbergen, G. Goossens, F. Catthoor, and H. J. de Man. Optimized synthesis of asynchronous control circuits from graph-theoretic specifications. *IEEE Transactions on Computer-Aided Design*, 11(11):1426–1438, November 1992.

391. P. Vanbekbergen, B. Lin, G. Goossens, and H. de Man. A generalized state assignment theory for transformations on signal transition graphs. *Journal of VLSI Signal Processing*, 7(1/2):101–115, February 1994.

392. W. S. VanScheik and R. F. Tinder. High speed externally asynchronous / internally clocked systems. *IEEE Transactions on Computers*, 46(7):824–829, July 1997.

393. V. I. Varshavsky, editor. *Self-Timed Control of Concurrent Processes: The Design of Aperiodic Logical Circuits in Computers and Discrete Systems*. Kluwer Academic Publishers, Boston, Dordrecht, The Netherlands, 1990.

394. H. J. M. Veendrick. The behavior of flip-flops used as synchronizers and prediction of their failure rate. *IEEE Journal of Solid-State Circuits*, 15(2):169–176, 1980.

395. P. Vingron. Coherent design of asynchronous circuits. *IEE Proceedings, Computers and Digital Techniques*, 130(6):190–202, 1983.

396. W. M. Waite. The production of completion signals by asynchronous, iterative networks. *IEEE Transactions on Computers*, 13(2):83–86, April 1964.

397. J. Walker and A. Cantoni. A new synchronizer design. *IEEE Transactions on Computers*, 45(11):1308–1311, November 1996.

398. D. T. Weih and M. R. Greenstreet. Verification of speed-independent data-path circuits. *IEE Proceedings, Computers and Digital Techniques*, 143(5):295–300, September 1996.

399. S. R. Whitaker and G. K. Maki. Pass-transistor asynchronous sequential circuits. *IEEE Journal of Solid-State Circuits*, 24(1):71–78, February 1989.

400. T. Williams, N. Patkar, and G. Shen. SPARC64: A 64-b 64-active-instruction out-of-order-execution MCM processor. *IEEE Journal of Solid-State Circuits*, 30(11):1215–1226, November 1995.

401. T. E. Williams. *Self-Timed Rings and Their Application to Division*. PhD thesis, Stanford University, June 1991.

402. T. E. Williams. Performance of iterative computation in self-timed rings. *Journal of VLSI Signal Processing*, 7(1/2):17–31, February 1994.

403. T. E. Williams and M. A. Horowitz. Bipolar circuit elements providing self-completion-indication. *IEEE Journal of Solid-State Circuits*, 25(1):309–312, January 1990.

404. T. E. Williams and M. A. Horowitz. A zero-overhead self-timed 160ns 54b CMOS divider. *IEEE Journal of Solid-State Circuits*, 26(11):1651–1661, November 1991.

405. A. S. Wojcik and K.-Y. Fang. On the design of three-valued asynchronous modules. *IEEE Transactions on Computers*, 29(10):889–898, October 1980.

406. J. V. Woods, P. Day, S. B. Furber, J. D. Garside, N. C. Paver, and S. Temple. AMULET1: An asynchronous ARM processor. *IEEE Transactions on Computers*, 46(4):385–398, April 1997.

407. S.-F. Wu and P. D. Fisher. Automating the design of asynchronous sequential logic circuits. *IEEE Journal of Solid-State Circuits*, 26(3):364–370, March 1991.

408. A. Xie and P. A. Beerel. Performance analysis of asynchronous circuits and systems using stochastic timed Petri nets. In A. Yakovlev, L. Gomes, and L. Lavagno, editors, *Hardware Design and Petri Nets*, pages 239–268. Kluwer Academic Publishers, Boston, March 2000.

409. A. Yakovlev, A. Petrov, and L. Lavagno. A low latency asynchronous arbitration circuit. *IEEE Transactions on VLSI Systems*, 2(3):372–377, September 1994.

410. T. Yamasaki, K. Shima, S. Komori, H. Takata, T. Tamura, F. Asai, T. Ohno, O. Tomisawa, and H. Terada. VLSI implementation of a variable-length pipeline scheme for data-driven processors. *IEEE Journal of Solid-State Circuits*, 24(4):933–937, August 1989.

411. O. Yenersoy. Synthesis of asynchronous machines using mixed-operation mode. *IEEE Transactions on Computers*, 28(4):325–329, 1979.

412. C. Ykman-Couvreur and B. Lin. Efficient state assignment framework for asynchronous state graphs. In *Proc. International Conference on Computer Design (ICCD)*, pages 692–697. IEEE Computer Society Press, Los Alamitos, CA, 1995.

413. C. Ykman-Couvreur and B. Lin. Optimised state assignment for asynchronous circuit synthesis. In *Asynchronous Design Methodologies*, pages 118–127. IEEE Computer Society Press, Los Alamitos, CA, May 1995.

414. T. Yoneda and H. Ryu. Timed trace theoretic verification using partial order reduction. In *Proc. International Symposium on Advanced Research in Asynchronous Circuits and Systems*, pages 108–121, April 1999.

415. T. Yoneda and B. Schlingloff. Efficient verification of parallel real-time systems. In C. Courcoubetis, editor, *Formal Methods in System Design*. Kluwer Academic Publishers, Boston, 1997.

416. T. Yoneda and T. Yoshikawa. Using partial orders for trace theoretic verification of asynchronous circuits. In *Proc. International Symposium on Advanced Research in Asynchronous Circuits and Systems*. IEEE Computer Society Press, Los Alamitos, CA, March 1996.

417. S. Yovine. KRONOS: A verification tool for real-time systems. *Springer International Journal of Software Tools for Technology Transfer*, 1(1/s2), October 1997.

418. K. Y. Yun. *Synthesis of Asynchronous Controllers for Heterogeneous Systems*. PhD thesis, Stanford University, August 1994.

419. K. Y. Yun, P. A. Beerel, V. Vakilotojar, A. E. Dooply, and J. Arceo. The design and verification of a high-performance low-control-overhead asynchronous differential equation solver. *IEEE Transactions on VLSI Systems*, 6(4):643–655, December 1998.

420. K. Y. Yun and D. L. Dill. Automatic synthesis of extended burst-mode circuits i: Specification and hazard-free implementation. *IEEE Transactions on Computer-Aided Design*, 18(2):101–117, February 1999.

421. K. Y. Yun and D. L. Dill. Automatic synthesis of extended burst-mode circuits II: Automatic synthesis. *IEEE Transactions on Computer-Aided Design*, 18(2):118–132, February 1999.

422. K. Y. Yun, D. L. Dill, and S. M. Nowick. Practical generalizations of asynchronous state machines. In *Proc. European Conference on Design Automation (EDAC)*, pages 525–530. IEEE Computer Society Press, Los Alamitos, CA, February 1993.

423. K. Y. Yun and A. E. Dooply. Pausible clocking-based heterogeneous systems. *IEEE Transactions on VLSI Systems*, 7(4):482–488, December 1999.

424. K. Y. Yun, B. Lin, D. L. Dill, and S. Devadas. BDD-based synthesis of extended burst-mode controllers. *IEEE Transactions on Computer-Aided Design*, 17(9):782–792, September 1998.

425. B. Zhou, T. Yoneda, and B.-G. Schlingloff. Conformance and mirroring for timed asynchronous circuits. In *Proc. Asia and South Pacific Design Automation Conference*, 2001.

Index

A

Absolute complement, 361
Absorbed, 167
Absorption, 188, 362
Acknowledge, 4, 57, 61
Actions, 116
Active, 6, 61
Adjacent, 87
Advance time, 271, 314
Adverse example, 280
AFSM, 89
 See also asynchronous finite state machine
Aliases, 42
Allowed sequences, 208
 See also trace
Always operator, 297
 timed, 303
AMULET, 323
Antisymmetric, 360, 363
Arbiter, 340
Arbitration, 340
 See also synchronization
Arc, 86
Architecture, 25
Assert statement, 59
Assign, 4, 59, 63
Associative, 188, 362

Asymmetric
 choice nets, 108
 confusion, 107
Asynchronous
 circuit design
  future?, 341
  history, 322
 finite state machines, 8, 88
 handshake, 2
 instruction-length decoder, 325
  See also RAPPID
 microprocessors, 323
  AMULET, 323
  Caltech, 323
  pager, 324
 timing, 2
Atomic gate, 225
 implementation, 225
Attribute, 27
Autofailure, 309
 manifestation, 308
Average-case
 logic optimization, 327
 performance analysis, 329

B

Base functions, 190
BCP algorithm, 134
 See also covering problem

393

bounding, 137
branching, 138
example, 149
reduction, 135
termination, 137
Binary
    decision diagram, 198
    relations, 362
Binate covering problem, 133
Bipartite graph, 88
Blocks, 155
BM machines, 92
    See also burst-mode
    edge-labeling functions, 92
    maximal set property, 92
    state diagram, 92
    unique entry point, 92
Boolean
    matching, 191
    matrix, 157
Bounded
    delay, 14
    gate and wire delay, 132
    response time, 300
    timing constraint, 118, 260
Bounding, 137
Branching, 138
Bridging fault model, 332
Buffer, 58
Bundled data, 61
Bundling constraint, 62
Burst-mode, 196
    See also BM machines
    generalized C-elements, 193
    hazard-free logic synthesis, 183
    hazards, 176
    multilevel logic synthesis, 188
    sequential hazards, 194
    state assignment, 183
    state machines, 91
    state minimization, 180
    synthesis, 171
    technology mapping, 189
    transition, 176

C

CALL module, 74
Caltech asynchronous microprocessor, 323
Candidate implicant, 233
Canonical
    DBM, 268
    trace structure, 310
Case statements, 31
Causal, 281
Chain, 363
Change diagrams, 111, 121

Channel, 3, 23
    active, 61
    init_channel, 25, 28
    package, 25, 348
    passive, 61
    port, 61
    probe, 35
    receive, 26
        parallel, 36
    send, 26
        parallel, 35
    type, 25
    VHDL modeling, 24
Characteristic marking, 102
Choke, 310
Circuit verification, 303
Class set, 145
Clause, 133
Closed, 148
    cover, 148
    system, 25
Closure, 145
    clause, 235
    constraint, 148
    section, 235
Column dominance, 136
Combinational
    hazards
        burst-mode, 176
        dynamic $0 \to 1$, 170
        dynamic $1 \to 0$, 170
        extended burst-mode, 177
        multiple-input change, 173
        single-input change, 169
        static 0-hazard, 169
        static 1-hazard, 169
    optimization, 228, 236
Communicating sequential processes, 21
Communication
    channels, 3
        See also channel
        basic structure, 24
    protocols, 4
        See also handshake
        basic structure, 58
Communitive, 362
Compatibility table, 141
Compatible
    conditionally, 142
    list, 144
    maximal, 140
    output, 141
    pairs, 140–141
    prime, 140, 145
    states, 13
    unconditionally, 141

Complement, 167
Complete
  circuit, 11, 208
  state code, 216
  state coding, 216
    algorithm, 223
    insertion point cost function, 220
    insertion points, 217
    state graph coloring, 219
    state signal insertion, 222
    transition points, 217
  sums, 167
Completely specified partition, 156
Component, 25
  declarations, 29
  instantiations, 29
Composition, 305
Compulsory transition, 94
Concurrent, 106
  statement section, 25
Conditional
  clause, 95
  input bursts, 95
Conditionally compatible, 142
Conflict, 107
  relation, 116, 119
Conformance, 20, 303, 308
Conformation equivalent, 308
Conforms, 303, 308
Confusion, 107
  asymmetric, 108
  symmetric, 108
Consensus, 167
Consistent
  signal sets, 305
  state assignment, 113
Constraint matrix, 133
Contain, 165
Context signals, 214, 240, 242
Conv_integer, 27
Correctness constraints, 231
Counterflow pipeline, 324
Cover, 148, 166, 230
Covering
  clause, 234
  constraint, 148, 231, 237
    violation, 241
  problem, 132
    See also BCP algorithm
    binate, 133
    context signals, 241–242
    prime implicant selection, 168
    speed-independent logic synthesis, 234
    state assignment, 161
    state minimization, 148
    unate, 133
  section, 235
Critical race, 155
CSC solver algorithm, 223
Cube, 165
  approximations, 315
Cycle, 87
  simple, 87
Cyclic core, 137

D

Data
  encoding
    bundled data, 61
    dual-rail, 58, 67
  hazards, 49
Data-driven processors, 323
DC-set, 165
Dead, 103
Deadlock, 34, 66
Declaration section, 25
Decomposition, 190
  See also technology mapping
  insertion points, 245
  speed-independent, 243
Delay, 26, 59
  elements, 11
  fault model, 332
  faults, 330
  problem of the first kind, 251
Delay-insensitive circuits, 11
  limits, 12
Delayed-reset domino, 342
DeMorgan's theorem, 188, 362
Deterministic, 32
Dichotomies, 156
Difference bound matrix, 268
Digraph
  acyclic, 88
  simple, 88
Direct transition, 155
Directed
  acyclic graph, 88
  don't cares, 93
  edge, 86
  graph, 86
Disabling rules, 118
Discrete-time, 265
  example, 265
  worst-case complexity, 265
Disengageable strong precedence arcs, 121
Disjoint, 361
  collection, 361
Distance 1 apart, 171
Distributive, 167, 188, 362
  state graphs, 214
Domain, 362

Dual-rail, 67
  protocol, 58
Duals, 362
Dummy timer, 268
Dynamic
  0 → 1 hazard
    multiple-input change, 175
    single-input change, 170
  0 → 1 transition, 169
  1 → 0 hazard
    multiple-input change, 175
    single-input change, 170
  1 → 0 transition, 169
  function hazard, 173
  hazard-free
    compatible, 181
    implicant, 185
    prime implicant, 185
  hazards
    generalized C-elements, 193
    multiple-input change, 175

E

Edges, 86
Empty set, 361
Enable module, 74
Enabled, 102, 118
  event, 118
Enabling event, 118
End
  cube, 177
  point, 172
  subcube, 178
  transitions, 217
Enters, 87
Entity, 25
Entrance constraint, 231, 237
  violation, 241
Enumerated types, 25
Environment, 4
Equilibrium, 112
Equipotential regions, 249
Equivalence
  class, 363
  relation, 363
Equivalent, 209
Essential
  hazard, 195
  primes, 168
  rows, 134
  variable, 134
Evans and Sutherland, 323
Events, 10, 116, 260
Eventually operator, 297
  timed, 303
Excitation
  cube, 238
  region, 213
  states, 212
Excited, 112
Expired, 118, 260
Extended
  burst-mode, 93
    See also XBM machines
    dynamic hazard problem, 179
    generalized C-elements, 193
    hazard-free logic synthesis, 183
    hazards, 177
    multilevel logic synthesis, 188
    sequential hazards, 194
    state assignment, 183
    state machines, 93
    state minimization, 180
    synthesis, 171
    technology mapping, 189
    transition, 177
  free-choice nets, 108
Extensions, 250

F

Failure
  exclusion, 310
  traces, 304
Failure-free, 307
Fault model
  bridging, 332
  delay, 332
  isochronic fork, 331
  stuck-at, 331
Faults, 330
Feedback delay requirement, 195
Finite state machine, 88
  Mealy machine, 89
  Moore machine, 89
Firing sequence, 102
Flow relation, 100
Floyd's algorithm, 269
Followed, 209
For loop, 33
Formal verification, 20
Four-phase handshaking, 5, 64
Fractional component, 262
Free-choice nets, 107
Function, 363
  hazard, 172
    dynamic, 173
    static, 173
Fundamental-mode, xv, 132, 195

G

Gate sharing, 237
Generalized

C-elements, 193, 226
    burst-mode, 193
    extended burst-mode, 193
    speed-independent, 226
    transition cube, 177
Glitch, 165
Globally
    asynchronous locally synchronous, 338
    synchronous locally asynchronous, 338
Good extension, 250
Graph, 85
    basic definitions, 85
    bipartite, 88
    connected, 87
    directed, 86
    order, 87
    size, 87
    strongly connected, 88
    undirected, 86
Graphical representations, 8, 85
Greatest
    lower bound, 364
    member, 364
Guard, 4, 59, 63
Guard_and, 18, 60
Guard_or, 59

H

Handshake, 2
    assign, 59
    guard, 59
    guard_and, 60
    guard_or, 59
    package, 58, 355
    protocols, 57
    vassign, 60
Handshaking expansion, 61
    dual-rail encoding, 67
    four-phase, 64
    lazy-active, 65
    probe, 71
    reshuffling, 65
    state variable insertion, 66
    two-phase, 62
    VHDL modeling, 58
Hazard, 13, 165
    burst-mode, 176
    dynamic $0 \to 1$, 170
    dynamic $1 \to 0$, 170
    dynamic function, 173
    extended burst-mode, 177
    function, 172
    generalized C-elements, 193
    preserving transformations, 188
    static 0, 169
    static 1, 169
    static function, 173
Hazard-free
    decomposition
        algorithm, 246
        speed-independent, 243
    logic synthesis
        atomic gate implementation, 225
        burst-mode, 183
        extended burst-mode, 183
        generalized C-element implementation, 226
        multiple-input change, 183
        single-cube algorithm, 238
        single-input change, 165
        speed-independent, 223
        standard C-implementation, 230
Hierarchical verification, 308
Hold time, 2, 195, 334
Huffman, xv
    circuits, 13, 131
    flow table, 8, 90
    school, 131

I

I-nets, 111
Idempotent, 167, 362
Identities, 361
Identity, 362
If-then-else statements, 31
ILLIAC II, xv, 322
Image, 363
Implicant, 166
    candidate, 233
    excitation region, 231
Implied
    state, 113
    states, 232
    value, 112
In, 28
    transition, 155
Incident
    from, 87
    on, 87
    to, 87
Incompatible, 141
Incompletely specified
    Boolean function, 165
    partition, 156
Infinite loop, 33
Initial
    marking, 100
    state, 116
Initially marked rules, 116
Init_channel, 28
Inout, 28
Input, 4

burst, 91
  conditional, 95
Insertion point, 217
  complete state coding, 217
  cost function, 220
  decomposition, 245
  filters, 246
Instruction statistics, 327
Interface nets, 121
Intersect, 361
Intersectible, 159
Intersection, 158, 165, 361
Into, 363
Inverse, 362
  delete, 305
IPCMOS, 345
Isochronic fork, 16
  fault model, 331

J

Joins, 87

K

K-bounded Petri net, 103
Karnaugh maps, 14

L

L0-live, 103
L1-live, 103
L2-live, 103
L3-live, 103
L4-live, 103
Labeled Petri net, 100
Lazy-active, 65
Least
  member, 364
  upper bound, 364
Leaves, 87
Level, 10
  expression, 118
  signaling, 5, 64
Library, 25
Linear
  inequalities, 268
  ordering, 363
Linear-time temporal logic, 296
Literal, 165
Live, 103, 111
  Petri net, 103
Liveness, 103
Lk-live, 104
Logic
  minimization, 131, 165
  optimization
    average-case, 327
  synthesis

burst-mode, 183
  combinational hazards, 169
  extended burst-mode, 183
  multiple-input change, 183
  prime implicant generation, 166
  prime implicant selection, 168
  single-input change, 165
  speed-independent, 223
  two-level, 165
Loop, 33
Lower bound, 364
  algorithm, 137
LTL formulas
  checking validity, 298

M

M-nets, 111, 120
Macromodules, xv, 322
Marked graph, 107
  component, 110
Marking, 9, 100
Matching/covering, 190
Matrix reduction method, 157
Maximal
  compatibles, 140, 143
  independent set, 137
  intersectible, 159
  member, 364
  set property
    BM machine, 92
    XBM machines, 94, 96
Mealy machine, 89
Mean time between failure, 335
Member, 360
Merge gate, 73
Metastable state, 333
Microprocessors
  asynchronous, 323
    AMULET, 323
    Caltech, 323
Minimal
  member, 364
  state, 214
MiniMIPS example, 36
  block diagram, 36
  channel-level diagram, 38
  decode block, 43
  dmem block, 47
  execute block, 45
  fetch block, 42
  imem block, 41
  instruction formats, 36
  optimized, 48
    decode block, 49
  structural VHDL code, 39
  write_back process, 52

INDEX   399

Minimum-transition-time
   state assignment, 155
Minterm, 165
Mirror, 310
Model checking, 20, 296
Modeling timing, 260
Monotonicity, 314
Moore machine, 89
Muller, xv, 322
   C-element, 7
   circuits, 16, 207
   school, 207
Multilevel logic synthesis
   burst-mode, 188
   extended burst-mode, 188
   multiple-input change, 188
Multiple-input change, 132
      See also burst-mode and extended burst-mode
   combinational hazards, 173
   dynamic hazards, 175
   hazard-free logic synthesis, 183
   multilevel logic synthesis, 188
   sequential hazards, 194
   state assignment, 183
   state minimization, 180
   static hazards, 174
   synthesis, 171
   technology mapping, 189
Mutual exclusion element, 337
   CMOS circuit, 338

N

n-ary relations, 362
Next state operator, 296
Nonadjacent, 87
Noncritical race, 155
Nondeterminism
   delay, 26, 59
   package, 25, 32, 347
   selection, 26, 32
Nondeterministic, 32
Nondisabling rules, 118
Normal flow table, 155, 173
Normalize, 271

O

OFF-set, 165
ON-set, 165
One-to-one, 363
Onto, 363
Open
   generalized transition cube, 177
   transition cube, 172
Ordered
   n-tuple, 362

   pair, 362
   partitions, 159
   triple, 362
Ordinary Petri net, 100
Others, 32
Out, 28
Output, 4
   burst, 91
   compatible, 141
   semi-modular, 212
   stuck-at fault model, 331

P

Packages, 25
Pair chart, 141
Parallel
   communication, 35
   composition
      syntax-directed translation, 78
Partial order, 315, 363
Partially ordered set, 280, 363
Partition, 155
   list, 156
   theory, 155
Partitioning, 190
Passive, 6, 61
Peephole optimizations, 78
      See also syntax-directed translation
   call module optimization, 78
   enable module optimization, 78
   merge module optimization, 78
   select module optimization, 78
Performance analysis, 329
   average-case, 329
Persistent, 111
Petri net, 9, 85, 100
      See also signal transition graph
   asymmetric
      choice nets, 108
   classifications, 107
   dead, 103
   enabled, 102
   firing sequence, 102
   flow relation, 100
   free-choice nets, 107
      extended, 108
   initial marking, 100
   k-bounded, 103
   L0-live, 103
   L1-live, 103
   L2-live, 103
   L3-live, 103
   L4-live, 103
   labeled, 100
   live, 103
   liveness, 109

Lk-live, 104
marked graph, 107
marking, 100
  characteristic, 102
  MG component, 110
  ordinary, 100
  places, 100
  postset, 100
  preset, 100
  reachability graph, 104
  reachable, 102
    markings, 102
  safe, 103
  safety, 109
  SM component, 110
  state machines, 107
  transition, 100
Philips pager, 324
Pipeline synchronization, xv, 345
Pipelining, 48
Places, 100
Port, 61
  declarations, 27
  map, 30
  modes, 28
  type, 28
POSET, 280
  algorithm, 280
  example, 281
  matrix, 280
  timing, 280
Possible traces, 304
Postset, 100
Power set, 361
Prefix-closed trace structure, 304
Premature firing, 331
Premax, 271
Preset, 100
Prime compatibles, 140, 145
  algorithm, 145
Prime implicant, 166
  essential, 168
  generation, 166
  selection, 168
Principle
  of abstraction, 360
  of duality, 362
  of extension, 360
Privileged cubes, 175, 184
Probe, 35
  handshaking expansion, 61, 71
Process statement, 26
Product, 165
Progress, 314
Project, 270
Propagation delay, 334

Propositional logic, 296
Protocol verification, 296

Q

Q-flop, 345
Q-modules, 345
Quantum synchronizers, 345
Quasi-delay insensitive, 16, 323
  See also speed-independent
Quiescent states, 213
Quotient set, 363

R

$\mathcal{R}$-related, 209
$\mathcal{R}$-sequence, 209
Racing, 155
Range, 362
RAPPID, xv, 325, 343
  microarchitecture, 327
  tag unit, 328
RAW hazard, 49
Reachability graph, 104
Reachable, 87, 102
  markings, 102
Recanonicalization, 269
Receive, 26
  parallel, 36
  syntax-directed translation, 76
Receptive, 304
Reduce algorithm, 135
Reduction, 135
Reflexive, 360, 363
Region function, 230
Regions, 262
  example, 262
  worst-case complexity, 265
Register locking, 49
Relative
  complement, 361
  timing, 316
Renaming, 305
Repetition, 31–32
  for loop, 33
  infinite loop, 33
  while loop, 33
Reply, 322
Request, 4, 57, 61
Required cubes, 174–175, 184
Reset region, 213
Reshuffling, 6, 65
Response time, 334
Restrict, 270
ρ-related to, 362
Row
  covers, 158
  dominance, 135

INDEX   401

includes, 158
length, 138
Rules, 116, 260

S

Safe, 111
   Petri net, 103
   substitute, 312
Safety failures, 314
Satisfied, 118, 260
Scan paths, 332
Selection, 26, 31
   case, 31
   deterministic, 31
   function, 32
   if-then-else, 31
   non-deterministic, 32
   syntax-directed translation, 76
Selector module, 73
Self-checking, 331
Self-reference, 401
Self-reseting domino, 342
Self-timed, 249
   See also speed-independent
Semi-modular, 212
   output, 212
Send, 26
   parallel, 35
   syntax-directed translation, 76
Sensitivity list, 35
Sequencing event, 116
Sequential
   composition
      syntax-directed translation, 77
   hazard, 194
   hazards
      burst-mode, 194
      extended burst-mode, 194
      multiple-input change, 194
Set, 360
   builder notation, 360
   region, 213
Setup time, 2, 195, 334
Shared row assignments, 197
Shortest path problem, 269
Signal
   assignment
      syntax-directed translation, 74
   in TEL structure, 116
   in VHDL, 25
   transition graph, 100, 111
      See also Petri net
Simple
   ordering, 363
   path, 87
Simply ordered set, 363

Single-cube algorithm, 238
Single-cycle transitions, 112
Single-input change, 132
   combinational hazards, 169
   hazard-free logic synthesis, 165
   state assignment, 154
   state minimization, 140
Siphon, 109
Skew-tolerant codes, 80
Specification, 4
Speed-independent, xv–16, 209, 322
   See also Muller circuits, quasi-delay
      insensitive, and self-timed
   atomic gate implementation, 225
   complete state coding, 216
   formal definition, 208
   generalized C-element implementation, 226
   hazard-free decomposition, 243
   hazard-free logic synthesis, 223
   limitations, 248
   output semi-modular, 212
   semi-modular, 212
   single-cube algorithm, 238
   standard C-implementation, 230
   totally sequential, 210
Stable, 155, 238
   state, 90
Standard C-implementation, 230
Start
   cube, 177
   point, 172
   subcube, 178
   transitions, 217
State, 208
   minimization
      burst-mode, 180
   assignment, 14, 131
      burst-mode, 183
      extended burst-mode, 183
      multiple-input change, 183
      outputs as state variables, 163
      single-input change, 154
   diagram, 89, 209
   graph, 112, 207, 210
      coloring, 219
   labeling function, 112
   machine, 85
      component, 110
   machines, 107
   minimization, 14, 131
      extended burst-mode, 180
      multiple-input change, 180
      single-input change, 140
   signal insertion, 222
   transitions, 112

variable, 12, 66
Static
　0 → 0 transition, 169
　1 → 1 transition, 169
　0-hazard
　　multiple-input change, 174
　　single-input change, 169
　1-hazard
　　multiple-input change, 174
　　single-input change, 169
　function hazard, 173
　hazards
　　multiple-input change, 174
Std_logic, 25, 58
　std_logic_vector, 25
STG properties
　live, 111
　persistent, 111
　safe, 111
　single-cycle transitions, 112
Stoppable clock, 336
Strong
　conformance, 314
　precedence arcs, 121
　until operator, 296–297
Strongly connected, 88
Structural
　modeling, 27
　pattern matching, 190
Stubborn sets, 315
Stuck-at fault model, 331
　output, 331
Subset, 360
Success traces, 304
Sum-of-products, 165
SUN counterflow pipeline, 324
Switching region, 213
Symmetric, 360, 363
　confusion, 107
　difference, 361
Synchronization, 321
　See also arbitration
　error, 335
　failure, 2, 333
　　eliminating the probability of, 336
　　probability of, 334
　　reducing the probability of, 335
　problem, 321, 332
Synchronous timing, 2
Syntax-directed translation, 73
　parallel composition, 78
　peephole optimizations, 78
　receive statement, 76
　selection statement, 76
　send statement, 76
　sequential composition, 77

　signal assignment, 74
　while loop, 74

T

Tag unit, 328
τ-partitions, 156
Technology mapping, 189
　See also decomposition
　burst-mode, 189
　extended burst-mode, 189
　multiple-input change, 189
TEL structures, 85, 116, 260
Temporal logic, 296
　linear-time, 296
　timed LTL, 300
Terminal class, 209
Terminating transition, 94
Termination, 137
Ternary relation, 362
Testing asynchronous circuits, 330
Time separation of events, 292
Time-quantified requirements, 300
Timed
　always operator, 303
　circuit, 17, 259, 289
　　tag unit, 328
　event/level structure, 9
　eventually operator, 303
　LTL, 300
　　interval operators, 303
　sequences, 261
　state space exploration, 262
　states, 260
　trace theory, 314
　trace, 314
　until operator, 300
Timers, 260
Timing failure, 314
Tokens, 9
Total state transition relation, 296
Totally sequential, 210
Trace, 100, 303
　See also allowed sequences
　structure, 303–304
　　canonical, 310
　　prefix-closed, 304
　theory, 303
Transit time, 249
Transition, 100
　compulsory, 94
　cube, 172
　cubes, 171
　dynamic 0 → 1, 169
　dynamic 1 → 0, 169
　labeling function, 111
　points, 217

complete state coding, 217
  restrictions, 218
  signaling, 5, 63
  static $0 \to 0$, 169
  static $1 \to 1$, 169
  subcubes, 175
  terminating, 94
Transitive, 360, 363
Trap, 110
Trigger signals, 214
Trigger
  cube, 239
  signals, 240
Two-level logic minimization, 165
  See also logic synthesis
Two-phase handshaking, 5, 63
Two-sided timing constraint, 329
Types, 25

U

U-v path, 87
Unacceptable variable, 134
Unate covering problem, 133
Unbalanced logic trees, 327
Unbounded gate delay model, 16, 207
Unconditionally compatible, 141
Unfoldings, 315
Unger, xiii
Union, 361
Unique
  entry point, 92
  state assignment, 113
  state code, 216
Universal set, 361
University of Utah, 323
Unspecified, 156
Until operator
  strong, 296
  timed, 300
  weak, 297
Untimed state, 260
Update
  poset algorithm, 280
  zone algorithm, 270
Upper bound, 364

V

Vacuous, 7, 65
Validation, 20
Vassign, 60
Verification, 296
  circuit, 303
  protocol, 296
Vertices, 85
VHDL constructs
  aliases, 42

architecture, 25
assert statement, 59
attribute, 27
buffer, 58
case, 31
comments, 25
component, 25
  declarations, 29
  instantiations, 29
concurrent statement section, 25
control structures, 31
declaration section, 25
entity, 25
enumerated types, 25
for loop, 33
if-then-else, 31
infinite loop, 33
library, 25
others, 32
packages, 25
port
  declarations, 27
  map, 30
  modes, 28
  type, 28
process statement, 26
repetition, 31
selection, 31
signals, 25
structural modeling, 27
types, 25
while loop, 33
Violating states, 240
  covering constraint, 241
  entrance constraint, 241
  generalized C-element implementation, 240
  standard C-implementation, 241
Violations
  unresolvable, 242

W

Wait statement, 35
Weak
  precedence arcs, 121
  until operator, 297
Well-ordered, 364
While loop, 33
  syntax-directed translation, 74
Wine shop example
  discrete-time, 265
  LTL, 297
  patron handshaking model
    3-bit dual-rail, 70
    dual-rail, 68
    four-phase bundled data, 64

404 INDEX

two wine shops, 72
two-phase bundled data, 63
POSETS, 281
problem specification, 1
regions, 262
shop handshaking model
  dual-rail, 68
  four-phase bundled data, 64
  lazy-active, 66
  reshuffled, 65
  two-phase bundled data, 63
STG models, 116
synchronization problem, 332
syntax-directed translation, 73
timed LTL, 303
timed, 260
two patron example, 96
  STG models, 116
  TEL structure model, 120
two wine shops, 71
VHDL channel model, 24
VHDL handshaking model, 58
VHDL structural model, 27
winery handshaking model
  3-bit dual-rail, 69
  dual-rail, 67
  four-phase bundled data, 64
  two wine shops, 72
  two-phase bundled data, 62
zones, 271

X

XBM machines, 93
  See also extended burst-mode
directed don't cares, 93
edge-labeling functions, 99
formal model, 98
maximal set property, 94, 96

Z

Zones, 267
  example, 271